The Theory of Partitions

George E. Andrews

Pennsylvania State University
University Park, Pennsylvania

PUBLISHED BY THE PRESS SYNDICATE OF THE UNIVERSITY OF CAMBRIDGE
The Pitt Building, Trumpington Street, Cambridge CB2 1RP

CAMBRIDGE UNIVERSITY PRESS
The Edinburgh Building, Cambridge CB2 2RU, United Kingdom
40 West 20th Street, New York, NY 10011-4211, USA
10 Stamford Road, Oakleigh, Melbourne 3166, Australia

Original hardcover edition first published 1976 by Addison-Wesley Publishing Company, Inc.
First paperback edition published 1998 by Cambridge University Press

Library of Congress Cataloging-in-Publication Data is available.

A catalog record for this book is available from the British Library.

ISBN 0-521-63766-X paperback

Transferred to digital printing 2003

To Joy

Amy, and

Katy

Editor's Statement

A large body of mathematics consists of facts that can be presented and described much like any other natural phenomenon. These facts, at times explicitly brought out as theorems, at other times concealed within a proof, make up most of the applications of mathematics, and are the most likely to survive changes of style and of interest.

This ENCYCLOPEDIA will attempt to present the factual body of all mathematics. Clarity of exposition, accessibility to the non-specialist, and a thorough bibliography are required of each author. Volumes will appear in no particular order, but will be organized into sections, each one comprising a recognizable branch of present-day mathematics. Numbers of volumes and sections will be reconsidered as times and needs change.

It is hoped that this enterprise will make mathematics more widely used where it is needed, and more accessible in fields in which it can be applied but where it has not yet penetrated because of insufficient information.

The theory of partitions is one of the very few branches of mathematics that can be appreciated by anyone who is endowed with little more than a lively interest in the subject. Its applications are found wherever discrete objects are to be counted or classified, whether in the molecular and the atomic studies of matter, in the theory of numbers, or in combinatorial problems from all sources.

Professor Andrews has written the first thorough survey of this many-sided field. The specialist will consult it for the more recondite results, the student will be challenged by many a deceptively simple fact, and the applied scientist may locate in it the missing identity he needs to organize his data.

Professor Turán's untimely death has left this book without a suitable introduction. It is fitting to dedicate it to the memory of one of the masters of number theory.

GIAN-CARLO ROTA

Contents

Preface to the Paperback Edition

In the past twenty years, the theory of partitions has blossomed. The object of this book, as appropriate in this series, is to provide the fundamentals in a form accessible to the nonspecialist, with references to the recent literature for those who wish to pursue a particular interest. The major changes I would have made in a total revision would have added greatly to the length of the book. This is a task of such magnitude that it will have to wait until a number of my other projects are completed.

In the light of this introduction, here are my comments on the chapters of the third printing of *The Theory of Partitions*. Chapter 1 contains the almost immutable basics. Chapter 2 is partially devoted to basic hypergeometric series. As such, it is a small introduction to the wonderful world of q that has been so beautifully chronicled by Gasper and Rahman (1990). Andrews (1986) provides a further survey of the interactions of partitions with q. For Chapter 3, it should be pointed out that O'Hara (1990) has shown how to prove the unimodality of Gaussian polynomials in a purely elementary (although not easy) manner. Chapters 3 through 5 are fairly current introductions to their topics.

The work in Chapter 6 has been greatly extended by Richard McIntosh in a series of papers: cf. McIntosh (1995). The material in Chapters 7, 8, and 9 has also been greatly extended in the past decade. Workers in partitions proper (for example, Alladi and Gordon) have made major advances. In the 1970s and 1980s, David Bressoud made spectacular progress in this area. For further references and an account of his central methods, see Bressoud (1980). In addition, the world of statistical mechanics has provided a flood of results, as well as important questions on this topic. The fountainhead of this work is Rodney Baxter (1982).

The introduction to Chapter 10 is reasonably up to date. However, there have been major recent discoveries for partition function congruences; the most notable have been found by Garavan, Kim, and Stanton (1990); Ono (1996); and Gordon and Ono (1997).

There have also been extensive discoveries for higher-dimensional partitions. Here the interested reader should consult surveys by Stanley (1986a, 1986b) and a related paper by Robbins (1991). As many of the conjectures in these papers are now theorums, these surveys are themselves becoming out of date, but they are good leads for what has been done.

The topics in Chapters 12 and 13 might well be augmented with an account of generalized Frobenius partitions (see Andrews [1984]).

Chapter 14 contains the appropriate computational rudiments. However, Zeilberger (1991) has pioneered exciting new computational work related to partitions and other combinatorial identities (cf. Zeilberger and Wilf [1990]).

The following list of references is designed only to provide the reader with leads into the literature. It does not in any way give adequate credit to the numerous contributions of scores of researchers during the past two decades.

References

Alladi, K. (1995). "The method of weighted words and applications to partitions," *Number Theory* (S. David, ed.). Cambridge University Press, Cambridge.

Andrews, G. E. (1984). "Generalized Frobenius partitions," *Mem. Amer. Math. Soc.* **301.**

Andrews, G. E. (1986). "q-Series: Their development and application in analysis, number theory, combinatorics, physics and computer algebra," *Mem. Amer. Math. Soc.* **66.**

Baxter, R. J. (1982). *Exactly Solved Models in Statistical Mechanics.* Academic Press, London and New York.

Bressoud, D. M. (1980a). "Analytic and combinatorial generalizations of the Rogers-Ramanujan identities," *Mem. Amer. Math. Soc.* **227**.

Bressoud, D. M. (1980b). "Extension of the partition sieve," *J. Number Th.* **12,** 76–100.

Bressoud, D. M. (1998). *Proofs and Confirmations.* Cambridge University Press, Cambridge.

Burge, W. H. (1981). "A correspondence between partitions related to generalizations of the Rogers-Ramanujan identities," *Discrete Math.* **34,** 9–15.

Garvan, F., Kim, D., and Stanton, D. (1990). "Cranks and t-cores," *Invent. Math.* **101,** 1–17.

Gasper, G, and Rahman, M. (1990). "Basic hypergeometric series," *Encycl. Math and Its Applications* **35** (G-C. Rota, ed.). Cambridge University Press, Cambridge.

Gordon, B., and Ono, K. (1997). "Divisibility properties of certain partition functions by powers of primes," *Ramunujan J.* **1,** 25–35.

McIntosh, R. J. (1995). "Some asymptotic formulae for q-hypergeometric series," *J. London Math. Soc. (2)* **51,** 120–136.

O'Hara, K. M. (1990). "Unimodality of Gaussian coefficients: A constructive proof," *J. Combinatorial Th.* **A53,** 29–52.

Ono, K. (1996). "On the parity of the partition function in arithmetic progressions," *J. Reine Angew. Math.* **472,** 1–15.

Robbins, D. P. (1991). "The story of 1, 2, 7, 42, 429, 7436, . . .," *Math. Intellig.* **13,** 12–19.

Stanley, R. P. (1986a). "A baker's dozen of conjectures concerning plane partitions," *Lecture Notes in Math.* **1234,** 285–293. Springer, Berlin.

Stanley, R. P. (1986b) "Symmetries of plane partitions," *J. Comb. Th.* **A43,** 103–113.

Zeilberger, D., and Wilf, H. S. (1990). "Rational functions certify combinatorial identities," *J. Amer. Math. Soc.* **3,** 147–158.

Zeilberger, D. (1991). "The method of creative telescoping," *J. Sym. Comp.* **11,** 195–204.

Preface

Let us begin by acknowledging that the word "partition" has numerous meanings in mathematics. Any time a division of some object into subobjects is undertaken, the word partition is likely to pop up. For the purposes of this book a "partition of n" is a nonincreasing finite sequence of positive integers whose sum is n. We shall extend this definition in Chapters 11, 12, and 13 when we consider higher-dimensional partitions, partitions of n-tuples, and partitions of sets, respectively. Compositions or ordered partitions (merely finite sequences of positive integers) will be considered in Chapter 4.

The theory of partitions has an interesting history. Certain special problems in partitions certainly date back to the Middle Ages; however, the first discoveries of any depth were made in the eighteenth century when L. Euler proved many beautiful and significant partition theorems. Euler indeed laid the foundations of the theory of partitions. Many of the other great mathematicians—Cayley, Gauss, Hardy, Jacobi, Lagrange, Legendre, Littlewood, Rademacher, Ramanujan, Schur, and Sylvester—have contributed to the development of the theory.

There have been almost no books devoted entirely to partitions. Generally the combinatorial and formal power series aspects of partitions have found a place in older books on elementary analysis (*Introductio in Analysin Infinitorum* by Euler, *Textbook of Algebra* by Chrystal), in encyclopedic surveys of number theory (*Niedere Zahlentheorie* by Bachman, *Introduction to the Theory of Numbers* by Hardy and Wright), and in combinatorial analysis books (*Combinatory Analysis* by MacMahon, *Introduction to Combinatorial Analysis* by Riordan, *Combinatorial Methods* by Percus, *Advanced Combinatorics* by Comtet). The asymptotic problems associated with partitions have, on the other hand, been treated in works on analytic or additive number theory (*Introduction to the Analytic Theory of Numbers* by Ayoub, *Modular Functions in Analytic Number Theory* by Knopp, *Topics from the Theory of Numbers* by Grosswald, *Additive Zahlentheorie* by Ostmann, *Topics in Analytic Number Theory* by Rademacher).

If one considers the applications of partitions in various branches of mathematics and statistics, one is struck by the interplay of combinatorial and asymptotic methods. We have tried to organize this book so that it adequately develops and interrelates both combinatorial and analytic methods.

Chapters 1–4 treat the elementary portions of the theory of partitions; of primary importance here is the use of generating functions.

Chapters 5 and 6 treat the asymptotic problems. Partition identities are dealt with in Chapters 7 through 9. Chapter 10 on partition function congruences returns to the analytic aspect of partitions. Chapters 11–13 treat several generalizations of partitions and Chapter 14 presents a brief discussion of the computational aspect of partitions.

There are three concluding sections of each chapter: A "Notes" section provides historical comment on the material covered; a "References" section provides a substantial but nonexhaustive list of relevant books and papers; and an "Examples" section provides statements of results not fully covered in the text. Those examples that occur with an asterisk are significant advances beyond the material presented in the text; the remainder form a reasonable set of exercises by which the reader may determine his grasp of the subject matter. References for the source of the examples occur in the related Notes section.

Many of the mathematical sciences have seen applications of partitions recently. Nonparametric statistics require restricted partitions like those in Chapter 3. Various permutation problems in probability and statistics are intimately linked with the Simon Newcomb problem of Chapter 4. Particle physics uses partition asymptotics and partition identities related to the work in Chapters 5–9. Group theory (through Young tableaux) is intimately connected with Chapter 12, and the relationship between partitions and combinatorial theory is explored in Chapter 13.

The material in this book has been developed over a period of years. My first acquaintance with partitions came from thrilling lectures delivered by my thesis adviser, the late Professor Hans Rademacher. Many of the topics herein have been presented in graduate courses at the Pennsylvania State University between 1964 and 1975, in seminars at MIT during the 1970–1971 academic year, at the University of Erlangen in the summer of 1975, and at the University of Wisconsin during the 1975–1976 academic year. I owe a great debt of gratitude to many people at these four universities. I wish to thank specially R. Askey, K. Baclawski, B. Berndt, and L. Carlitz, who contributed many valuable suggestions and comments during the preparation of this book.

Finally I thank my wife, Joy, who has throughout this project been both a help and an inspiration to me.

George E. Andrews

CHAPTER 1

The Elementary Theory of Partitions

1.1 Introduction

In this book we shall study in depth the fundamental additive decomposition process: the representation of positive integers by sums of other positive integers.

DEFINITION 1.1. A *partition* of a positive integer n is a finite nonincreasing sequence of positive integers $\lambda_1, \lambda_2, \ldots, \lambda_r$ such that $\sum_{i=1}^{r} \lambda_i = n$. The λ_i are called the *parts* of the partition.

Many times the partition $(\lambda_1, \lambda_2, \ldots, \lambda_r)$ will be denoted by λ, and we shall write $\lambda \vdash n$ to denote "λ is a partition of n." Sometimes it is useful to use a notation that makes explicit the number of times that a particular integer occurs as a part. Thus if $\lambda = (\lambda_1, \lambda_2, \ldots, \lambda_r) \vdash n$, we sometimes write

$$\lambda = (1^{f_1} 2^{f_2} 3^{f_3} \cdots)$$

where exactly f_i of the λ_j are equal to i. Note now that $\sum_{i \geqslant 1} f_i i = n$.

Numerous types of partition problems will concern us in this book; however, among the most important and fundamental is the question of enumerating various sets of partitions.

DEFINITION 1.2. The partition function $p(n)$ is the number of partitions of n.

Remark. Obviously $p(n) = 0$ when n is negative. We shall set $p(0) = 1$ with the observation that the empty sequence forms the only partition of zero. The following list presents the next six values of $p(n)$ and tabulates the actual partitions.

$p(1) = 1$: $1 = (1)$;
$p(2) = 2$: $2 = (2)$, $1 + 1 = (1^2)$;
$p(3) = 3$: $3 = (3)$, $2 + 1 = (12)$, $1 + 1 + 1 = (1^3)$;
$p(4) = 5$: $4 = (4)$, $3 + 1 = (13)$, $2 + 2 = (2^2)$,
$2 + 1 + 1 = (1^2 2)$, $1 + 1 + 1 + 1 = (1^4)$;

ENCYCLOPEDIA OF MATHEMATICS and Its Applications, Gian-Carlo Rota (ed.). 2, George E. Andrews, The Theory of Partitions

$p(5) = 7$: $5 = (5)$, $4 + 1 = (14)$, $3 + 2 = (23)$,
$3 + 1 + 1 = (1^3 3)$, $2 + 2 + 1 = (12^2)$,
$2 + 1 + 1 + 1 = (1^3 2)$, $1 + 1 + 1 + 1 + 1 = (1^5)$;
$p(6) = 11$: $6 = (6)$, $5 + 1 = (15)$, $4 + 2 = (24)$,
$4 + 1 + 1 = (1^2 4)$, $3 + 3 = (3^2)$, $3 + 2 + 1 = (123)$,
$3 + 1 + 1 + 1 = (1^3 3)$, $2 + 2 + 2 = (2^3)$,
$2 + 2 + 1 + 1 = (1^2 2^2)$, $2 + 1 + 1 + 1 + 1 = (1^4 2)$,
$1 + 1 + 1 + 1 + 1 + 1 = (1^6)$.

The partition function increases quite rapidly with n. For example, $p(10) = 42$, $p(20) = 627$, $p(50) = 204226$, $p(100) = 190569292$, and $p(200) = 3972999029388$.

Many times we are interested in problems in which our concern does not extend to all partitions of n but only to a particular subset of the partitions of n.

DEFINITION 1.3. Let $\mathscr{S}$ denote the set of all partitions.

DEFINITION 1.4. Let $p(S, n)$ denote the number of partitions of n that belong to a subset S of the set $\mathscr{S}$ of all partitions.

For example, we might consider $\mathscr{O}$ the set of all partitions with odd parts and $\mathscr{D}$ the set of all partitions with distinct parts. Below we tabulate partitions related to $\mathscr{O}$ and to $\mathscr{D}$.

$p(\mathscr{O}, 1) = 1$: $1 = (1)$,
$p(\mathscr{O}, 2) = 1$: $1 + 1 = (1^2)$,
$p(\mathscr{O}, 3) = 2$: $3 = (3)$, $1 + 1 + 1 = (1^3)$,
$p(\mathscr{O}, 4) = 2$: $3 + 1 = (13)$, $1 + 1 + 1 + 1 = (1^4)$,
$p(\mathscr{O}, 5) = 3$: $5 = (5)$, $3 + 1 + 1 = (1^2 3)$,
$1 + 1 + 1 + 1 + 1 = (1^5)$,
$p(\mathscr{O}, 6) = 4$: $5 + 1 = (15)$, $3 + 3 = (3^2)$,
$3 + 1 + 1 + 1 = (1^3 3)$,
$1 + 1 + 1 + 1 + 1 + 1 = (1^6)$,
$p(\mathscr{O}, 7) = 5$: $7 = (7)$, $5 + 1 + 1 = (1^2 5)$, $3 + 3 + 1 = (13^2)$,
$3 + 1 + 1 + 1 + 1 = (1^4 3)$,
$1 + 1 + 1 + 1 + 1 + 1 + 1 = (1^7)$.

$p(\mathscr{D}, 1) = 1$: $1 = (1)$,
$p(\mathscr{D}, 2) = 1$: $2 = (2)$,
$p(\mathscr{D}, 3) = 2$: $3 = (3)$, $2 + 1 = (12)$,
$p(\mathscr{D}, 4) = 2$: $4 = (4)$, $3 + 1 = (13)$,
$p(\mathscr{D}, 5) = 3$: $5 = (5)$, $4 + 1 = (14)$, $3 + 2 = (23)$,
$p(\mathscr{D}, 6) = 4$: $6 = (6)$, $5 + 1 = (15)$, $4 + 2 = (24)$,
$3 + 2 + 1 = (123)$,
$p(\mathscr{D}, 7) = 5$: $7 = (7)$, $6 + 1 = (16)$, $5 + 2 = (25)$,
$4 + 3 = (34)$, $4 + 2 + 1 = (124)$.

We point out the rather curious fact that $p(\mathcal{O}, n) = p(\mathcal{D}, n)$ for $n \leqslant 7$, although there is little apparent relationship between the various partitions listed (see Corollary 1.2).

In this chapter, we shall present two of the most elemental tools for treating partitions: (1) infinite product generating functions; (2) graphical representation of partitions.

1.2 Infinite Product Generating Functions of One Variable

DEFINITION 1.5. The generating function $f(q)$ for the sequence $a_0, a_1, a_2, a_3, \ldots$ is the power series $f(q) = \sum_{n \geq 0} a_n q^n$.

Remark. For many of the problems we shall encounter, it suffices to consider $f(q)$ as a "formal power series" in q. With such an approach many of the manipulations of series and products in what follows may be justified almost trivially. On the other hand, much asymptotic work (see Chapter 6) requires that the generating functions be analytic functions of the complex variable q. In actual fact, both approaches have their special merits (recently, E. Bender (1974) has discussed the circumstances in which we may pass from one to the other). Generally we shall state our theorems on generating functions with explicit convergence conditions. For the most part we shall be dealing with absolutely convergent infinite series and infinite products; consequently, various rearrangements of series and interchanges of summation will be justified analytically from this simple fact.

DEFINITION 1.6. Let H be a set of positive integers. We let "H" denote the set of all partitions whose parts lie in H. Consequently, $p(\text{"}H\text{"}, n)$ is the number of partitions of n that have all their parts in H.

Thus if H_0 is the set of all odd positive integers, then "H_0" $= \mathcal{O}$.

$$p(\text{"}H_0\text{"}, n) = p(\mathcal{O}, n).$$

DEFINITION 1.7. Let H be a set of positive integers. We let "H"$(\leqslant d)$ denote the set of all partitions in which no part appears more than d times and each part is in H.

Thus if N is the set of all positive integers, then $p(\text{"}N\text{"}(\leqslant 1), n) = p(\mathcal{D}, n)$.

THEOREM 1.1. *Let H be a set of positive integers, and let*

$$f(q) = \sum_{n \geqslant 0} p(\text{"}H\text{"}, n) q^n, \tag{1.2.1}$$

$$f_d(q) = \sum_{n \geqslant 0} p(\text{"}H\text{"}(\leqslant d), n) q^n. \tag{1.2.2}$$

Then for $|q| < 1$

$$f(q) = \prod_{n \in H} (1 - q^n)^{-1}, \tag{1.2.3}$$

$$f_d(q) = \prod_{n \in H} (1 + q^n + \cdots + q^{dn})$$

$$= \prod_{n \in H} (1 - q^{(d+1)n})(1 - q^n)^{-1}. \tag{1.2.4}$$

Remark. The equivalence of the two forms for $f_d(q)$ follows from the simple formula for the sum of a finite geometric series:

$$1 + x + x^2 + \cdots + x^r = \frac{1 - x^{r+1}}{1 - x}.$$

Proof. We shall proceed in a formal manner to prove (1.2.3) and (1.2.4); at the conclusion of our proof we shall sketch how to justify our steps analytically. Let us index the elements of H, so that $H = \{h_1, h_2, h_3, h_4, \ldots\}$. Then

$$\prod_{n \in H} (1 - q^n)^{-1} = \prod_{n \in H} (1 + q^n + q^{2n} + q^{3n} + \cdots)$$

$$= (1 + q^{h_1} + q^{2h_1} + q^{3h_1} + \cdots)$$

$$\times (1 + q^{h_2} + q^{2h_2} + q^{3h_2} + \cdots)$$

$$\times (1 + q^{h_3} + q^{2h_3} + q^{3h_3} + \cdots)$$

$$\cdots$$

$$= \sum_{a_1 \geq 0} \sum_{a_2 \geq 0} \sum_{a_3 \geq 0} \cdots q^{a_1 h_1 + a_2 h_2 + a_3 h_3 + \cdots}$$

and we observe that the exponent of q is just the partition $(h_1^{a_1} h_2^{a_2} h_3^{a_3} \cdots)$. Hence q^N will occur in the foregoing summation once for each partition of n into parts taken from H. Therefore

$$\prod_{n \in H} (1 - q^n)^{-1} = \sum_{n \geq 0} p(\text{“}H\text{”}, n) q^n.$$

The proof of (1.2.4) is identical with that of (1.2.3) except that the infinite geometric series is replaced by the finite geometric series:

$$\prod_{n \in H} (1 + q^n + q^{2n} + \cdots + q^{dn})$$

$$= \sum_{d \geq a_1 \geq 0} \sum_{d \geq a_2 \geq 0} \sum_{d \geq a_3 \geq 0} \cdots q^{a_1 h_1 + a_2 h_2 + a_3 h_3 + \cdots}$$

$$= \sum_{n \geq 0} p(\text{“}H\text{”}(\leq d), n) q^n.$$

If we are to view the foregoing procedures as operations with convergent infinite products, then the multiplication of infinitely many series together requires some justification. The simplest procedure is to truncate the infinite product to $\prod_{i=1}^{n}(1-q^{h_i})^{-1}$. This truncated product will generate those partitions whose parts are among $h_1, h_2, \ldots, h_n$. The multiplication is now perfectly valid since only a finite number of absolutely convergent series are involved. Now assume q is real and $0 < q < 1$; then if $M = h_n$,

$$\sum_{j=0}^{M} p(\text{“}H\text{”}, j)q^j \leqslant \prod_{i=1}^{n}(1-q^{h_i})^{-1} \leqslant \prod_{i=1}^{\infty}(1-q^{h_i})^{-1} < \infty.$$

Thus the sequence of partial sums $\sum_{j=0}^{M} p(\text{“}H\text{”}, j)q^j$ is a bounded increasing sequence and must therefore converge. On the other hand

$$\sum_{j=0}^{\infty} p(\text{“}H\text{”}, j)q^j \geqslant \prod_{i=1}^{n}(1-q^{h_i})^{-1} \to \prod_{i=1}^{\infty}(1-q^{h_i})^{-1} \qquad \text{as} \quad n \to \infty.$$

Therefore

$$\sum_{j=0}^{\infty} p(\text{“}H\text{”}, j)q^j = \prod_{i=1}^{\infty}(1-q^{h_i})^{-1} = \prod_{n\in H}(1-q^n)^{-1}.$$

Similar justification can be given for the proof of (1.2.4). ■

COROLLARY 1.2 (Euler). *$p(\mathcal{O}, n) = p(\mathcal{D}, n)$ for all n.*

Proof. By Theorem 1.1,

$$\sum_{n\geqslant 0} p(\mathcal{O}, n)q^n = \prod_{n=1}^{\infty}(1-q^{2n-1})^{-1}$$

and

$$\sum_{n\geqslant 0} p(\mathcal{D}, n)q^n = \prod_{n=1}^{\infty}(1+q^n).$$

Now

$$\prod_{n=1}^{\infty}(1+q^n) = \prod_{n=1}^{\infty}\frac{(1-q^{2n})}{(1-q^n)} = \prod_{n=1}^{\infty}\frac{1}{1-q^{2n-1}}. \tag{1.2.5}$$

Hence

$$\sum_{n\geqslant 0} p(\mathcal{O}, n)q^n = \sum_{n\geqslant 0} p(\mathcal{D}, n)q^n,$$

and since a power series expansion of a function is unique, we see that $p(\mathscr{C}, n) = p(\mathscr{D}, n)$ for all n. ■

COROLLARY 1.3 (Glaisher). *Let N_d denote the set of those positive integers not divisible by d. Then*

$$p(\text{"}N_{d+1}\text{"}, n) = p(\text{"}N\text{"}(\leqslant d), n)$$

for all n.

Proof. By Theorem 1.1,

$$\sum_{n\geqslant 0} p(\text{"}N\text{"}(\leqslant d), n)q^n = \prod_{n=1}^{\infty} \frac{(1 - q^{(d+1)n})}{(1 - q^n)}$$

$$= \prod_{\substack{n=1 \\ (d+1)\nmid n}}^{\infty} \frac{1}{(1 - q^n)}$$

$$= \sum_{n\geqslant 0} p(\text{"}N_{d+1}\text{"}, n)q^n,$$

and the result follows as before. ■

There are numerous results of the type typified by Corollaries 1.2 and 1.3. We shall run into such results again in Chapters 7 and 8, where much deeper theorems of a similar nature will be discussed.

1.3 Graphical Representation of Partitions

Another effective elementary device for studying partitions is the graphical representation. To each partition λ is associated its *graphical representation* $\mathscr{G}_\lambda$ (or Ferrers graph), which formally is the set of points with integral coordinates (i, j) in the plane such that if $\lambda = (\lambda_1, \lambda_2, \ldots, \lambda_n)$, then $(i, j) \in \mathscr{G}_\lambda$ if and only if $0 \geqslant i \geqslant -n + 1$, $0 \leqslant j \leqslant \lambda_{|i|+1} - 1$. Rather than dwell on this formal definition, we shall, by means of a few examples, fully explain the graphical representation.

The graphical representation of the partition $8 + 6 + 6 + 5 + 1$ is

```
· · · · · · · ·
· · · · · ·
· · · · · ·
· · · · ·
·
```

The graphical representation of the partition $7 + 3 + 3 + 2 + 1 + 1$ is

. . . .
. . .
. . .
. .
.
.

Note that the ith row of the graphical representation of $(\lambda_1, \lambda_2, \ldots, \lambda_n)$ contains λ_i points (or dots, or nodes).

We remark that there are several equivalent ways of forming the graphical representation. Some authors use unit squares instead of points, so that the graphical representation of $8 + 6 + 6 + 5 + 1$ becomes

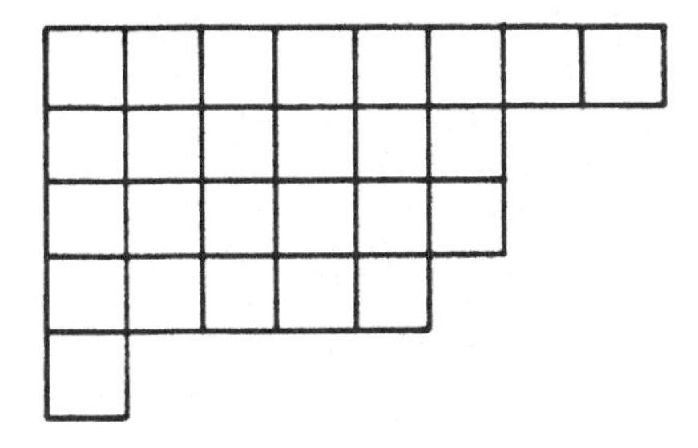

Such a representation is extremely useful when we consider applications of partitions to plane partitions or Young tableaux (see Chapter 11).

Other authors prefer the representation to be upside down (they would say right side up); for example, in the case of $8 + 6 + 6 + 5 + 1$

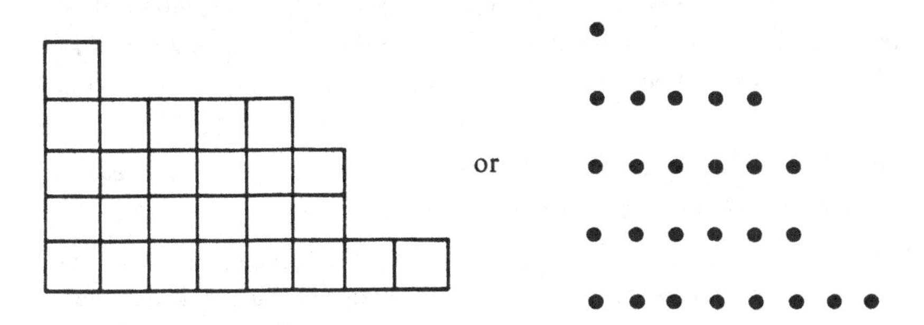

Since most of the classical texts on partitions use the first representation shown in this section, we shall also.

Definition 1.8. If $\lambda = (\lambda_1, \ldots, \lambda_n)$ is a partition, we may define a new partition $\lambda' = (\lambda_1', \ldots, \lambda_m')$ by choosing λ_i' as the number of parts of λ that are $\geqslant i$. The partition λ' is called the *conjugate* of λ.

While the formal definition of conjugate is not too revealing, we may better understand the conjugate by using graphical representation. From the definition, we see that the conjugate of the partition $8 + 6 + 6 + 5 + 1$ is $5 + 4 + 4 + 4 + 4 + 3 + 1 + 1$. The graphical representation of $8 + 6 + 6 + 5 + 1$ is

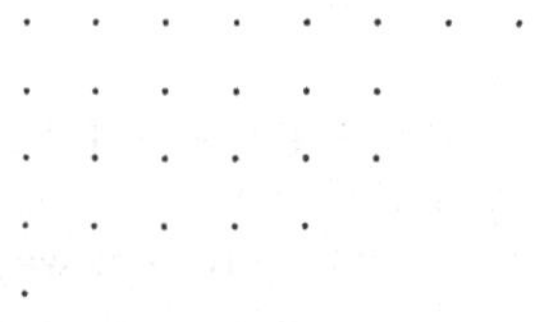

```
. . . . . . . .
. . . . . .
. . . . . .
. . . . .
.
```

and the conjugate of this partition is obtained by counting the dots in successive columns; that is, the graphical representation of the conjugate is obtained by reflecting the graph in the main diagonal. Thus the graph of the conjugate partition is

```
. . . . .
. . . .
. . . .
. . . .
. . . .
. . .
.
.
```

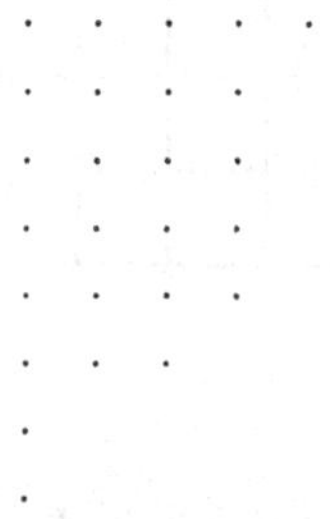

Notice that not only does the graphical representation provide a simple method by which to obtain the conjugate of λ, but it also shows directly that the conjugate partition λ' is a partition of the same integer as λ is; that is, $\Sigma\lambda_i = \Sigma\lambda_i'$. Furthermore, it is clear that conjugation is an involution of the partitions of any integer, in that the conjugate of the conjugate of λ is again λ.

Let us now prove some theorems on partitions, using graphical representation.

THEOREM 1.4. *The number of partitions of n with at most m parts equals the number of partitions of n in which no part exceeds m.*

Proof. We may set up a one-to-one correspondence between the two classes of partitions under consideration by merely mapping each partition onto its conjugate. The mapping is certainly one-to-one, and by considering the graphical representation we see that under conjugation the condition "at most m parts" is transformed into "no part exceeds m" and vice versa. ■

As an example, let us consider the partitions of 6, first into at most three parts and then into parts none of which exceeds 3. We shall list conjugates opposite each other.

$$\begin{array}{ll}
6 & 1+1+1+1+1+1 \\
5+1 & 2+1+1+1+1 \\
4+2 & 2+2+1+1 \\
4+1+1 & 3+1+1+1 \\
3+3 & 2+2+2 \\
3+2+1 & 3+2+1 \\
2+2+2 & 3+3
\end{array}$$

Theorem 1.4 is quite useful and shows how a graphical representation can be used directly to obtain important information. More subtle uses of this technique can be seen in the following two theorems.

THEOREM 1.5. *The number of partitions of $a - c$ into exactly $b - 1$ parts, none exceeding c, equals the number of partitions of $a - b$ into $c - 1$ parts, none exceeding b.*

Proof. Let us consider the graphical representation of a typical partition of the first type mentioned in the theorem. We transform the partition as follows: first we adjoin a new top row of c nodes; then we delete the first column (which now has b nodes); and then we take the conjugate:

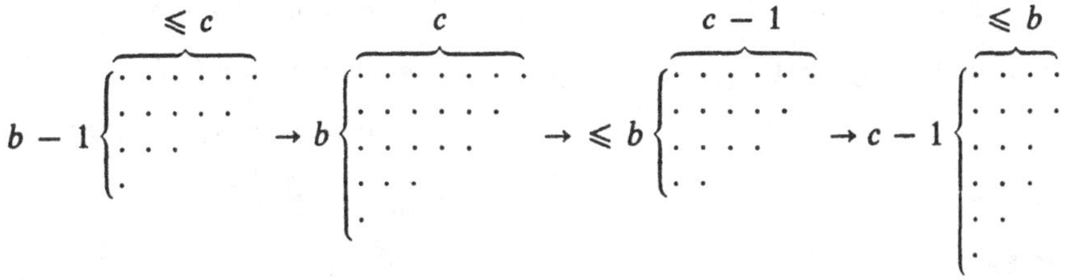

$$b-1\left\{\overbrace{\begin{array}{l}\cdot\ \cdot\ \cdot\ \cdot\ \cdot\ \cdot\\ \cdot\ \cdot\ \cdot\ \cdot\ \cdot\\ \cdot\ \cdot\ \cdot\\ \cdot\end{array}}^{\leqslant c}\right. \rightarrow b\left\{\overbrace{\begin{array}{l}\cdot\ \cdot\ \cdot\ \cdot\ \cdot\ \cdot\ \cdot\\ \cdot\ \cdot\ \cdot\ \cdot\ \cdot\ \cdot\\ \cdot\ \cdot\ \cdot\ \cdot\ \cdot\\ \cdot\ \cdot\ \cdot\\ \cdot\end{array}}^{c}\right. \rightarrow \leqslant b\left\{\overbrace{\begin{array}{l}\cdot\ \cdot\ \cdot\ \cdot\ \cdot\ \cdot\\ \cdot\ \cdot\ \cdot\ \cdot\ \cdot\\ \cdot\ \cdot\ \cdot\ \cdot\\ \cdot\ \cdot\end{array}}^{c-1}\right. \rightarrow c-1\left\{\overbrace{\begin{array}{l}\cdot\ \cdot\ \cdot\ \cdot\\ \cdot\ \cdot\ \cdot\ \cdot\\ \cdot\ \cdot\ \cdot\\ \cdot\ \cdot\ \cdot\\ \cdot\ \cdot\\ \cdot\end{array}}^{\leqslant b}\right.$$

We see immediately that this composite transformation provides a one-to-one correspondence between the two types of partitions considered, and consequently the theorem is established. ∎

As an example, let us consider the case in which $a = 14$, $b = 5$, $c = 4$.

$$\begin{array}{lcl}
4+4+1+1 & \rightarrow & 3+3+3 \\
4+3+2+1 & \rightarrow & 4+3+2 \\
4+2+2+2 & \rightarrow & 5+2+2 \\
3+3+3+1 & \rightarrow & 4+4+1 \\
3+3+2+2 & \rightarrow & 5+3+1
\end{array}$$

We conclude this chapter with one of the truly remarkable achievements of nineteenth-century American mathematics: F. Franklin's proof of Euler's pentagonal number theorem. Franklin's accomplishment was to prove Legendre's combinatorial interpretation of Euler's theorem. We shall actually state the pentagonal number theorem as Corollary 1.7, and we shall show how useful the theorem is computationally in Corollary 1.8.

THEOREM 1.6. *Let $p_e(\mathscr{D}, n)$ (resp. $p_o(\mathscr{D}, n)$) denote the number of partitions of n into an even (resp. odd) number of distinct parts. Then*

$$p_e(\mathscr{D}, n) - p_o(\mathscr{D}, n) = \begin{cases} (-1)^m & \text{if } n = \frac{1}{2}m(3m \pm 1), \\ 0 & \text{otherwise.} \end{cases}$$

Proof. We shall attempt to establish a one-to-one correspondence between the partitions enumerated by $p_e(\mathscr{D}, n)$ and those enumerated by $p_o(\mathscr{D}, n)$. For most integers n our attempt will be successful; however, whenever n is one of the pentagonal numbers $\frac{1}{2}m(3m \pm 1)$, a single exceptional case will arise.

To begin with, we note that each partition $\lambda = (\lambda_1, \ldots, \lambda_r)$ of n has a smallest part $s(\lambda) = \lambda_r$; also, we observe that the largest part λ_1 of $\lambda = (\lambda_1, \lambda_2, \ldots, \lambda_r)$ is the first of a sequence of, say, $\sigma(\lambda)$ consecutive integers that are parts of λ (formally $\sigma(\lambda)$ is the largest j such that $\lambda_j = \lambda_1 - j + 1$). Graphically the parameters $s(\lambda)$ and $\sigma(\lambda)$ are easily described:

$\lambda = (76432)$ $\qquad$ $\lambda = (8765)$

$\sigma(\lambda) = 2$ $\qquad$ $\sigma(\lambda) = 4$

$s(\lambda) = 2$ $\qquad$ $s(\lambda) = 5$

We transform partitions as follows.

Case 1. $s(\lambda) \leqslant \sigma(\lambda)$. In this event, we add one to each of the $s(\lambda)$ largest parts of λ and we delete the smallest part. Thus

$$\lambda = (76432) \rightarrow \lambda' = (8743);$$

that is

Case 2. $s(\lambda) > \sigma(\lambda)$. In this event, we subtract one from each of the $\sigma(\lambda)$ largest parts of λ and insert a new smallest part of size $\sigma(\lambda)$. Thus

$$\lambda = (8743) \to (76432);$$

that is

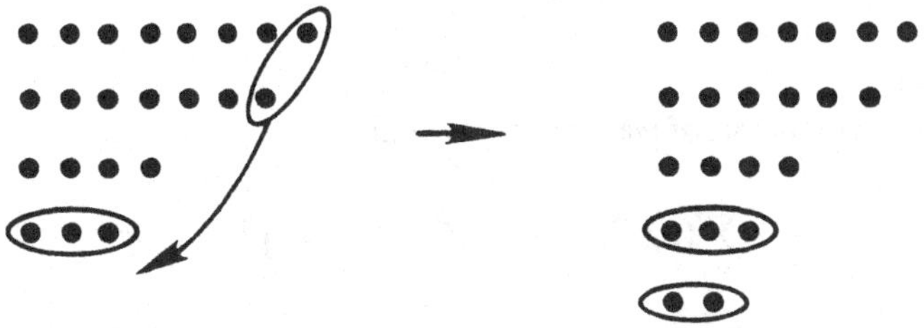

The foregoing procedure in either case changes the parity of the number of parts of the partition, and noting that exactly one case is applicable to any partition λ, we see directly that the mapping establishes a one-to-one correspondence. However, there are certain partitions for which the mapping will not work. The example $\lambda = (8765)$ is a case in point. Case 2 should be applicable to it; however, the image partition is *no longer* one with distinct parts. Indeed, Case 2 breaks down in precisely those cases when the partition has r parts, $\sigma(\lambda) = r$ and $s(\lambda) = r + 1$, in which case the number being partitioned is

$$(r+1) + (r+2) + \cdots + 2r = \tfrac{1}{2}r(3r+1).$$

On the other hand, Case 1 breaks down in precisely those cases when the partition has r parts, $\sigma(\lambda) = r$ and $s(\lambda) = r$, in which case the number being partitioned is

$$r + (r+1) + \cdots + (2r-1) = \tfrac{1}{2}r(3r-1).$$

Consequently, if n is not a pentagonal number, $p_e(\mathscr{D}, n) = p_o(\mathscr{D}, n)$; if $n = \frac{1}{2}r(3r \pm 1)$, $p_e(\mathscr{D}, n) = p_o(\mathscr{D}, n) + (-1)^r$. ■

Corollary 1.7 (Euler's pentagonal number theorem).

$$\prod_{n=1}^{\infty}(1-q^n) = 1 + \sum_{m=1}^{\infty}(-1)^m q^{\frac{1}{2}m(3m-1)}(1+q^m)$$

$$= \sum_{m=-\infty}^{\infty}(-1)^m q^{\frac{1}{2}m(3m-1)}. \qquad (1.3.1)$$

Proof. Clearly

$$\sum_{m=-\infty}^{\infty}(-1)^m q^{\frac{1}{2}m(3m-1)} = 1 + \sum_{m=1}^{\infty}(-1)^m q^{\frac{1}{2}m(3m-1)} + \sum_{m=-1}^{-\infty}(-1)^m q^{\frac{1}{2}m(3m-1)}$$

$$= 1 + \sum_{m=1}^{\infty}(-1)^m q^{\frac{1}{2}m(3m-1)} + \sum_{m=1}^{\infty}(-1)^m q^{\frac{1}{2}m(3m+1)}$$

$$= 1 + \sum_{m=1}^{\infty} (-1)^m q^{\frac{1}{2}m(3m-1)}(1 + q^m)$$

$$= 1 + \sum_{n=1}^{\infty} (p_e(\mathscr{D}, n) - p_o(\mathscr{D}, n))q^n,$$

by Theorem 1.6.

To complete the proof we must show that

$$1 + \sum_{n=1}^{\infty} (p_e(\mathscr{D}, n) - p_o(\mathscr{D}, n))q^n = \prod_{n=1}^{\infty} (1 - q^n).$$

Now

$$\prod_{n=1}^{\infty} (1 - q^n) = \sum_{a_1=0}^{1} \sum_{a_2=0}^{1} \sum_{a_3=0}^{1} \cdots (-1)^{a_1+a_2+a_3+\cdots} q^{a_1 \cdot 1 + a_2 \cdot 2 + a_3 \cdot 3 \ldots},$$

as in the proof of (1.2.4) in Theorem 1.1. Note now that each partition with distinct parts is counted with a weight $(-1)^{a_1+a_2+a_3+\cdots}$, which is $+1$ if the partition has an even number of parts and -1 if the partition has an odd number of parts. Consequently

$$\prod_{n=1}^{\infty} (1 - q^n) = \sum_{a_1=0}^{1} \sum_{a_2=0}^{1} \sum_{a_3=0}^{1} \cdots (-1)^{a_1+a_2+a_3+\cdots} q^{a_1 \cdot 1 + a_2 \cdot 2 + a_3 \cdot 3 + \cdots}$$

$$= 1 + \sum_{n=1} (p_e(\mathscr{D}, n) - p_o(\mathscr{D}, n))q^n,$$

and so we have the desired result. ∎

COROLLARY 1.8 (Euler). *If $n > 0$, then*

$$p(n) - p(n-1) - p(n-2) + p(n-5) + p(n-7)$$
$$+ \cdots + (-1)^m p(n - \tfrac{1}{2}m(3m-1))$$
$$+ (-1)^m p(n - \tfrac{1}{2}m(3m+1)) + \cdots = 0, \tag{1.3.2}$$

where we recall that $p(M) = 0$ for all negative M.

Proof. Let a_n denote the left-hand side of (1.3.2). Then clearly

$$\sum_{n=0}^{\infty} a_n q^n = \sum_{n=0}^{\infty} p(n)q^n \cdot \left[1 + \sum_{m=1}^{\infty} (-1)^m q^{\frac{1}{2}m(3m-1)}(1 + q^m)\right]$$

$$= \prod_{n=1}^{\infty} (1 - q^n)^{-1} \cdot \prod_{n=1}^{\infty} (1 - q^n)$$

$$= 1$$

where the penultimate equation follows immediately by (1.2.3) and Corollary 1.7. Hence, $a_n = 0$ for $n > 0$. ■

Corollary 1.8 provides an extremely efficient algorithm for computing $p(n)$ that we shall discuss further in Chapter 14.

Examples

1. (Subbarao) The number of partitions of n in which each part appears two, three, or five times equals the number of partitions of n into parts congruent to 2, 3, 6, 9, or 10 modulo 12.

2. The number of partitions of n in which only odd parts may be repeated equals the number of partitions of n in which no part appears more than three times.

3. The number of partitions of n in which only parts $\not\equiv 0 \pmod{2^m}$ may be repeated equals the number of partitions of n in which no part appears more than $2^{m+1} - 1$ times.

4. (Ramanujan) The number of partitions of n with unique smallest part and largest part at most twice the smallest part equals the number of partitions of n in which the largest part is odd and the smallest part is larger than half the largest part.

5. Let $P_1(r; n)$ denote the number of partitions of n into parts that are either even and not congruent to $4r - 2 \pmod{4r}$ or odd and congruent to $2r - 1$ or $4r - 1 \pmod{4r}$. Let $P_2(r; n)$ denote the number of partitions of n in which only even parts may be repeated and all odd parts are congruent to $2r - 1$ modulo $2r$. Then $P_1(r; n) = P_2(r; n)$.

Comment on Examples 6–7. P. A. MacMahon introduced what he termed "modular" partitions. Given the positive integers k and n, there exist (by the Euclidean algorithm) $h \geqslant 0$ and $0 < j \leqslant k$ such that

$$n = kh + j.$$

The "modular" partitions are a modification of the Ferrers graph so that n is represented by a row of h k's and one j. Thus the representation of $8 + 8 + 7 + 7 + 6 + 5 + 2$ to the modulus 2 is

$$\begin{array}{llll} 2 & 2 & 2 & 2 \\ 2 & 2 & 2 & 2 \\ 2 & 2 & 2 & 1 \\ 2 & 2 & 2 & 1 \\ 2 & 2 & 2 & \\ 2 & 2 & 1 & \\ 2 & & & \end{array}$$

Note that the ordinary Ferrers graph is just the modular representation with modulus 1.

6. Let $W_1(r, m, n)$ denote the number of partitions of n into m parts, each larger than 1, with exactly r odd parts, each distinct. Let $W_2(r, m, n)$ denote the number of partitions of n with $2m$ as largest part and exactly r

odd parts, each distinct. Then $W_1(r, m, n) = W_2(r, m, n)$ (use modular representations of the partitions in question with modulus 2).

7. Let $P_3(r; n)$ denote the number of partitions $(\lambda_1 \lambda_2 \cdots \lambda_s)$ of n such that if λ_i is odd, then $\lambda_i - \lambda_{i+1} \geqslant 2r - 1$ $(1 \leqslant i \leqslant s, \lambda_{s+1} = 0)$. Then $P_2(r; n) = P_3(r; n)$ (see Exercise 5 for $P_2(r; n)$) (use modular representation of the partitions in question with modulus 2).

8. (Sylvester) The number of partitions of n with distinct odd parts equals the number of partitions of n that are self-conjugate (i.e., identical with their conjugate).

9. (MacMahon) Let $M_1(n)$ denote the number of partitions of n into parts, each larger than 1, such that consecutive integers do not both appear as parts. Let $M_2(n)$ denote the number of partitions of n in which no part appears exactly once. Then $M_1(n) = M_2(n)$.

10. (MacMahon) Let $M_3(N)$ denote the number of partitions of n into parts not congruent to 1 or 5 modulo 6. Then $M_2(n) = M_3(n)$ (see Example 9 for $M_2(n)$).

11. (Euler) The absolute value of excess of the number of partitions n with an odd number of parts over the number of those with an even number of parts equals the number of partitions of n into distinct odd parts.

Notes

Many books on number theory or combinatorics present material of the type chosen for this chapter, for example, Andrews (1971), Comtet (1974), Hardy and Wright (1960), MacMahon (1916), Ostmann (1956), and Riordan (1958). Euler's contributions are primarily found in Euler (1748). Glaisher's theorem (Corollary 1.3) appears in Glaisher (1883). Theorem 1.5 can be found in Sylvester's monumental paper (Sylvester, 1882–1884; see also G. W. Starcher, 1930). Franklin's proof of the pentagonal number theorem appears in Franklin (1881); see also Subbarao (1971a), Andrews (1972). Recurrence relations (related to Corollary 1.8) and recent surveys are reviewed in Sections P56 and P02, respectively, of LeVeque (1974).

Example 1. Subbarao (1971b).
Example 4. Andrews (1967b).
Examples 5, 7. Andrews (1970).
Example 6. Andrews (1974), MacMahon (1923).
Examples 9, 10. Andrews (1967a).
Example 11. Sylvester (1882–1884).

References

Andrews, G. E. (1967a). "A generalization of a partition theorem of MacMahon," *J. Combinatorial Theory* 3, 100–101.

Andrews, G. E. (1967b). "Enumerative proofs of certain q-identities," *Glasgow Math. J.* **8**, 33–40.

Andrews, G. E. (1970). "Note on a partition theorem." *Glasgow Math. J.* **11**, 108–109.

Andrews, G. E. (1971). *Number Theory*. Saunders, Philadelphia.

Andrews, G. E. (1972). "Two theorems of Gauss and allied identities proved arithmetically," *Pacific J. Math.* **41**, 563–578.

Andrews, G. E. (1974). "Applications of basic hypergeometric functions," *S.I.A.M. Rev.* **16**, 441–484.

Bender, E. A. (1974). "A lifting theorem for formal power series," *Proc. Amer. Math. Soc.* **42**, 16–22.

Comtet, L. (1974). *Advanced Combinatorics*. D. Reidel, Dordrecht.

Euler, L. (1748). *Introductio in analysin infinitorum*, Chapter 16. Marcum-Michaelem Bousquet, Lausannae.

Franklin, F. (1881). "Sur le développement du produit infini $(1-x)(1-x^2)(1-x^3)\ldots$," *Comptes Rendus* **82**, 448–450.

Glaisher, J. W. L. (1883). "A theorem in partitions," *Messenger of Math.* **12**, 158–170.

Hardy, G. H., and Wright, E. M. (1960). *An Introduction to the Theory of Numbers*, 4th ed. Oxford Univ. Press, London and New York.

Hickerson, D. R. (1973). "Identities relating the number of partitions into an even and odd number of parts," *J. Combinatorial Theory* **A15**, 351–353.

Hickerson, D. R. (1974). "A partition identity of the Euler type," *Amer. Math. Monthly* **81**, 627–629.

Knutson, D. (1972). "A lemma on partitions," *Amer. Math. Monthly* **79**, 1111–1112.

LeVeque, W. J. (1974). *Reviews in Number Theory*, Vol. 4. Amer. Math. Soc., Providence, R.I.

MacMahon, P. A. (1916). *Combinatory Analysis*, Vol. 2. Cambridge Univ. Press, London and New York (reprinted by Chelsea, New York, 1960).

MacMahon, P. A. (1923). "The theory of modular partitions," *Proc. Cambridge Phil. Soc.* **21**, 197–204.

Moore, E. (1974a). "Generalized Euler-type partition identities," *J. Combinatorial Theory* **A17**, 78–83.

Moore, E. (1974b). "Partitions with parts appearing a specified number of times," *Proc. Amer. Math. Soc.* **46**, 205–210.

Netto, E. (1927). *Lehrbuch der Kombinatorik*, 2nd ed. Teubner, Leipzig (reprinted by Chelsea, New York, 1958).

Ostmann, H. H. (1956). *Additive Zahlentheorie*, 2 vols. Springer, Berlin.

Rademacher, H. (1973). *Topics in Analytic Number Theory*. Springer, Berlin.

Riordan, J. (1958). *An Introduction to Combinatorial Analysis*. Wiley, New York.

Starcher, G. W. (1930). "On identities arising from solutions of q-difference equations and some interpretations in number theory," *Amer. J. Math.* **53**, 801–816.

Subbarao, M. V. (1971a). "Combinatorial proofs of some identities," *Proc. Washington State Univ. Conf. Number Theory* pp. 80–91.

Subbarao, M. V. (1971b). "On a partition theorem of MacMahon-Andrews," *Proc. Amer. Math. Soc.* **27**, 449–450.

Sylvester, J. J. (1882–1884). "A constructive theory of partitions, arranged in three acts, an interact and an exodion," *Amer. J. Math.* **5**, 251–330; **6**, 334–336 (or pp. 1–83 of *The Collected Mathematical Papers of J. J. Sylvester*, Vol. 4. Cambridge Univ. Press, London and New York, 1912; reprinted by Chelsea, New York, 1974).

CHAPTER 2

Infinite Series Generating Functions

2.1 Introduction

In this chapter we shall consider the fruitful interaction between partition theorems and certain elementary methods for manipulating infinite series and products. The convergence questions that arose in Chapter 1 arise again; as before, however, they pose few if any difficulties in our work.

We note that now we shall consider $p(S, m, n)$, the number of partitions of n that lie in a set of partitions S and that have m parts. This immediately leads to the two-variable generating function

$$f_s(z; q) = \sum_{m=0}^{\infty} \sum_{n=0}^{\infty} p(S, m, n) z^m q^n$$
$$= \sum_{\lambda \in S} z^{\#(\lambda)} q^{\sigma(\lambda)}$$

where if $\lambda = (\lambda_1, \ldots, \lambda_r)$, $\#(\lambda) = r$, $\sigma(\lambda) = \lambda_1 + \cdots + \lambda_r$. The double series above converges absolutely for $|z| < |q|^{-1} > 1$. This may be easily seen by proving (in the manner in which Theorem 1.1 is proved) that

$$\sum_{m=0}^{\infty} \sum_{n=0}^{\infty} p(\mathscr{S}, m, n) z^m q^n = \prod_{n=1}^{\infty} (1 - zq^n)^{-1}$$

where $\mathscr{S}$ is the set of all partitions.

In fact, the proof of Theorem 1.1 with slight modification may be employed to prove that

$$\sum_{m=0}^{\infty} \sum_{n=0}^{\infty} p(\text{“}H\text{”}, m, n) z^m q^n = \prod_{n \in H} (1 - zq^n)^{-1}, \tag{2.1.1}$$

$$\sum_{m=0}^{\infty} \sum_{n=0}^{\infty} p(\text{“}H\text{”}(\leqslant d), m, n) z^m q^n = \prod_{n \in H} (1 - z^{d+1} q^{(d+1)n})(1 - zq^n)^{-1}. \tag{2.1.2}$$

ENCYCLOPEDIA OF MATHEMATICS and Its Applications, Gian-Carlo Rota (ed.). 2, George E. Andrews, The Theory of Partitions

Other parameters besides the number of parts $\#(\lambda)$ of a partition λ will interest us from time to time; so we shall have occasion to consider other types of partition generating functions of several variables.

The preceding comments suggest the interest of considering infinite series and products in two (or more) variables. In the following section, we shall develop an elementary technique for proving many series and product identities. We shall obtain several classical theorems of great importance, such as Jacobi's triple product identity. As will become clear in Section 2.3, the results of Section 2.2 are quite useful in treating partition identities. It is possible, however, to skip Section 2.2 and read Section 2.3, referring back only for the statements of theorems. For the reader who needs series transformations to attack a partition problem, the first six examples at the end of this chapter form a good test of the techniques used in Section 2.2.

2.2 Elementary Series-Product Identities

We begin with a theorem due to Cauchy; as we shall see, this result provides the tool for doing everything else in this section.

THEOREM 2.1. *If* $|q| < 1$, $|t| < 1$, *then*

$$1 + \sum_{n=1}^{\infty} \frac{(1-a)(1-aq)\cdots(1-aq^{n-1})t^n}{(1-q)(1-q^2)\cdots(1-q^n)} = \prod_{n=0}^{\infty} \frac{(1-atq^n)}{(1-tq^n)}. \qquad (2.2.1)$$

Remark. We shall try always to state our theorems with as little notational disguise as possible. However, for the proofs, it seems only sensible to use the following standard abbreviations

$$(a)_n = (a;q)_n = (1-a)(1-aq)\cdots(1-aq^{n-1}),$$

$$(a)_\infty = (a;q)_\infty = \lim_{n\to\infty}(a;q)_n,$$

$$(a)_0 = 1.$$

We may define $(a)_n$ for all real numbers n by

$$(a)_n = (a)_\infty/(aq^n)_\infty.$$

The series in (2.2.1) is an example of a basic hypergeometric series. The study of basic series (or q-series, or Eulerian series) is an extensive branch of analysis and we shall only touch upon it in this book. Most of the theorems of this section may be viewed as elementary results in the theory of basic hypergeometric series. Theorem 2.1 has become known as the "q-analog of the binomial series," for if we write $a = q^\alpha$ where α is a nonnegative integer, then (2.2.1) formally tends to

$$1 + \sum_{n=1}^{\infty} \binom{\alpha + n - 1}{n} t^n = (1 - t)^{-\alpha}, \quad \text{as} \quad q \to 1^-.$$

Proof. Let us consider

$$F(t) = \prod_{n=0}^{\infty} \frac{(1 - atq^n)}{(1 - tq^n)} = \sum_{n=0}^{\infty} A_n t^n \tag{2.2.2}$$

where $A_n = A_n(a, q)$. We note that the A_n exist since the infinite product is uniformly convergent for fixed a and q inside $|t| \leqslant 1 - \varepsilon$, and therefore it defines a function of t analytic inside $|t| < 1$.

Now

$$\begin{aligned}(1 - t)F(t) &= (1 - at) \prod_{n=1}^{\infty} \frac{(1 - atq^n)}{(1 - tq^n)} \\ &= (1 - at) \prod_{n=0}^{\infty} \frac{(1 - atq^{n+1})}{(1 - tq^{n+1})} = (1 - at)F(tq).\end{aligned} \tag{2.2.3}$$

Clearly $A_0 = F(0) = 1$, and by comparing coefficients of t^n in the extremes of (2.2.3) we see that

$$A_n - A_{n-1} = q^n A_n - aq^{n-1} A_{n-1},$$

or

$$A_n = \frac{(1 - aq^{n-1})}{(1 - q^n)} A_{n-1}. \tag{2.2.4}$$

Iterating (2.2.4) we see that

$$\begin{aligned}A_n &= \frac{(1 - aq^{n-1})(1 - aq^{n-2}) \cdots (1 - a)A_0}{(1 - q^n)(1 - q^{n-1}) \cdots (1 - q)} \\ &= \frac{(a)_n}{(q)_n}.\end{aligned}$$

Substituting this value for A_n into (2.2.2), we obtain the theorem. ■

Euler found the two following special cases of Theorem 2.1. Each of these identities is directly related to partitions in Example 17 at the end of this chapter.

COROLLARY 2.2 (Euler). *For* $|t| < 1$, $|q| < 1$,

$$1 + \sum_{n=1}^{\infty} \frac{t^n}{(1-q)(1-q^2)\cdots(1-q^n)} = \prod_{n=0}^{\infty} (1 - tq^n)^{-1}, \tag{2.2.5}$$

$$1 + \sum_{n=1}^{\infty} \frac{t^n q^{\frac{1}{2}n(n-1)}}{(1-q)(1-q^2)\cdots(1-q^n)} = \prod_{n=0}^{\infty} (1 + tq^n). \tag{2.2.6}$$

Proof. Equation (2.2.5) follows immediately by setting $a = 0$ in (2.2.1). To obtain (2.2.6) we replace a by a/b and t by bz in (2.2.1); hence for $|bz| < 1$

$$1 + \sum_{n=1}^{\infty} \frac{(b-a)(b-aq)\cdots(b-aq^{n-1})z^n}{(1-q)(1-q^2)\cdots(1-q^n)} = \prod_{n=0}^{\infty} \frac{(1-azq^n)}{(1-bzq^n)}. \tag{2.2.7}$$

Now set $b = 0$, $a = -1$ in (2.2.7) and we derive (2.2.6) directly. ∎

The following result is Heine's fundamental transformation, and it is instrumental in proving each of the succeeding four corollaries.

COROLLARY 2.3 (Heine). *For* $|q| < 1$, $|t| < 1$, $|b| < 1$

$$1 + \sum_{n=1}^{\infty} \frac{(1-a)(1-aq)\cdots(1-aq^{n-1})(1-b)(1-bq)\cdots(1-bq^{n-1})t^n}{(1-q)(1-q^2)\cdots(1-q^n)(1-c)(1-cq)\cdots(1-cq^{n-1})}$$

$$= \prod_{m=0}^{\infty} \frac{(1-bq^m)(1-atq^m)}{(1-cq^m)(1-tq^m)}$$

$$\times \left\{ 1 + \sum_{n=1}^{\infty} \frac{(1-c/b)(1-cq/b)\cdots(1-cq^{n-1}/b)\times}{(1-q)(1-q^2)\cdots(1-q^n)\times} \right.$$

$$\left. \frac{\times(1-t)(1-tq)\cdots(1-tq^{n-1})b^n}{\times(1-at)(1-atq)\cdots(1-atq^{n-1})} \right\}.$$

Proof.

$$\sum_{n=0}^{\infty} \frac{(a)_n(b)_n t^n}{(q)_n(c)_n} = \frac{(b)_\infty}{(c)_\infty} \sum_{n=0}^{\infty} \frac{(a)_n t^n}{(q)_n} \cdot \frac{(cq^n)_\infty}{(bq^n)_\infty}$$

$$= \frac{(b)_\infty}{(c)_\infty} \sum_{n=0}^{\infty} \sum_{m=0}^{\infty} \frac{(a)_n t^n}{(q)_n} \cdot \frac{(c/b)_m b^m q^{nm}}{(q)_m}$$

$$= \frac{(b)_\infty}{(c)_\infty} \sum_{m=0}^{\infty} \frac{(c/b)_m b^m}{(q)_m} \frac{(atq^m)_\infty}{(tq^m)_\infty}$$

$$= \frac{(b)_\infty (at)_\infty}{(c)_\infty (t)_\infty} \sum_{m=0}^{\infty} \frac{(c/b)_m (t)_m b^m}{(q)_m (at)_m}.$$ ∎

COROLLARY 2.4 (Heine). *If* $|c| < |ab|$, $|q| < 1$,

$$1 + \sum_{n=1}^{\infty} \frac{(1-a)(1-aq)\cdots(1-aq^{n-1})(1-b)(1-bq)\cdots(1-bq^{n-1})(c/ab)^n}{(1-q)(1-q^2)\cdots(1-q^n)(1-c)(1-cq)\cdots(1-cq^{n-1})}$$

$$= \prod_{m=0}^{\infty} \frac{(1-cq^m/a)(1-cq^m/b)}{(1-cq^m)(1-cq^m/ab)}.$$

Proof. By Corollary 2.3,

$$\sum_{n=0}^{\infty} \frac{(a)_n(b)_n(c/ab)^n}{(q)_n(c)_n} = \frac{(b)_\infty(c/b)_\infty}{(c)_\infty(c/ab)_\infty} \sum_{n=0}^{\infty} \frac{(c/ab)_n b^n}{(q)_n}$$

$$= \frac{(b)_\infty(c/b)_\infty}{(c)_\infty(c/ab)_\infty} \cdot \frac{(c/a)_\infty}{(b)_\infty} = \frac{(c/a)_\infty(c/b)_\infty}{(c)_\infty(c/ab)_\infty}. \qquad \blacksquare$$

COROLLARY 2.5 (Bailey). *If* $|q| < \min(1, |b|)$, *then*

$$1 + \sum_{n=1}^{\infty} \frac{(1-a)(1-aq)\cdots(1-aq^{n-1})(1-b)(1-bq)\cdots(1-bq^{n-1})(-q/b)^n}{(1-q)(1-q^2)\cdots(1-q^n)(1-aq/b)(1-aq^2/b)\cdots(1-aq^n/b)}$$

$$= \prod_{m=0}^{\infty} \frac{(1-aq^{2m+1})(1+q^{m+1})(1-aq^{2m+2}/b^2)}{(1-aq^{m+1}/b)(1+q^{m+1}/b)}.$$

Proof. By Corollary 2.3 (interchanging a and b)

$$\sum_{n=0}^{\infty} \frac{(b)_n(a)_n(-q/b)^n}{(q)_n(aq/b)_n} = \frac{(a)_\infty(-q)_\infty}{(aq/b)_\infty(-q/b)_\infty} \sum_{m=0}^{\infty} \frac{(q/b)_m(-q/b)_m a^m}{(q)_m(-q)_m}$$

$$= \frac{(a)_\infty(-q)_\infty}{(aq/b)_\infty(-q/b)_\infty} \sum_{m=0}^{\infty} \frac{(q^2/b^2;q^2)_m a^m}{(q^2;q^2)_m}$$

$$= \frac{(a)_\infty(-q)_\infty(aq^2/b^2;q^2)_\infty}{(aq/b)_\infty(-q/b)_\infty(a;q^2)_\infty}$$

$$= \frac{(aq;q^2)_\infty(-q)_\infty(aq^2/b^2;q^2)_\infty}{(aq/b)_\infty(-q/b)_\infty}. \qquad \blacksquare$$

We remark that Corollary 2.4 is commonly referred to as the "q-analog of Gauss's theorem," while Corollary 2.5 is the "q-analog of Kummer's theorem."

COROLLARY 2.6. *If* $|q| < 1$,

$$1 + \sum_{n=1}^{\infty} \frac{q^{n^2-n}z^n}{(1-q)(1-q^2)\cdots(1-q^n)(1-z)(1-zq)\cdots(1-zq^{n-1})}$$

$$= \prod_{m=0}^{\infty} (1-zq^m)^{-1}, \tag{2.2.8}$$

$$1+\sum_{n=1}^{\infty}\frac{q^{n^2}}{(1-q)^2(1-q^2)^2\cdots(1-q^n)^2}=\prod_{m=1}^{\infty}(1-q^m)^{-1}. \tag{2.2.9}$$

Remark. Equation (2.2.8) is due to Cauchy, and Eq. (2.2.9) is due to Euler.

Proof. First we note that (2.2.9) is obtained from (2.2.8) by setting $z = q$. In Corollary 2.4, set $a = \alpha^{-1}$, $b = \beta^{-1}$, $c = z$. Hence

$$1+\sum_{n=1}^{\infty}\frac{(\alpha-1)(\alpha-q)\cdots(\alpha-q^{n-1})(\beta-1)(\beta-q)\cdots(\beta-q^{n-1})z^n}{(q)_n(z)_n}$$

$$=\frac{(z\alpha)_\infty(z\beta)_\infty}{(z)_\infty(z\alpha\beta)_\infty},$$

and if we set $\alpha = \beta = 0$ in this identity, we obtain (2.2.8). ∎

COROLLARY 2.7. *If* $|q| < 1$,

$$1+\sum_{n=1}^{\infty}\frac{(1-a)(1-aq)\cdots(1-aq^{n-1})q^{n(n+1)/2}}{(1-q)(1-q^2)\cdots(1-q^n)}=\prod_{m=1}^{\infty}(1-aq^{2m-1})(1+q^m).$$

Proof. Set $b = \beta^{-1}$ in Corollary 2.5. Hence

$$\sum_{n=0}^{\infty}\frac{(a)_n(\beta-1)(\beta-q)\cdots(\beta-q^{n-1})(-q)^n}{(q)_n(aq\beta)_n}=\frac{(aq;q^2)_\infty(-q)_\infty(aq^2\beta^2;q^2)_\infty}{(aq\beta)_\infty(-q\beta)_\infty}.$$

Now set $\beta = 0$ in this identity and we obtain the desired result. ∎

The next result, Jacobi's triple product identity, may be viewed as a corollary of Corollary 2.2; however, it is so important that we label it a theorem.

THEOREM 2.8. *For* $z \neq 0$, $|q| < 1$,

$$\sum_{n=-\infty}^{\infty}z^nq^{n^2}=\prod_{n=0}^{\infty}(1-q^{2n+2})(1+zq^{2n+1})(1+z^{-1}q^{2n+1}). \tag{2.2.10}$$

Proof. For $|z| > |q|$, $|q| < 1$,

$$\prod_{n=0}^{\infty}(1+zq^{2n+1})=\sum_{m=0}^{\infty}\frac{z^mq^{m^2}}{(q^2;q^2)_m} \qquad \text{(by (2.2.6))}$$

$$=\frac{1}{(q^2;q^2)_\infty}\sum_{m=0}^{\infty}z^mq^{m^2}(q^{2m+2};q^2)_\infty$$

$$=\frac{1}{(q^2;q^2)_\infty}\sum_{m=-\infty}^{\infty}z^mq^{m^2}(q^{2m+2};q^2)_\infty$$

(since $(q^{2m+2};q^2)_\infty$ vanishes for m negative)

$$= \frac{1}{(q^2;q^2)_\infty} \sum_{m=-\infty}^{\infty} z^m q^{m^2} \sum_{r=0}^{\infty} \frac{(-1)^r q^{r^2+2mr+r}}{(q^2;q^2)_r}$$

$$= \frac{1}{(q^2;q^2)_\infty} \sum_{r=0}^{\infty} \frac{(-1)^r z^{-r} q^r}{(q^2;q^2)_r} \sum_{m=-\infty}^{\infty} q^{(m+r)^2} z^{m+r}$$

$$= \frac{1}{(q^2;q^2)_\infty} \sum_{r=0}^{\infty} \frac{(-q/z)^r}{(q^2;q^2)_r} \sum_{m=-\infty}^{\infty} q^{m^2} z^m$$

$$= \frac{1}{(q^2;q^2)_\infty (-q/z;q^2)_\infty} \sum_{m=-\infty}^{\infty} q^{m^2} z^m.$$

This is the desired result. Note that absolute convergence pertains everywhere only so long as $|z| > |q|$, $|q| < 1$. However, the full result of the theorem follows either by invoking analytic continuation, or by observing that the entire argument may be carried out again with z^{-1} replacing z. ∎

COROLLARY 2.9. *For* $|q| < 1$,

$$\sum_{n=-\infty}^{\infty} (-1)^n q^{(2k+1)n(n+1)/2-in}$$

$$= \sum_{n=0}^{\infty} (-1)^n q^{(2k+1)n(n+1)/2-in}(1 - q^{(2n+1)i})$$

$$= \prod_{n=0}^{\infty} (1 - q^{(2k+1)(n+1)})(1 - q^{(2k+1)n+i})(1 - q^{(2k+1)(n+1)-i}). \tag{2.2.11}$$

Proof. Replace q by $q^{k+\frac{1}{2}}$ and then set $z = -q^{k+\frac{1}{2}-i}$ in (2.2.10). This substitution immediately yields the equality of the extremes in (2.2.11). Now

$$\sum_{n=0}^{\infty} (-1)^n q^{(2k+1)n(n+1)/2-in}(1 - q^{(2n+1)i})$$

$$= \sum_{n=0}^{\infty} (-1)^n q^{(2k+1)n(n+1)/2-in} + \sum_{n=1}^{\infty} (-1)^n q^{(2k+1)n(n-1)/2+in}$$

$$= \sum_{n=0}^{\infty} (-1)^n q^{(2k+1)n(n+1)/2-in} + \sum_{n=-1}^{-\infty} (-1)^n q^{(2k+1)n(n+1)/2-in}$$

$$= \sum_{n=-\infty}^{\infty} (-1)^n q^{(2k+1)n(n+1)/2-in}. \qquad ∎$$

We remark that Corollary 2.9 reduces to Corollary 1.7 when $k = i = 1$ once we observe that

$$\prod_{n=0}^{\infty}(1-q^{3n+3})(1-q^{3n+1})(1-q^{3n+2}) = \prod_{n=1}^{\infty}(1-q^n).$$

COROLLARY 2.10 (Gauss)

$$\sum_{n=-\infty}^{\infty}(-1)^n q^{n^2} = \prod_{m=1}^{\infty}\frac{(1-q^m)}{(1+q^m)}, \tag{2.2.12}$$

$$\sum_{n=0}^{\infty} q^{n(n+1)/2} = \prod_{m=1}^{\infty}\frac{(1-q^{2m})}{(1-q^{2m-1})}. \tag{2.2.13}$$

Proof. By (2.2.10) with $z = -1$,

$$\sum_{n=-\infty}^{\infty}(-1)^n q^{n^2} = (q^2;q^2)_\infty(q;q^2)_\infty(q;q^2)_\infty$$

$$= (q)_\infty(q;q^2)_\infty = (q)_\infty/(-q)_\infty$$

where the final equation follows from (1.2.5). Next

$$\sum_{n=0}^{\infty} q^{n(n+1)/2} = \tfrac{1}{2}\sum_{n=-\infty}^{\infty} q^{n(n+1)/2}$$

$$= \tfrac{1}{2}(q)_\infty(-q)_\infty(-1)_\infty$$

$$= (q)_\infty(-q)_\infty(-q)_\infty = (q^2;q^2)_\infty(-q)_\infty = (q^2;q^2)_\infty/(q;q^2)_\infty$$

where again the final equation follows from (1.2.5). ■

So far this section seems filled with much mathematics and little commentary. It has been the hope that the power of Theorem 2.1 and simple series manipulation would be fully appreciated if numerous significant results followed in rapid-fire order. The reader will have a chance to practice the techniques involved in the many examples at the end of this chapter.

2.3 Applications to Partitions

We shall prove four theorems on partitions utilizing either the actual results or the methods of Section 2.2. We conclude with an examination of "Durfee squares," which allows us to obtain (2.2.9) from purely combinatorial considerations. We begin with an interpretation of Corollary 2.9.

THEOREM 2.11. *Let* $\mathscr{D}(k,i)$ *denote all those partitions with distinct parts in which each part is congruent to* 0, $\pm i$ (modulo $2k+1$). *Let* $p_e(\mathscr{D}(k,i),n)$ (resp. $p_o(\mathscr{D}(k,i),n)$) *denote the number of partitions of* n *taken from* $\mathscr{D}(k,i)$ *with an even* (resp. *odd*) *number of parts. Then*

$$p_e(\mathscr{D}(k, i), n) - p_o(\mathscr{D}(k, i), n)$$
$$= \begin{cases} (-1)^m & \text{if } n = (k \mp \tfrac{1}{2})m(m+1) \pm im, \\ 0 & \text{otherwise.} \end{cases}$$

Proof. The proof is exactly like the proof of Corollary 1.7 done backwards. Here we read the partition-theoretic result by comparing coefficients on both sides of (2.2.11). ■

THEOREM 2.12 (Sylvester). *Let $A_k(n)$ denote the number of partitions of n into odd parts (repetitions allowed) such that exactly k different parts occur. Let $B_k(n)$ denote the number of partitions $\lambda = (\lambda_1, \ldots, \lambda_r)$ of n such that the sequence $(\lambda_1, \ldots, \lambda_r)$ is composed of exactly k noncontiguous sequences of one or more consecutive integers. Then $A_k(n) = B_k(n)$ for all k and n.*

Remarks. First of all, we note that this theorem of Sylvester is a refinement of Euler's theorem (Corollary 1.2), in that $p(\mathscr{O}, n) = \sum_{k=0}^{\infty} A_k(n)$ and $p(\mathscr{D}, n) = \sum_{k=0}^{\infty} B_k(n)$. Sylvester obtained a purely graphical proof of this result. Our proof (due to Ramamani and Venkatachaliengar) illustrates nicely the way in which combinatorics and formal series analysis can interact to facilitate a proof.

The exact meaning of $A_k(n)$ and $B_k(n)$ should be crystal clear from the following example: $A_3(14) = 7$, since the relevant partitions are $(1^2 39)$, $(1^2 57)$, $(13^2 7)$, $(1^4 37)$, (135^2), $(1^3 3^2 5)$, $(1^6 35)$; and $B_3(14) = 7$, since the relevant partitions are (1, 3, 10), (149), (248), (158), (257), (1247), (1346).

Proof. We note that the method of proof of (1.2.3) in Theorem 1.1 may be extended to show that

$$\begin{aligned}
\sum_{k=0}^{\infty} \sum_{n=0}^{\infty} A_k(n) a^k q^n &= \prod_{j=1}^{\infty} (1 + aq^{2j-1} + aq^{2(2j-1)} + aq^{3(2j-1)} + \cdots) \\
&= \prod_{j=1}^{\infty} \left(1 + \frac{aq^{2j-1}}{1 - q^{2j-1}}\right) \\
&= \prod_{j=1}^{\infty} \frac{(1 - (1-a)q^{2j-1})}{(1 - q^{2j-1})} \\
&= \frac{((1-a)q; q^2)_\infty}{(q; q^2)_\infty} = ((1-a)q; q^2)_\infty (-q)_\infty \qquad (2.3.1)
\end{aligned}$$

where the final equation follows from Euler's theorem (1.2.5).

It is not a simple matter to find directly the two-variable generating functions for $B_k(n)$. However, if we examine the conjugates of the partitions enumerated by $B_k(n)$, things simplify greatly. Actually the conjugate partitions λ' fall into two distinct classes: (1) if 1 is a part of λ, then λ' is a partition of n in which the largest part is unique, every positive integer smaller than the largest part appears, and exactly $k-1$ of these parts appear more than once; (2) if 1 is not a part of λ, then λ' is a partition of n in which the largest part is repeated, every positive integer smaller than the largest part appears, and exactly k parts appear more than once.

Consequently

$$\begin{aligned}
\sum_{k=0}^{\infty}\sum_{n=0}^{\infty} B_k(n)a^k q^n &= 1 + \sum_{N=1}^{\infty} aq^N \prod_{j=1}^{N-1}(q^j + aq^{2j} + aq^{3j} + \cdots) \\
&\quad + \sum_{N=1}^{\infty}(aq^{2N} + aq^{3N} + \cdots)\prod_{j=1}^{N-1}(q^j + aq^{2j} + aq^{3j} + \cdots) \\
&= 1 + \sum_{N=1}^{\infty} \frac{aq^N}{1-q^N} \prod_{j=1}^{N-1} q^j\left(1 + \frac{aq^j}{1-q^j}\right) \\
&= 1 + \sum_{N=1}^{\infty} \frac{(1-(1-a))q^{N(N+1)/2}((1-a)q)_{N-1}}{(q)_N} \\
&= \sum_{N=0}^{\infty} \frac{((1-a))_N q^{N(N+1)/2}}{(q)_N} \\
&= ((1-a)q; q^2)_\infty(-q)_\infty \qquad \text{(by Corollary 2.7)} \\
&= \sum_{k=0}^{\infty}\sum_{n=0}^{\infty} A_k(n)a^k q^n \qquad \text{(by (2.3.1))}.
\end{aligned}$$

Comparing coefficients of $a^k q^n$ in the extremes of the foregoing string of equations, we deduce that $A_k(n) = B_k(n)$ for all k and n. ■

Now we shall consider an analytic approach to Theorem 1.5; generally, analytic proofs require less ingenuity and provide less insight than combinatorial proofs.

Analytic Proof of Theorem 1.5. If we let $p_{b,c}(a-c)$ denote the number of the first type of partition described in Theorem 1.5, then the theorem asserts that $p_{b,c}(a-c) = p_{c,b}(a-b)$. Now

$$\begin{aligned}
\sum_{a=0}^{\infty}\sum_{b=0}^{a}\sum_{c=0}^{a} p_{b,c}(a)x^b y^c q^a &= 1 + \sum_{n=1}^{\infty} xy^n \prod_{j=1}^{n}(1 + xq^j + x^2q^{2j} + x^3q^{3j} + \cdots) \\
&= 1 + \sum_{n=1}^{\infty} \frac{xy^n}{(xq)_n}.
\end{aligned}$$

Therefore if

$$f(x, y) = \sum_{a=0}^{\infty} \sum_{b=0}^{\infty} \sum_{c=0}^{\infty} p_{b,c}(a - c)x^b y^c q^a$$

$$= 1 + \sum_{n=1}^{\infty} \frac{xy^n q^n}{(xq)_n},$$

then we need only show $f(x, y) = f(y, x)$ to obtain the desired result:

$$f(x, y) - 1 + x = \sum_{n=0}^{\infty} \frac{xy^n q^n}{(xq)_n}$$

$$= x \sum_{n=0}^{\infty} \frac{(0)_n (q)_n (yq)^n}{(q)_n (xq)_n}$$

$$= x \frac{(q)_\infty}{(xq)_\infty (yq)_\infty} \sum_{n=0}^{\infty} \frac{(x)_n (yq)_n q^n}{(q)_n} \qquad \text{(by Corollary 2.3)}$$

$$= x \frac{(q)_\infty}{(xq)_\infty (yq)_\infty} \sum_{n=0}^{\infty} \frac{(yq)_n (x)_n q^n}{(q)_n (0)_n}$$

$$= x \frac{(q)_\infty}{(xq)_\infty (yq)_\infty} \frac{(x)_\infty (yq^2)_\infty}{(q)_\infty} \sum_{n=0}^{\infty} \frac{(q)_n x^n}{(q)_n (yq^2)_n}$$

(by Corollary 2.3)

$$= (1 - x) \sum_{n=0}^{\infty} \frac{x^{n+1}}{(yq)_{n+1}}.$$

Therefore

$$f(x, y) = (1 - x) \sum_{n=0}^{\infty} \frac{x^n}{(yq)_n} = \sum_{n=0}^{\infty} \frac{x^n}{(yq)_n} - \sum_{n=0}^{\infty} \frac{x^{n+1}}{(yq)_n}$$

$$= 1 + \sum_{n=1}^{\infty} \frac{x^n}{(yq)_n} - \sum_{n=1}^{\infty} \frac{x^n(1 - yq^n)}{(yq)_n}$$

$$= 1 + \sum_{n=1}^{\infty} \frac{yx^n q^n}{(yq)_n} = f(y, x). \qquad \blacksquare$$

As a third example we consider a second refinement of Euler's theorem (Corollary 1.2) due to N. J. Fine.

THEOREM 2.13. *The number of partitions of n into distinct parts with largest part k equals the number of partitions of n into odd parts such that 2k + 1 equals the largest part plus twice the number of parts.*

Proof. Combinatorial reasoning of the type used in the previous three proofs shows that Fine's theorem is equivalent to the following assertion:

$$\sum_{j=0}^{\infty} \frac{t^{j+1}q^{2j+1}}{(tq;q^2)_{j+1}} = tq\sum_{j=0}^{\infty}(-q)_j q^j t^j.$$

Now

$$tq\sum_{j=0}^{\infty}(-q)_j q^j t^j$$

$$= tq\sum_{j=0}^{\infty}\frac{(q^2;q^2)_j q^j t^j}{(q)_j}$$

$$= tq(q^2;q^2)_\infty \sum_{j=0}^{\infty}\frac{t^j q^j}{(q)_j}\frac{1}{(q^{2j+2};q^2)_\infty}$$

$$= tq(q^2;q^2)_\infty \sum_{j=0}^{\infty}\frac{t^j q^j}{(q)_j}\sum_{m=0}^{\infty}\frac{q^{2jm+2m}}{(q^2;q^2)_m} \qquad \text{(by (2.2.5))}$$

$$= tq(q^2;q^2)_\infty \sum_{m=0}^{\infty}\frac{q^{2m}}{(q^2;q^2)_m}\sum_{j=0}^{\infty}\frac{t^j q^{j(2m+1)}}{(q)_j}$$

$$= tq(q^2;q^2)_\infty \sum_{m=0}^{\infty}\frac{q^{2m}}{(q^2;q^2)_m(tq^{2m+1})_\infty} \qquad \text{(by (2.2.5))}$$

$$= \frac{tq(q^2;q^2)_\infty}{(tq)_\infty}\sum_{m=0}^{\infty}\frac{(tq;q^2)_m(tq^2;q^2)_m q^{2m}}{(q^2;q^2)_m}$$

$$= \frac{tq(q^2;q^2)_\infty}{(tq)_\infty}\frac{(tq^2;q^2)_\infty(tq^3;q^2)_\infty}{(q^2;q^2)_\infty}\sum_{m=0}^{\infty}\frac{(q^2;q^2)_m t^m q^{2m}}{(q^2;q^2)_m(tq^3;q^2)_m}$$

(by Corollary 2.3)

$$= \sum_{m=0}^{\infty}\frac{t^{m+1}q^{2m+1}}{(tq;q^2)_{m+1}}. \qquad \blacksquare$$

We conclude this chapter with a look at a property of partitions called the Durfee square, and we utilize it to provide a new proof of (2.2.9).

Combinatorial Proof of Eq. (2.2.9). To each partition $\lambda = (\lambda_1, \lambda_2, \ldots, \lambda_r)$ we may assign a parameter $d(\lambda)$ as the number of λ_j such that $\lambda_j \geqslant j$. Let us see what $d(\lambda)$ measures in the graphical representation of λ. Suppose $\lambda = (124^2 57^2)$, then $d(\lambda) = 4$, the graphical representation is

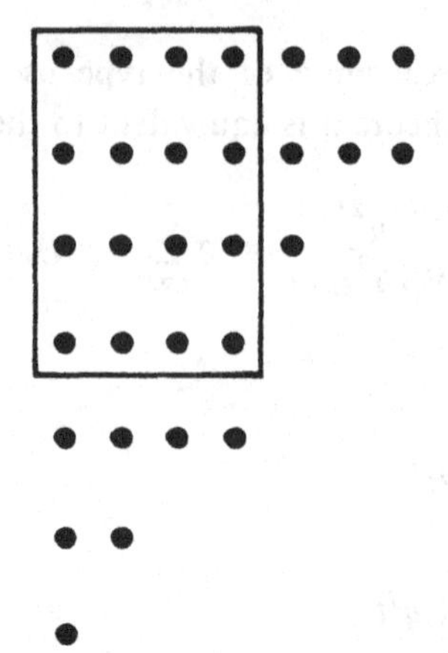

and as we have indicated, $d(\lambda)$ measures the largest square of nodes contained in the partition λ. This square is called the *Durfee square* (after W. P. Durfee), and $d(\lambda)$ is called the *side of the Durfee square.* It is clear from the graphical representation that if $\lambda \vdash n$ and $d(\lambda) = s$, then the partition λ may be uniquely written as $(s^s) + \lambda' + \lambda''$ where (s^s) counts the nodes in the Durfee square, λ' represents the nodes below the Durfee square (and is therefore some partition all of whose parts are $\leqslant s$), and λ'' represents the conjugate of the nodes to the right of the Durfee square and so λ'' is also some partition whose parts are $\leqslant s$. In the foregoing example the partition $\lambda = (124^2 57^2)$ is uniquely written as $(4^4) + (124) + (2^2 3)$. Since partitions with parts $\leqslant s$ are generated by

$$\frac{1}{(1-q)(1-q^2)\cdots(1-q^s)} = \frac{1}{(q)_s}$$

(Theorem 1.1), we see that the set of all partitions with Durfee square of side s is generated by

$$q^{s^2}\frac{1}{(q)_s}\cdot\frac{1}{(q)_s} = \frac{q^{s^2}}{(q)_s^{\,2}}.$$

Therefore

$$\frac{1}{(q)_\infty} = \sum_{n=0}^{\infty} p(n)q^n = \sum_{s=0}^{\infty}\frac{q^{s^2}}{(q)_s^{\,2}}. \qquad \blacksquare$$

Examples

1. The following generalization of Corollary 2.3 is valid for each integer $k \geqslant 1$,

$$\sum_{n=0}^{\infty}\frac{(a;q^k)_n(b)_{kn}t^n}{(q^k;q^k)_n(c)_{kn}} = \frac{(b)_\infty(at;q^k)_\infty}{(c)_\infty(t;q^k)_\infty}\sum_{n=0}^{\infty}\frac{(c/b)_n(t;q^k)_n b^n}{(q)_n(at;q^k)_n}.$$

2. $$\sum_{n=0}^{\infty} \frac{(b)_{2n}t^{2n}}{(q^2;q^2)_n} = \frac{(-tb)_\infty}{(-t)_\infty} \sum_{m=0}^{\infty} \frac{(b)_m t^m}{(q)_m(-tb)_m}.$$

3. $$\sum_{n=0}^{\infty} \frac{(t;q^2)_n b^n}{(q)_n} = \frac{(btq;q^2)_\infty}{(bq;q^2)_\infty} \sum_{m=0}^{\infty} \frac{(t;q^2)_m b^m}{(q^2;q^2)_m(btq;q^2)_m}.$$

4. $$\sum_{n=0}^{\infty} \frac{(a)_n(b;q^2)_n t^n}{(q)_n(atb;q^2)_n} = \frac{(at;q^2)_\infty(bt;q^2)_\infty}{(t;q^2)_\infty(abt;q^2)_\infty} \sum_{m=0}^{\infty} \frac{(a;q^2)_m(b;q^2)_m(tq)^m}{(q^2;q^2)_m(bt;q^2)_m}.$$

5. $$(1+a)\sum_{n=0}^{\infty} \frac{(-a)^n}{(-q)_n} = \sum_{n=0}^{\infty} \frac{q^{n^2}a^n}{(-q)_n(-aq)_n}.$$

6. $$\sum_{m=0}^{\infty} \frac{q^{m^2}x^m}{(y;q^2)_{m+1}} = \sum_{m=0}^{\infty} (-xq/y;q^2)_m y^m.$$

7. The identity used in the proof of Theorem 2.13 is a special case of the one in Example 4.

8. $$\sum_{m=0}^{\infty} (-a^{-1})_m(aq)^m - \sum_{m=0}^{\infty} (a^{-1})_m(-aq)^m = 2\sum_{m=0}^{\infty} \frac{(aq)^{2m+1}}{(q;q^2)_{m+1}}.$$

9. The identity in Example 8 may be used to prove another theorem of N. J. Fine:

$$U_{2r+1}(n) = V_{2r+1}(n) + V_{2r}(n)$$

where $U_{2r+1}(n)$ is the number of partitions of n with odd parts and largest part $2r+1$, and where $V_s(n)$ is the number of partitions of n into distinct parts such that the largest part minus the number of parts equals s.

10. It is possible to prove that

$$1 - x\sum_{n=1}^{\infty} (xq)_{n-1}(xq)^n = \sum_{n=0}^{\infty} \frac{(-1)^n x^{2n} q^{n(n+1)/2}}{(xq)_n}$$

$$= 1 + \sum_{n=1}^{\infty} (-1)^n (q^{\frac{1}{2}n(3n-1)}x^{3n-1} + q^{\frac{1}{2}n(3n+1)}x^{3n})$$

by showing that each of the series $s(x, q)$ in question satisfies

$$s(x,q) = 1 - x^2q - q^2x^3s(xq,q), \qquad s(0,q) = 1.$$

11. Euler's pentagonal number theorem is a corollary of Example 10.

12. Example 10 can be proved in a purely combinatorial manner. The approach must consider not only what number n is being partitioned, but also m, the sum of the largest part and the number of parts. The invariance of m in the transformations of the proof of Theorem 1.6 suffices to treat the second equation in Example 10.

In his last letter to G. H. Hardy, Ramanujan considered several families of functions which he called "mock-theta functions." Four of the fifth-order mock-theta functions are

$$f_0(q) = \sum_{n=0}^{\infty} \frac{q^{n^2}}{(-q)_n}, \qquad \phi_0(q) = \sum_{n=0}^{\infty} q^{n^2}(-q; q^2)_n,$$

$$\psi_0(q) = \sum_{n=0}^{\infty} q^{(n+1)(n+2)/2}(-q)_n, \quad F_0(q) = \sum_{n=0}^{\infty} \frac{q^{2n^2}}{(q; q^2)_n}.$$

Exercises 13 and 14 provide several relations among these functions:

13. $$\psi_0(q) = q\left(\sum_{n=0}^{\infty} q^{n(n+1)}\right) \sum_{n=0}^{\infty} \frac{q^{4n^2+4n}}{(q^4; q^4)_n} + F_0(q^2) - 1.$$

14. $$f_0(q) - \left(\sum_{n=-\infty}^{\infty} (-1)^n q^{n^2}\right) \sum_{n=0}^{\infty} \frac{q^{n^2}}{(q)_n} = 2\phi_0(-q^2).$$

15. $$\sum_{n=0}^{\infty} \frac{z^n}{(-z)_{n+1}} = \sum_{n=0}^{\infty} (q; q^2)_n z^{2n}.$$

16. If in Theorem 2.1 we replace q by q^2 and then replace t by tq^2 and a by $-aq$, then the resulting identity may be deduced from Example 6 of Chapter 1.

17. If t is replaced by tq in Corollary 2.2, the resulting two equations have simple combinatorial proofs. Equation (2.2.5) may be proved by the consideration of partitions of n with m parts; Eq. (2.2.6) may be proved by the consideration of partitions of n with m distinct parts.

18. Let us say a partition λ is *Sylvestered* if the smallest part to appear an even number of times (0 is an even number) is even. For each set of positive integers T, define $S(T; m, n)$ as the number of Sylvestered partitions of n into m parts such that no elements of T are repeated. Then

$$\sum_{m;n \geq 0} S(T; m, n)x^m q^n = \frac{\prod_{k \in T}(1 - x^2 q^{2k})}{\prod_{j=1}^{\infty}(1 - xq^j)} \sum_{n=1}^{\infty} \frac{x^{2n-1}q^{n(2n-1)}}{(-xq)_{2n}}.$$

19. The number of non-Sylvestered partitions of n with an odd number of parts equals the number of non-Sylvestered partitions of n with an even number of parts.

20. $$\sum_{m,n \geq 0} (-1)^m S(T; m, n) q^n = \prod_{k \in T} (1 - q^{2k}) \left(\frac{1}{(-q)_\infty} - 1\right).$$

Notes

The origin of Theorem 2.1 is uncertain; it has been attributed to Euler (Hardy, 1940, p. 223). Heine (1847) was the first to study systematically this type of series, and so Theorem 2.1 is often attributed to him; however, Cauchy proved this result in 1843 (see Cauchy, 1893, p. 45). Corollary 2.2 is truly due to Euler (1748). Corollaries 2.3 and 2.4 are due to Heine (1847). W. N. Bailey (1941) and J. A. Daum (1942) independently discovered Corollary 2.5. Corollary 2.6 is due to Cauchy (1893). Corollary 2.7 appears to be due to V. A. Lebesgue (1840). Corollary 2.8 is the celebrated Jacobi triple product identity (Jacobi, 1829); the proof given here was found independently by

Andrews (1965) and P. K. Menon (1965). Equations (2.2.12) and (2.2.13) are generally attributed either to Jacobi (1829) or to Gauss (1866). Theorem 2.12 is an extension of Euler's theorem (Corollary 1.2) due to Sylvester (1884–1886); the proof we give is in essence due to V. Ramamani and K. Venkatachaliengar (1972; see also Andrews, 1974). Theorem 2.13 is an unpublished result due to N. J. Fine (1954; see also Andrews, 1966b). The use of Durfee squares has been studied extensively in Andrews (1971b, 1972a).

The relationship between series and product identities has not been extensively treated in books. Certain elementary problems are broached in Andrews (1971a; see also 1972b). The advanced theory of these series may be found in Andrews (1974), Bailey (1935), Hahn (1949, 1950), and Slater (1966). Much of the recent literature is reviewed in Section P60 of LeVeque (1974).

Examples 1–3. Andrews (1966c).

Example 4. Andrews (1966b).

Example 5. Andrews (1966a, c).

Examples 6, 7. Andrews (1966b).

Example 8. This identity is originally due to N. J. Fine (1954), and it appears in Andrews (1966b).

Example 9. Andrews (1966b), Fine (1948).

Example 10. Rogers (1916), Andrews (1966b), Fine (1954).

Example 12. Subbarao (1971), Andrews (1972a).

Examples 13, 14. Watson (1936), Andrews (1966a).

Example 15. Carlitz (1967).

Example 16. Andrews (1974).

Example 18. Andrews (1970). In this paper, Sylvestered partitions are called "flushed" in accordance with Sylvester's original work. Since Sylvester was one of the all-time champion coiners of new mathematical terms, it seems appropriate to call flushed partitions "Sylvestered."

Examples 19, 20. Andrews (1970).

References

Andrews, G. E. (1965). "A simple proof of Jacobi's triple product identity," *Proc. Amer. Math. Soc.* **16**, 333–334.

Andrews, G. E. (1966a). "On basic hypergeometric series, mock theta functions, and partitions, I," *Quart. J. Math. Oxford Ser.* **17**, 64–80.

Andrews, G. E. (1966b). "On basic hypergeometric series, mock theta functions, and partitions, II," *Quart. J. Math. Oxford Ser.* **17**, 132–143.

Andrews, G. E. (1966c). "q-Identities of Auluck, Carlitz and Rogers," *Duke Math. J.* **33**, 575–582.

Andrews, G. E. (1966d). "On generalizations of Euler's partition theorem," *Michigan Math. J.* **13**, 491–498.

Andrews, G. E. (1970). "On a partition problem of J. J. Sylvester," *J. London Math. Soc.* (2) **2**, 571–576.

Andrews, G. E. (1971a). *Number Theory.* Saunders, Philadelphia.

Andrews, G. E. (1971b). "Generalization of the Durfee square," *J. London Math. Soc.* (2) **3**, 563–570.

Andrews, G. E. (1972a). "Two theorems of Gauss and allied identities proved arithmetically," *Pacific J. Math.* **41**, 563–578.

Andrews, G. E. (1972b). "Partition identities," *Advances in Math.* **9**, 10–51.

Andrews, G. E. (1974). "Applications of basic hypergeometric functions," *S.I.A.M. Rev.* **16**, 441–484.

Bailey, W. N. (1935). *Generalized Hypergeometric Series*. Cambridge Univ. Press, London and New York (reprinted by Hafner, New York, 1964).

Bailey, W. N. (1941). "A note on certain q-identities," *Quart. J. Math. Oxford Ser.* **12**, 173–175.

Carlitz, L. (1967). "Problem 66-9: A pair of identities," *S.I.A.M. Rev.* **9**, 254–256.

Cauchy, A. (1893). *Oeuvres*, Ser. 1, Vol. 8. Gauthier-Villars, Paris.

Daum, J. A. (1942). "The basic analog of Kummer's theorem," *Bull. Amer. Math. Soc.* **48**, 711–713.

Euler, L. (1748). *Introductio in analysin infinitorum*, Chapter 16. Marcum-Michaelem Bousquet, Lausannae.

Fine, N. J. (1948). "Some new results on partitions," *Proc. Nat. Acad. Sci. USA* **34**, 616–618.

Fine, N. J. (1954). "Some Basic Hypergeometric Series and Applications," unpublished monograph.

Gauss, C. F. (1866). *Werke*, Vol. 3. Konigliche Gesellschaft der Wissenschaften, Göttingen.

Hahn, W. (1949). "Beiträge zur Theorie der Heineschen Reihen, Die 24 Integrale der hypergeometrischen q-Differenzengleichung, Das q-Analogon der Laplace Transformation," *Math. Nachr.* **2**, 340–379.

Hahn, W. (1950). "Uber die hoheren Heineschen Reihen und eine einheitliche Theorie der sogenannten speziellen Funktionen," *Math. Nachr.* **3**, 257–294.

Hardy, G. H. (1940). *Ramanujan*. Cambridge Univ. Press, London and New York (reprinted by Chelsea, New York).

Hardy, G. H., and Wright, E. M. (1960). *An Introduction to the Theory of Numbers*, 4th ed. Oxford Univ. Press, London and New York.

Heine, E. (1847). "Untersuchungen über die Reihe ...," *J. Reine Angew. Math.* **34**, 285–328.

Jacobi, C. G. J. (1829). *Fundamenta nova theoriae functionum ellipticarum*. Regiomonti, fratrum Bornträger (reprinted in *Gesammelte Werke*, Vol. 1, pp. 49–239; Reimer, Berlin, 1881).

Lebesgue, V. A. (1840). "Sommation de quelques séries," *J. Math. Pures Appl.* **5**, 42–71.

LeVeque, W. J. (1974). *Reviews in Number Theory*, Vol. 4. Amer. Math. Soc., Providence, R.I.

Menon, P. K. (1965). "On Ramanujan's continued fraction and related identities," *J. London Math. Soc.* **40**, 49–54.

Ramamani, V., and Venkatachaliengar, K. (1972). "On a partition theorem of Sylvester," *Michigan Math. J.* **19**, 137–140.

Rogers, L. J. (1916). "On two theorems of combinatory analysis and some allied identities," *Proc. London Math. Soc.* (2) **16**, 315–336.

Slater, L. J. (1966). *Generalized Hypergeometric Functions*. Cambridge Univ. Press, London and New York.

Subbarao, M. V. (1971). "Combinatorial proofs of some identities," *Proc. Washington State Univ. Conf. Number Theory*, pp. 80–91.

Sylvester, J. J. (1884–1886). "A constructive theory of partitions arranged in three acts, an interact, and an exodion," *Amer. J. Math.* **5**, 251–330; **6**, 334–336 (or pp. 1–83 of the *Collected Papers of J. J. Sylvester*, Vol. 4, Cambridge Univ. Press, London and New York, 1912; reprinted by Chelsea, New York, 1974).

Watson, G. N. (1936). "The mock theta functions (2)," *Proc. London Math. Soc.* (2), **42**, 274–304.

CHAPTER 3

Restricted Partitions and Permutations

3.1 Introduction

In Chapters 1 and 2, we studied various partition problems utilizing infinite series and infinite products. In the application of partitions (such as in statistics), we are often interested in restricted partitions, that is, partitions in which the largest part is, say, $\leqslant N$ and the number of parts is $\leqslant M$. In this chapter, we shall examine restricted partitions. This undertaking will naturally lead us to the Gaussian polynomials and from there we shall be led to questions concerning permutations.

3.2 The Generating Function for Restricted Partitions

Let $p(N, M, n)$ denote the number of partitions of n into at most M parts, each $\leqslant N$. Clearly

$$p(N, M, n) = 0 \qquad \text{if} \quad n > MN,$$

$$p(N, M, NM) = 1.$$

Therefore the generating function

$$G(N, M; q) = \sum_{n \geqslant 0} p(N, M, n) q^n$$

is a polynomial in q of degree NM.

THEOREM 3.1. *For* $M, N \geqslant 0$

$$G(N, M; q) = \frac{(1 - q^{N+M})(1 - q^{N+M-1}) \cdots (1 - q^{M+1})}{(1 - q^N)(1 - q^{N-1}) \cdots (1 - q)}$$

$$= \frac{(q)_{N+M}}{(q)_N (q)_M}. \tag{3.2.1}$$

ENCYCLOPEDIA OF MATHEMATICS and Its Applications, Gian-Carlo Rota (ed.). 2, George E. Andrews, The Theory of Partitions

Proof. Let $g(N, M; q)$ denote the right-hand side of (3.2.1); then

$$g(N, 0; q) = g(0, M; q) = 1 \tag{3.2.2}$$

and

$$\begin{aligned} g(N, M; q) - g(N, M-1; q) &= \frac{(q)_{N+M-1}}{(q)_N(q)_M}[(1-q^{N+M}) - (1-q^M)] \\ &= \frac{(q)_{N+M-1}}{(q)_N(q)_M} q^M(1-q^N) = q^M \frac{(q)_{N+M-1}}{(q)_{N-1}(q)_M} \\ &= q^M g(N-1, M; q). \end{aligned} \tag{3.2.3}$$

Note that (3.2.2) and (3.2.3) uniquely define $g(N, M; q)$ for all nonnegative integers M and N (a fact easily proved by a double mathematical induction on N and M).

On the other hand,

$$p(N, 0, n) = p(0, M, n) = \begin{cases} 1 & \text{if } N = M = n = 0, \\ 0 & \text{otherwise,} \end{cases} \tag{3.2.4}$$

since the empty partition of 0 is the only partition in which no part is positive and is also the only partition in which the number of parts is nonpositive.

Equation (3.2.4) means that

$$G(N, 0; q) = G(0, M; q) = 1. \tag{3.2.5}$$

Furthermore, $p(N, M, n) - p(N, M-1, n)$ enumerates the number of partitions of n into *exactly* M parts, each $\leqslant N$. We transform each of these partitions by deleting every 1 that is a part, and subtracting 1 from each part larger than 1. The resulting partitions of $n - M$ have at most M parts and each part is $\leqslant N - 1$. Since the foregoing transformation is clearly reversible, it establishes a bijection between the partitions enumerated by $p(N, M, n) - p(N, M-1, n)$ and those enumerated by $p(N-1, M, n-M)$. Therefore

$$p(N, M, n) - p(N, M-1, n) = p(N-1, M, n-M), \tag{3.2.6}$$

and translating (3.2.6) into a generating function identity, we obtain

$$G(N, M; q) - G(N, M-1; q) = q^M G(N-1, M; q). \tag{3.2.7}$$

Thus since $g(N, M; q)$ and $G(N, M; q)$ satisfy the same initial conditions ((3.2.2) and (3.2.5) resp.) and the same defining recurrence ((3.2.3) and (3.2.7) resp.), they must be identical. Therefore

$$G(N, M; q) = g(N, M; q) = \frac{(q)_{N+M}}{(q)_N(q)_M}. \qquad \blacksquare$$

3.3 Properties of Gaussian Polynomials

The polynomials $G(N, M; q)$ appearing in Theorem 3.1 were first studied by Gauss and have come to be known as Gaussian polynomials. In this section we shall derive a number of useful formulas for these polynomials.

DEFINITION 3.1. The *Gaussian polynomial* $\left[\begin{smallmatrix} n \\ m \end{smallmatrix}\right]$ is defined by

$$\begin{bmatrix} n \\ m \end{bmatrix} = \begin{cases} (q)_n(q)_m^{-1}(q)_{n-m}^{-1} & \text{if } 0 \leqslant m \leqslant n, \\ 0 & \text{otherwise.} \end{cases}$$

Note that

$$\begin{bmatrix} N \\ M \end{bmatrix} = G(N - M, M; q)$$

by Theorem 3.1. We shall, however, not make use of this fact in this section.

THEOREM 3.2. *Let $0 \leqslant m \leqslant n$ be integers. The Gaussian polynomial $\left[\begin{smallmatrix} n \\ m \end{smallmatrix}\right]$ is a polynomial of degree $m(n - m)$ in q that satisfies the following relations.*

$$\begin{bmatrix} n \\ 0 \end{bmatrix} = \begin{bmatrix} n \\ n \end{bmatrix} = 1; \tag{3.3.1}$$

$$\begin{bmatrix} n \\ m \end{bmatrix} = \begin{bmatrix} n \\ n - m \end{bmatrix}; \tag{3.3.2}$$

$$\begin{bmatrix} n \\ m \end{bmatrix} = \begin{bmatrix} n - 1 \\ m \end{bmatrix} + q^{n-m} \begin{bmatrix} n - 1 \\ m - 1 \end{bmatrix}; \tag{3.3.3}$$

$$\begin{bmatrix} n \\ m \end{bmatrix} = \begin{bmatrix} n - 1 \\ m - 1 \end{bmatrix} + q^{m} \begin{bmatrix} n - 1 \\ m \end{bmatrix}; \tag{3.3.4}$$

$$\lim_{q \to 1} \begin{bmatrix} n \\ m \end{bmatrix} = \frac{n!}{m!(n - m)!} = \binom{n}{m}. \tag{3.3.5}$$

Proof. Equations (3.3.1) and (3.3.2) are obvious from Definition 3.1. As for (3.3.3), we see that

$$\begin{aligned} \begin{bmatrix} n \\ m \end{bmatrix} - \begin{bmatrix} n - 1 \\ m \end{bmatrix} &= \frac{(q)_{n-1}}{(q)_m(q)_{n-m}} [(1 - q^n) - (1 - q^{n-m})] \\ &= \frac{(q)_{n-1}q^{n-m}(1 - q^m)}{(q)_m(q)_{n-m}} \\ &= \frac{q^{n-m}(q)_{n-1}}{(q)_{m-1}(q)_{n-m}} = q^{n-m} \begin{bmatrix} n - 1 \\ m - 1 \end{bmatrix}. \end{aligned}$$

Equation (3.3.4) follows by replacing m by $n - m$ in (3.3.3) and applying (3.3.2).

The fact that $\begin{bmatrix} n \\ m \end{bmatrix}$ is a polynomial in q of degree $m(n - m)$ follows by induction on n using (3.3.1) and (3.3.4).

Finally

$$\lim_{q\to 1} \begin{bmatrix} n \\ m \end{bmatrix} = \lim_{q\to 1} \frac{(1 - q^n)}{(1 - q^m)} \frac{(1 - q^{n-1})}{(1 - q^{m-1})} \cdots \frac{(1 - q^{n-m+1})}{(1 - q)}$$

$$= \frac{n}{m} \frac{n-1}{m-1} \cdots \frac{n-m+1}{1}$$

$$= \frac{n!}{m!(n-m)!} = \binom{n}{m}. \quad ■$$

Many times the Gaussian polynomials arise due to their relationship with certain finite products.

THEOREM 3.3

$$(z)_N = \sum_{j=0}^{N} \begin{bmatrix} N \\ j \end{bmatrix} (-1)^j z^j q^{j(j-1)/2}; \tag{3.3.6}$$

$$(z)_N^{-1} = \sum_{j=0}^{\infty} \begin{bmatrix} N + j - 1 \\ j \end{bmatrix} z^j. \tag{3.3.7}$$

Proof. By Theorem 2.1,

$$(z)_N = \frac{(z)_\infty}{(zq^N)_\infty} = \sum_{j=0}^{\infty} \frac{(q^{-N})_j z^j q^{Nj}}{(q)_j}$$

$$= \sum_{j=0}^{\infty} \frac{(1 - q^{-N})(1 - q^{-N+1}) \cdots (1 - q^{-N+j-1}) z^j q^{Nj}}{(q)_j}$$

$$= \sum_{j=0}^{N} \frac{(-1)^j q^{-Nj + j(j-1)/2}(1 - q^N)(1 - q^{N-1}) \cdots (1 - q^{N-j+1}) z^j q^{Nj}}{(q)_j}$$

$$= \sum_{j=0}^{N} \frac{(q)_N}{(q)_j (q)_{N-j}} (-1)^j z^j q^{j(j-1)/2} = \sum_{j=0}^{N} \begin{bmatrix} N \\ j \end{bmatrix} (-1)^j z^j q^{j(j-1)/2},$$

which is (3.3.6).

Again by Theorem 2.1,

$$(z)_N^{-1} = \frac{(zq^N)_\infty}{(z)_\infty} = \sum_{j=0}^{\infty} \frac{(q^N)_j z^j}{(q)_j}$$

$$= \sum_{j=0}^{\infty} \frac{(q)_{N+j-1}}{(q)_j (q)_{N-1}} z^j = \sum_{j=0}^{\infty} \begin{bmatrix} N + j - 1 \\ j \end{bmatrix} z^j. \quad ■$$

There are many other formulas related to the Gaussian polynomials. We shall conclude this section with a few of the most useful ones.

THEOREM 3.4

$$\sum_{j=0}^{m}(-1)^j\begin{bmatrix}m\\j\end{bmatrix}=\begin{cases}(q;q^2)_n & \textit{if} \quad m=2n,\\ 0 & \textit{if} \quad m \textit{ is odd};\end{cases} \tag{3.3.8}$$

$$\begin{bmatrix}n+m+1\\m+1\end{bmatrix}=\sum_{j=0}^{n}q^j\begin{bmatrix}m+j\\m\end{bmatrix} \quad \textit{for} \quad m,n\geqslant 0; \tag{3.3.9}$$

$$\sum_{k=0}^{h}\begin{bmatrix}n\\k\end{bmatrix}\begin{bmatrix}m\\h-k\end{bmatrix}q^{(n-k)(h-k)}=\begin{bmatrix}m+n\\h\end{bmatrix}; \tag{3.3.10}$$

$$\sum_{r\geqslant 0}\begin{bmatrix}M-m\\r\end{bmatrix}\begin{bmatrix}N+m\\m+r\end{bmatrix}\begin{bmatrix}m+n+r\\M+N\end{bmatrix}q^{(N-r)(M-r-m)}=\begin{bmatrix}m+n\\M\end{bmatrix}\begin{bmatrix}n\\N\end{bmatrix}. \tag{3.3.11}$$

Proof. We begin with (3.3.8), an identity important in the evaluation of Gaussian sums. Let $f(m)$ denote the left-hand side of (3.3.8); then

$$\begin{aligned}\sum_{m=0}^{\infty}\frac{f(m)z^m}{(q)_m}&=\sum_{m=0}^{\infty}\sum_{j=0}^{m}(-1)^j\frac{z^m}{(q)_j(q)_{m-j}}\\&=\sum_{j=0}^{\infty}\sum_{m=j}^{\infty}\frac{(-1)^jz^m}{(q)_j(q)_{m-j}}\\&=\sum_{j=0}^{\infty}\sum_{m=0}^{\infty}\frac{(-1)^jz^{m+j}}{(q)_j(q)_m}\\&=\sum_{j=0}^{\infty}\frac{(-1)^jz^j}{(q)_j}\sum_{m=0}^{\infty}\frac{z^m}{(q)_m}\\&=(-z)_\infty^{-1}(z)_\infty^{-1} \qquad \text{(by (2.2.5))}\\&=(z^2;q^2)_\infty^{-1}\\&=\sum_{n=0}^{\infty}\frac{z^{2n}}{(q^2;q^2)_n} \qquad \text{(by (2.2.5)).}\end{aligned}$$

Equation (3.3.8) now follows by comparing coefficients of z^m in the extremes of the foregoing string of equations.

We may prove (3.3.9) by induction on n. If $n=0$, the equation reduces to $1=1$. Assuming the result true for a specific n, we see that by (3.3.4)

$$\begin{bmatrix} n+m+2 \\ m+1 \end{bmatrix} = \begin{bmatrix} n+m+1 \\ m+1 \end{bmatrix} + q^{n+1} \begin{bmatrix} n+m+1 \\ m \end{bmatrix}$$
$$= \sum_{j=0}^{n} q^j \begin{bmatrix} m+j \\ m \end{bmatrix} + q^{n+1} \begin{bmatrix} m+n+1 \\ m \end{bmatrix}$$
$$= \sum_{j=0}^{n+1} q^j \begin{bmatrix} m+j \\ m \end{bmatrix}.$$

To obtain (3.3.10) (the q-analog of the Chu–Vandermonde summation) we compare coefficients of z^h on both sides of the identity

$$\sum_{h=0}^{m+n} \begin{bmatrix} m+n \\ h \end{bmatrix} z^h(-1)^h q^{h(h-1)/2}$$
$$= (z)_{m+n} = (z)_n (zq^n)_m$$
$$= \sum_{k=0}^{n} \begin{bmatrix} n \\ k \end{bmatrix} z^k(-1)^k q^{k(k-1)/2} \sum_{i=0}^{m} \begin{bmatrix} m \\ i \end{bmatrix} (-1)^i z^i q^{ni+i(i-1)/2}$$
$$= \sum_{h \geq 0} z^h(-1)^h q^{h(h-1)/2} \sum_{k \geq 0} \begin{bmatrix} n \\ k \end{bmatrix} \begin{bmatrix} m \\ h-k \end{bmatrix} q^{(n-k)(h-k)}.$$

Instead of proving (3.3.11), we prove the more general

$$\sum_{n \geq 0} \frac{(a)_n (b)_n (q^{-N})_n q^n}{(q)_n (c)_n (abq^{1-N}c^{-1})_n} = \frac{(c/a)_N (c/b)_N}{(c)_N (c/ab)_N}. \tag{3.3.12}$$

Equation (3.3.12) was first proved by F. H. Jackson and is called the q-analog of Saalschutz's theorem; Eq. (3.3.12) reduces to (3.3.11) if we make the substitutions $a = q^{-M+m}$, $b = q^{m+n+1}$, $c = q^{m+1}$, and then simplify.

To obtain (3.3.12), we must utilize an identity that is easily deduced from Corollary 2.3.

$$\sum_{n=0}^{\infty} \frac{(a)_n (b)_n t^n}{(q)_n (c)_n} = \frac{(b)_\infty (at)_\infty}{(c)(t)_\infty} \sum_{n=0}^{\infty} \frac{(c/b)_n (t)_n b^n}{(q)_n (at)_n} \quad \text{(by Corollary 2.3)}$$
$$= \frac{(b)_\infty (at)_\infty}{(c)_\infty (t)_\infty} \sum_{n=0}^{\infty} \frac{(t)_n (c/b)_n b^n}{(q)_n (at)_n}$$
$$= \frac{(b)_\infty (at)_\infty}{(c)_\infty (t)_\infty} \cdot \frac{(c/b)_\infty (bt)_\infty}{(at)_\infty (b)_\infty} \sum_{n=0}^{\infty} \frac{(abt/c)_n (b)_n (c/b)^n}{(q)_n (bt)_n}$$
$$\text{(by Corollary 2.3)}$$
$$= \frac{(c/b)_\infty (bt)_\infty}{(c)_\infty (t)_\infty} \sum_{n=0}^{\infty} \frac{(b)_n (abt/c)_n (c/b)^n}{(q)_n (bt)_n}$$

$$= \frac{(c/b)_\infty (bt)_\infty}{(c)_\infty (t)_\infty} \cdot \frac{(abt/c)_\infty (c)_\infty}{(bt)_\infty (c/b)_\infty} \sum_{n=0}^{\infty} \frac{(c/a)_n (c/b)_n (abt/c)^n}{(q)_n (c)_n}$$

$$= \frac{(abt/c)_\infty}{(t)_\infty} \sum_{n=0}^{\infty} \frac{(c/a)_n (c/b)_n (abt/c)^n}{(q)_n (c)_n}. \tag{3.3.13}$$

Let us multiply the extremes of this string of equations by $(t)_\infty/(abt/c)_\infty$ and then compare coefficients of t^N on each side. Thus

$$\sum_{n=0}^{\infty} \frac{(a)_n (b)_n (c/ab)_{N-n}}{(q)_n (c)_n (q)_{N-n}} \left(\frac{ab}{c}\right)^{N-n} = \frac{(c/a)_N (c/b)_N a^N b^N c^{-N}}{(q)_N (c)_N},$$

Multiplying both sides of (3.3.14) by $(q)_N a^{-N} b^{-N} c^N/(c/ab)_N$ and simplifying, we obtain (3.3.12). ■

3.4 Permutations and Gaussian Multinomial Coefficients

The Gaussian polynomial is often called the "q-binomial coefficient" due to (3.3.5). In this section we shall use Gaussian polynomials to study certain types of permutation problems, and we shall be naturally led to the treatment of q-multinomial coefficients:

DEFINITION 3.2. For $m_1, \ldots, m_r \geqslant 0$, we define the *Gaussian multinomial coefficient* (or *q-multinomial coefficient*) by

$$\begin{bmatrix} m_1 + m_2 + \cdots + m_r \\ m_1, m_2, \ldots, m_r \end{bmatrix} = \frac{(q)_{m_1+m_2+\ldots+m_r}}{(q)_{m_1}(q)_{m_2}\cdots(q)_{m_r}}. \tag{3.4.1}$$

Note that

$$\begin{bmatrix} n \\ m \end{bmatrix} = \begin{bmatrix} n \\ m, n-m \end{bmatrix}.$$

Now when $r = 2$, Theorem 3.1 tells us that

$$\begin{bmatrix} m_1 + m_2 \\ m_1, m_2 \end{bmatrix}$$

is the generating function for $p(m_1, m_2, n)$, the number of partitions of n with at most m_2 parts, each $\leqslant m_1$.

We shall now consider a different type of mathematical object with the same enumerative function $p(m_1, m_2, n)$.

DEFINITION 3.3. A *multiset* is a set with possibly repeated elements.

To be quite correct we might define a multiset as an ordered pair (M, f) where M is a set and f is a function from M to the nonnegative integers; for each $m \in M$, $f(m)$ would be called the *multiplicity* of m. When M is a finite set, say $\{m_1, m_2, \ldots, m_r\}$, we shall write

$$(M, f) = \{m_1^{f(m_1)} m_2^{f(m_2)} \cdots m_r^{f(m_r)}\}.$$

Let us begin by considering permutations of multisets (a permutation of (M, f) is a word in which each letter belongs to M and for each $m \in M$ the total number of appearances of m in the word is $f(m)$). Thus 3 2 1 2 2 3 2 1 1 2 is a permutation of the multiset $\{1^3\, 2^5\, 3^2\}$.

DEFINITION 3.4. We let $\mathrm{inv}(m_1, m_2, \ldots, m_r; n)$ denote the number of permutations $\xi_1\xi_2 \cdots \xi_{m_1+m_2+\cdots+m_r}$ of $\{1^{m_1}2^{m_2} \cdots r^{m_r}\}$ in which there are exactly n pairs (ξ_i, ξ_j) such that $i < j$ and $\xi_i > \xi_j$.

THEOREM 3.5. $\mathrm{inv}(m_1, m_2; n) = p(m_1, m_2, n)$.

Proof. We shall define a bijection between the permutations enumerated by $\mathrm{inv}(m_1, m_2; n)$ and the partitions enumerated by $p(m_1, m_2, n)$.

Let us use the "box" graphical representation for the Ferrers graph of a partition with each part $\leqslant$ 11 and at most seven parts (here $8 + 6 + 6 + 1 + 1$).

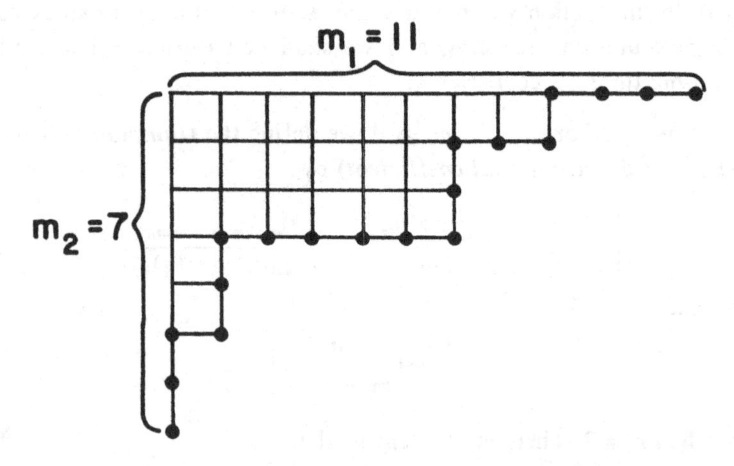

We follow the path indicated by the dots, starting with the upper right node and moving to the left and downward: if the path moves vertically, we write a 2 and if horizontally we write a 1. Hence the sequence corresponding to this graph is 1 1 1 2 1 1 2 2 1 1 1 1 1 2 2 1 2 2.

Notice that the number of 1's to the right of the first 2 tells us the largest part of our partition; the number of 1's to the right of our second 2 tells us the second part of our partition, and in general the number of 1's to the right of the ith 2 tells us the ith part of our partition. Clearly the above relationship between partitions and permutations establishes a bijection between the permutations of $\{1^{m_1}2^{m_2}\}$ with n inversions and the partitions of n with at most m_2 parts, each $\leqslant m_1$. Hence, $\mathrm{inv}(m_1, m_2; n) = p(m_1, m_2, n)$. ∎

Theorem 3.5, together with Theorem 3.1, yields the special case in which $r = 2$ of the following general result:

THEOREM 3.6. *For* $r \geqslant 1$,

$$\sum_{n \geqslant 0} \operatorname{inv}(m_1, m_2, \ldots, m_r; n) q^n = \begin{bmatrix} m_1 + m_2 + \cdots + m_r \\ m_1, m_2, \ldots, m_r \end{bmatrix}. \tag{3.4.2}$$

Remark. This theorem is due to P. A. MacMahon.

Proof. We proceed by induction on r. When $r = 1$, the result in Eq. (3.4.2) clearly reduces to $1 = 1$. When $r = 2$, we see that

$$\sum_{n \geqslant 0} \operatorname{inv}(m_1, m_2; n) q^n = \sum_{n \geqslant 0} p(m_1, m_2; n) q^n = \begin{bmatrix} m_1 + m_2 \\ m_1, m_2 \end{bmatrix},$$

by Theorems 3.5 and 3.1.

Now in general

$$\operatorname{inv}(m_1, \ldots, m_r; n) = \sum_{j=0}^{n} \operatorname{inv}(m_1 + \cdots + m_{r-1}, m_r; j) \cdot \operatorname{inv}(m_1, \ldots, m_{r-1}; n - j). \tag{3.4.3}$$

To see (3.4.3), let us examine those permutations in which exactly j of the n inverted pairs have r as first element. The easiest way to construct all such permutations is as follows: first, take a permutation of $\{1^{m_1 + m_2 + \cdots + m_{r-1}} 2^{m_r}\}$ with j inversions; next replace each 2 by an r and then replace the $m_1 + m_2 + \cdots + m_{r-1}$ appearances of 1 by a permutation of $\{1^{m_1} 2^{m_2} \cdots (r-1)^{m_{r-1}}\}$ with $n - j$ inversions. Since the two choices of permutations were totally independent, we see that the number of permutations of $\{1^{m_1} 2^{m_2} \cdots r^{m_r}\}$ with r appearing in exactly j of the inverted pairs is just

$$\operatorname{inv}(m_1 + \cdots + m_{r-1}, m_r; j) \operatorname{inv}(m_1, \ldots, m_{r-1}; n - j).$$

Summing on all $j \geqslant 0$, we obtain (3.4.3).

It is now a simple matter to prove (3.4.2). It is true for $r = 1$ and 2, and assuming it true for r, we have by (3.4.3)

$$\sum_{n \geqslant 0} \operatorname{inv}(m_1, \ldots, m_{r+1}; n) q^n$$

$$= \sum_{n \geqslant 0} \left[\sum_{j=0}^{n} \operatorname{inv}(m_1 + \cdots + m_r, m_{r+1}; j) \operatorname{inv}(m_1, \ldots, m_r; n - j) \right] q^n$$

$$= \sum_{j=0}^{\infty} \operatorname{inv}(m_1 + \cdots + m_r, m_{r+1}; j) q^j \sum_{k=0}^{\infty} \operatorname{inv}(m_1, \ldots, m_r; k) q^k$$

$$= \frac{(q)_{m_1+\cdots+m_{r+1}}}{(q)_{m_{r+1}}(q)_{m_1+\cdots+m_r}} \cdot \frac{(q)_{m_1+\cdots+m_r}}{(q)_{m_1}(q)_{m_2}\cdots(q)_{m_r}}$$

$$= \frac{(q)_{m_1+\cdots+m_{r+1}}}{(q)_{m_1}(q)_{m_2}\cdots(q)_{m_{r+1}}} = \begin{bmatrix} m_1 + m_2 + \cdots + m_{r+1} \\ m_1, m_2, \ldots, m_{r+1} \end{bmatrix},$$

and the desired result follows by mathematical induction. ∎

Let us now introduce a further parameter related to multiset permutations.

DEFINITION 3.5. We let $\text{ind}(m_1, m_2, \ldots, m_r; n)$ denote the number of permutations $\xi_1\xi_2\cdots\xi_{m_1+\cdots+m_r}$ of $\{1^{m_1}2^{m_2}\cdots r^{m_r}\}$ for which

$$\sum_{i=1}^{m_1+\cdots+m_r-1} \chi(\xi_i) = n$$

where $\chi(\xi_i) = i$ if $\xi_i > \xi_{i+1}$ and $\chi(\xi_i) = 0$ otherwise. The sum $\sum_{i=1}^{m_1+\cdots+m_r-1} \chi(\xi_i)$ is called the *greater index* of the permutation.

Thus the greater index of 1 1 1 2 1 1 2 2 1 1 1 1 1 2 2 1 2 is $0+0+0+4+0+0+0+8+0+0+0+0+0+0+15+0=27$.

We now obtain a second surprising result of MacMahon:

THEOREM 3.7. *For* $r \geqslant 1$

$$\sum_{n\geqslant 0} \text{ind}(m_1, m_2, \ldots, m_r; n)q^n = \begin{bmatrix} m_1 + m_2 + \cdots + m_r \\ m_1, m_2, \ldots, m_r \end{bmatrix}. \tag{3.4.4}$$

Proof. We shall actually prove that

$$\frac{\sum_{n\geqslant 0} \text{ind}(m_1, m_2, \ldots, m_r; n)q^n}{(q)_{m_1+m_2+\cdots+m_r}} = \frac{1}{(q)_{m_1}(q)_{m_2}\cdots(q)_{m_r}}. \tag{3.4.5}$$

We begin by considering the right-hand side of (3.4.5). We recall from Theorem 1.1 that $(q)_{m_i}^{-1}$ is the generating function for partitions in which no part exceeds m_i, and by Theorem 1.4 it is also the generating function for partitions with at most m_i parts. Now each partition with at most m_i (positive) parts corresponds to a partition with exactly m_i nonnegative parts (just fill in the necessary number of zeros). Therefore the coefficient of q^N in

$$\frac{1}{(q)_{m_1}(q)_{m_2}\cdots(q)_{m_r}}$$

is the number of arrays

$$
\begin{aligned}
&a_1 \geqslant a_2 \geqslant \cdots \geqslant a_{m_1} \geqslant 0\\
&b_1 \geqslant b_2 \geqslant \cdots \geqslant b_{m_2} \geqslant 0\\
&\vdots\\
&x_1 \geqslant x_2 \geqslant \cdots \geqslant x_{m_r} \geqslant 0
\end{aligned} \tag{3.4.6}
$$

where $\Sigma a_i + \Sigma b_i + \cdots + \Sigma x_i = N$.

Now we define a one-to-one correspondence between such arrays and ordered pairs (π, σ) where σ is a permutation of $\{1^{m_1}2^{m_2}\cdots r^{m_r}\}$ and π is a partition with exactly $m_1 + m_2 + \cdots + m_r$ nonnegative parts. For convenience we let $M = m_1 + m_2 + \cdots + m_r$.

The correspondence is constructed thus: we take the partition π: $A_1 + A_2 + \cdots + A_M$ and write directly below it the permutation $\sigma = \xi_1\xi_2\cdots\xi_M$:

$$
\begin{gathered}
A_1A_2A_3\cdots A_M\\
\xi_1\xi_2\xi_3\cdots\xi_M.
\end{gathered}
$$

We now form an array resembling (3.4.6) by placing each A_i (starting with A_1) in the ξ_ith row. This construction is inadequate as it stands since we have, for example, when $\pi = (2^3 34)$ and σ is a permutation of $\{1^3 2^2\}$

$$
\begin{matrix} 4&3&2&2&2\\ 1&2&1&1&2 \end{matrix} \;\rightarrow\; \begin{matrix} 4&2&2\\ 3&2& \end{matrix}
$$

$$
\begin{matrix} 4&3&2&2&2\\ 1&2&1&2&1 \end{matrix} \;\rightarrow\; \begin{matrix} 4&2&2\\ 3&2& \end{matrix}
$$

$$
\begin{matrix} 4&3&2&2&2\\ 1&2&2&1&1 \end{matrix} \;\rightarrow\; \begin{matrix} 4&2&2\\ 3&2& \end{matrix}
$$

However, we may make the mapping a bijection by the specification that whenever a fall occurs in the permutation (i.e., $\xi_i > \xi_{i+1}$) we must also have $A_i > A_{i+1}$. Examining the above three pairs $\binom{\pi}{\sigma}$, we find that the first pair is the only one that fits this requirement. Clearly the mapping is now one-to-one, since we may take any array of type (3.4.6) with the specified inequalities and successively insert the second row into the first, the third row into the first, and so on, always making our insertion as far to the right as possible consistent with the nonincreasing order; this way of inserting guarantees that the "$>$" will always appear whenever an element of one row appears just to the left of one of a lower-numbered row.

For example, let us work our insertion process on

$$
\begin{matrix} 5&3&2&2&0\\ 3'&3'&2'&1'&\\ 2''&1''&1''&& \end{matrix} \;\rightarrow\; \begin{matrix} 5\;3\;3'\;3'\;2\;2\;2'\;1'\;0\\ 2''\;1''\;1'' \end{matrix} \;\rightarrow\; 5\;3\;3'\;3'\;2\;2\;2'\;2''\;1'\;1''\;1''\;0.
$$

The appropriate permutation is 1 1 2 2 1 1 2 3 2 3 3 1, and the mapping we originally constructed yields

$$\begin{matrix} 5\,3\,3\,3\,2\,2\,2\,2\,1\,1\,1\,0 \\ 1\,1\,2\,2\,1\,1\,2\,3\,2\,3\,3\,1 \end{matrix} \rightarrow \begin{matrix} 5 & 3 & 2 & 2 & 0 \\ 3 & 3 & 2 & 1 & \\ 2 & 1 & 1 & & \end{matrix}$$

as desired.

To prove that we have a partition $\pi = (A_1 A_2 \cdots A_M)$ with strict inequality at the appropriate terms (specified by the falls in the permutation $\sigma = \xi_1 \xi_2 \cdots \xi_M$), we start with an arbitrary partition

$$\pi_0 = (a_1 a_2 \cdots a_M),$$

and we define

$$\pi = (a_1 + \phi_1, a_2 + \phi_2, a_3 + \phi_3, \ldots, a_M + \phi_M)$$

where ϕ_i is the number of falls in $\{\xi_i \xi_{i+1} \cdots \xi_M\}$. Note that $\phi_1 + \phi_2 + \cdots + \phi_M = \chi(\xi_1) + \chi(\xi_2) + \cdots + \chi(\xi_M)$, the greater index of σ, since on the left-hand side the fall $\xi_i > \xi_{i+1}$ is counted exactly i times.

Thus our mapping actually provides a bijection between arrays (3.4.6) whose total sum is N and ordered pairs $\binom{\pi_0}{\sigma}$ where σ is a permutation of $\{1^{m_1} 2^{m_2} \cdots r^{m_r}\}$ with greater index $g(\sigma)$ and π_0 is a partition of $N - g$ with at most M parts.

Therefore

$$\sum_\sigma q^g \cdot \frac{1}{(q)_{m_1+m_2+\cdots+m_r}} = \frac{1}{(q)_{m_1}(q)_{m_2}\cdots(q)_{m_r}}. \tag{3.4.7}$$

But $\sum_\sigma q^g = \sum_{n \geq 0} \operatorname{ind}(m_1, m_2, \ldots, m_r; n) q^n$, and so (3.4.5) is established, as desired. ■

To make our correspondence clear we exhibit in full the case in which $N = 3$, $m_1 = 3$, $m_2 = 2$.

Array (3.4.6)	$\binom{\pi}{\sigma}$	π_0	σ
0 0 0 3 0	3 0 0 0 0 2 1 1 1 2	2 0 0 0 0	2 1 1 1 2
3 0 0 0 0	3 0 0 0 0 1 1 1 2 2	3 0 0 0 0	1 1 1 2 2
2 1 0 0 0	2 1 0 0 0 1 1 1 2 2	2 1 0 0 0	1 1 1 2 2

2 0 0	2 1 0 0 0	1 0 0 0 0	1 2 1 1 2
1 0	1 2 1 1 2		
1 0 0	2 1 0 0 0	1 1 0 0 0	2 1 1 1 2
2 0	2 1 1 1 2		
0 0 0	2 1 0 0 0	1 0 0 0 0	2 2 1 1 1
2 1	2 2 1 1 1		
1 1 1	1 1 1 0 0	1 1 1 0 0	1 1 1 2 2
0 0	1 1 1 2 2		
1 1 0	1 1 1 0 0	0 0 0 0 0	1 1 2 1 2
1 0	1 1 2 1 2		
1 0 0	1 1 1 0 0	0 0 0 0 0	1 2 2 1 1
1 1	1 2 2 1 1		

COROLLARY 3.8

$$\operatorname{ind}(m_1, m_2, \ldots, m_r; n) = \operatorname{inv}(m_1, m_2, \ldots, m_r; n).$$

Proof. Compare coefficients of q^n in the identity

$$\sum_{n=0}^{\infty} \operatorname{inv}(m_1, m_2, \ldots, m_r; n) q^n = \begin{bmatrix} m_1 + m_2 + \cdots + m_r \\ m_1, m_2, \ldots, m_r \end{bmatrix}$$

$$= \sum_{n=0}^{\infty} \operatorname{ind}(m_1, m_2, \ldots, m_r; n) q^n,$$

which is valid in light of Theorems 3.6 and 3.7. ■

The subject matter of this section has been greatly extended by D. Foata and R. P. Stanley. We note in passing that Foata has provided a purely combinatorial proof of Corollary 3.8.

3.5 The Unimodal Property

In many applications, we are interested in the distribution of values of functions like $p(m_1, m_2; n)$. We shall prove a simple result that will allow us to provide relevant information on such questions in numerous cases.

DEFINITION 3.6. A polynomial $p(q) = a_0 + a_1 q + \cdots + a_n q^n$ is called *reciprocal* if for each i, $a_i = a_{n-i}$, equivalently $q^n p(q^{-1}) = p(q)$.

DEFINITION 3.7. A polynomial $p(q) = a_0 + a_1 q + \cdots + a_n q^n$ is called *unimodal* if there exists m such that

$$a_0 \leqslant a_1 \leqslant a_2 \leqslant \cdots \leqslant a_m \geqslant a_{m+1} \geqslant a_{m+2} \geqslant \cdots \geqslant a_n.$$

THEOREM 3.9. *Let $p(q)$ and $r(q)$ be reciprocal, unimodal polynomials with nonnegative coefficients; then $p(q)r(q)$ is also a reciprocal, unimodal polynomial with nonnegative coefficients.*

Proof. Let $p(q) = a_0 + a_1 q + \cdots + a_n q^n$, $r(q) = b_0 + b_1 q + \cdots + b_m q^m$, and let

$$s(q) = p(q)r(q) = c_0 + c_1 q + \cdots + c_{n+m} q^{n+m}$$

where

$$c_i = \sum_{j=-\infty}^{\infty} a_j b_{i-j} \tag{3.5.1}$$

with the convention that $a_j = 0$ if $j < 0$ or $j > n$ and $b_j = 0$ if $j < 0$ or $j > m$.

Now by (3.5.1) all the c_i are nonnegative. Since

$$\begin{aligned} q^{n+m}s(q^{-1}) &= q^n p(q^{-1}) q^m r(q^{-1}) \\ &= p(q)r(q) = s(q), \end{aligned}$$

we see that $s(q)$ is reciprocal.

Finally, to prove that $s(q)$ is unimodal, we begin by noting that $a_j - a_{j-1} \geqslant 0$ for all $j \leqslant n/2$ and $b_j - b_{j-1} \geqslant 0$ for all $j \leqslant m/2$; this is because each of $p(q)$ and $r(q)$ is both unimodal and reciprocal. Now

$$\begin{aligned} 2(c_j - c_{j-1}) &= \sum_{i=-\infty}^{\infty} a_i b_{j-i} + \sum_{i=-\infty}^{\infty} a_{n-i+1} b_{j-n+i-1} \\ &\quad - \sum_{i=-\infty}^{\infty} a_{i-1} b_{j-i} - \sum_{i=-\infty}^{\infty} a_{n-i} b_{j-n+i-1} \\ &= \sum_{i=-\infty}^{\infty} (a_i - a_{i-1}) b_{j-i} + \sum_{i=-\infty}^{\infty} b_{j-n+i-1}(a_{n-i+1} - a_{n-i}) \\ &= \sum_{i=-\infty}^{\infty} (a_i - a_{i-1})(b_{j-i} - b_{j-n+i-1}) \qquad (\text{since } a_i = a_{n-i}) \\ &= \sum_{i=0}^{n+1} (a_i - a_{i-1})(b_{j-i} - b_{j-n+i-1}) \\ &= \sum_{i=0}^{\frac{1}{2}(n+1)} (a_i - a_{i-1})(b_{j-i} - b_{j-n+i-1}) \\ &\quad + \sum_{i=0}^{\frac{1}{2}(n+1)} (a_{n+1-i} - a_{n-i})(b_{j-n-1+i} - b_{j-i}) \end{aligned}$$

(note that if $(n+1)/2$ is an integer, $a_{(n+1)/2} - a_{(n-1)/2} = 0$)

$$= 2\sum_{i=0}^{n/2}(a_i - a_{i-1})(b_{j-i} - b_{j-n+i-1}).$$

Hence

$$c_j - c_{j-1} = \sum_{i=0}^{n/2}(a_i - a_{i-1})(b_{j-i} - b_{j-n+i-1}). \qquad (3.5.2)$$

Since $0 \leqslant i \leqslant n/2$, we see that $a_i - a_{i-1} \geqslant 0$. If $j - i \leqslant m/2$, then since $n + 1 > 2i$, we see that $m/2 \geqslant j - i > j - n + i - 1$; hence $b_{j-i} - b_{j-n+i-1} \geqslant 0$ in this case. If $j - i > m/2$, then for $0 \leqslant j \leqslant (m+n)/2$, we see that $m + n + 1 > 2j$, and so $m/2 > m - j + i > j - n + i - 1$; therefore in this second case $b_{j-i} - b_{j-n+i-1} = b_{m-j+i} - b_{j-n+i-1} \geqslant 0$. Hence, in any case, both factors of each term on the right-hand side of (3.5.2) are nonnegative provided that $0 \leqslant j \leqslant (m+n)/2$. Therefore, since $s(q)$ is reciprocal, we see that it is also unimodal. ∎

THEOREM 3.10. *For all* $N, M, n \geqslant 0$

$$p(N, M, n) = p(M, N, n); \qquad (3.5.3)$$

$$p(N, M, n) = p(N, M, NM - n); \qquad (3.5.4)$$

$$p(N, M, n) - p(N, M, n-1) \geqslant 0 \quad \text{for} \quad 0 < n \leqslant NM/2. \qquad (3.5.5)$$

Proof. By Theorem 3.1,

$$\sum_{n=0}^{\infty} p(N, M, n)q^n = G(N, M; q) = \begin{bmatrix} N+M \\ M \end{bmatrix} = \frac{(q)_{N+M}}{(q)_N(q)_M}.$$

Equation (3.5.3) follows from the fact that

$$\begin{bmatrix} N+M \\ M \end{bmatrix}$$

is symmetric in N and M.

The degree of $G(N, M; q)$ is just

$$\binom{N+M+1}{2} - \binom{N+1}{2} - \binom{M+1}{2} = NM,$$

and

$q^{NM}G(N, M; q^{-1})$

$$= q^{NM}\frac{(1-q^{-1})(1-q^{-2})\cdots(1-q^{-N-M})}{(1-q^{-1})(1-q^{-2})\cdots(1-q^{-N})(1-q^{-1})(1-q^{2})\cdots(1-q^{-M})}$$

$$= \frac{q^{NM - \binom{N+M+1}{2} + \binom{N+1}{2} + \binom{M+1}{2}}(q)_{N+M}}{(q)_N(q)_M}$$

$$= \frac{(q)_{N+M}}{(q)_N(q)_M} = G(N, M; q).$$

Hence $G(N, M; q)$ is a reciprocal polynomial of degree NM; therefore (3.5.4) is true.

Now (3.5.5) is the unimodal property and unfortunately its proof (at least with current knowledge) lies outside of the theory of partitions proper. It has been proved through invariant theory, where it is shown that $p(N, M; n) - p(N, M; n - 1)$ is the number of linearly independent semi-invariants of degree M, weight n, and extent not exceeding N (I. J. Schur, *Vorlesungen über Invariantentheorie*, Satz 2.22, p. 76, *Grundlehren der Mathematischen Wissenschaften*, Vol. 143). To my knowledge no simple combinatorial proof of (3.5.5) is known. ∎

THEOREM 3.11. *For all* $m_1, m_2, \ldots, m_r, n \geqslant 0$

$$\operatorname{ind}(m_1, \ldots, m_r; n) = \operatorname{ind}(m_{i_1}, \ldots, m_{i_r}; n)$$
$$= \operatorname{inv}(m_1, \ldots, m_r; n) = \operatorname{inv}(m_{i_1}, \ldots, m_{i_r}; n) \tag{3.5.6}$$

where $\{i_1, \ldots, i_r\}$ *is a permutation of* $\{1, 2, \ldots, r\}$;

$$\operatorname{ind}(m_1, \ldots, m_r; n) = \operatorname{ind}(m_1, \ldots, m_r; S - n)$$
$$= \operatorname{inv}(m_1, \ldots, m_r; n) = \operatorname{inv}(m_1, \ldots, m_r; S - n) \tag{3.5.7}$$

where $S = \sum_{1 \leqslant i < j \leqslant r} m_i m_j$ *is the second elementary symmetric function of the* m_i;

$$\operatorname{ind}(m_1, \ldots, m_r; n) - \operatorname{ind}(m_1, \ldots, m_r; n - 1)$$
$$= \operatorname{inv}(m_1, \ldots, m_r; n) - \operatorname{inv}(m_1, \ldots, m_r; n - 1) \geqslant 0 \tag{3.5.8}$$

for $0 < n \leqslant S/2$.

Proof. Since

$$\sum_{n \geqslant 0} \operatorname{ind}(m_1, \ldots, m_r; n) q^n = \sum_{n \geqslant 0} \operatorname{inv}(m_1, \ldots, m_r; n) q^n$$

$$= \begin{bmatrix} m_1 + \cdots + m_r \\ m_1, \ldots, m_r \end{bmatrix} = \frac{(q)_{m_1 + m_2 + \cdots + m_r}}{(q)_{m_1}(q)_{m_2} \cdots (q)_{m_r}}$$

$$= \begin{bmatrix} m_1 + m_2 \\ m_1 \end{bmatrix} \begin{bmatrix} m_1 + m_2 + m_3 \\ m_1 + m_2 \end{bmatrix} \cdots \begin{bmatrix} m_1 + \cdots + m_r \\ m_1 + \cdots + m_{r-1} \end{bmatrix},$$

we see that

$$\begin{bmatrix} m_1 + \cdots + m_r \\ m_1, \ldots, m_r \end{bmatrix}$$

is symmetric in the m_i (whence (3.5.6)) and is the product of unimodal reciprocal polynomials (by Theorem 3.10) and is therefore unimodal and reciprocal by Theorem 3.9. Equations (3.5.7) and (3.5.8) immediately follow once we observe that the degree of

$$\begin{bmatrix} m_1 + m_2 + \cdots + m_r \\ m_1, m_2, \ldots, m_r \end{bmatrix}$$

is just

$$\binom{m_1 + \cdots + m_r + 1}{2} - \binom{m_1 + 1}{2} - \binom{m_2 + 1}{2} - \cdots - \binom{m_r + 1}{2}$$

$$= \sum_{1 \leqslant i < j \leqslant r} m_i m_j = S. \qquad \blacksquare$$

Examples

1. The following finite form of Jacobi's triple product identity may be deduced from Eq. (3.3.6):

$$\sum_{j=-n}^{n} q^{j^2} x^j \begin{bmatrix} 2n \\ n+j \end{bmatrix}_{q^2} = (-x^{-1}q; q^2)_n(-xq; q^2)_n,$$

where $\left[{m \atop r}\right]_{q^2}$ is the ordinary Gaussian polynomial with q replaced by q^2.

2. Jacobi's triple product identity (Theorem 2.8) may be deduced from Example 1.

In Examples 3–9, the polynomials $H_n(t)$ are the Rogers–Szegö polynomials

$$H_n(t) = \sum_{j=0}^{n} \begin{bmatrix} n \\ j \end{bmatrix} t^j.$$

3. $\displaystyle\sum_{n=0}^{\infty} \frac{H_n(t)x^n}{(q)_n} = \frac{1}{(x)_\infty (xt)_\infty}.$

4. Identity (3.3.8) is deduced from Example 3 in the case in which $t = -1$.

5. $H_n(q^{\frac{1}{2}}) = (-q^{\frac{1}{2}}; q^{\frac{1}{2}})_n$.

6. $H_{n+1}(t) = (1 + t)H_n(t) - (1 - q^n)tH_{n-1}(t)$.

7. Utilizing Example 6 and the recurrence formula for the Gaussian polynomials (3.3.3), we may prove by induction on m that

$$H_m(t)H_n(t) = \sum_{r=0}^{m} \begin{bmatrix} m \\ r \end{bmatrix} \begin{bmatrix} n \\ r \end{bmatrix} (q)_r t^r H_{m+n-2r}(t).$$

8. From Examples 3 and 7 it follows that

$$\sum_{m,n \geq 0} \frac{H_{m+n}(t) y^m z^n}{(q)_m (q)_n} = \frac{(tyz)_\infty}{(y)_\infty (z)_\infty (ty)_\infty (tz)_\infty}.$$

9. From Example 8, we may deduce that

$$\sum_{n=0}^{\infty} \frac{H_n(t) H_n(s) x^n}{(q)_n} = \frac{(stx^2)_\infty}{(x)_\infty (tx)_\infty (sx)_\infty (tsx)_\infty}.$$

10. Let $[x]$ denote the greatest integer function, then we may prove

$$\sum_{j \geq 0} q^{j^2+aj} \begin{bmatrix} n+1-a-j \\ j \end{bmatrix} = \sum_{h=-\infty}^{\infty} (-1)^h q^{\frac{1}{2}h(5h+1)-2ah} \begin{bmatrix} n+1 \\ [\frac{1}{2}(n+1-5h)+a] \end{bmatrix}$$

(where $a = 0$ or 1), by showing that the assertion is true for $n = 0$ and 1 and that each side satisfies the recurrence $f_n = f_{n-1} + q^n f_{n-2}$.

11. The Rogers–Ramanujan identities may be deduced from Example 10, namely, for $a = 0$ or 1,

$$\sum_{j \geq 0} \frac{q^{j^2+aj}}{(q)_j} = \prod_{j \geq 0} \frac{1}{(1-q^{5j+a+1})(1-q^{5j+4-a})}.$$

12. Let D_n denote the polynomial defined in Example 10 when $a = 0$. These polynomials satisfy the identities

$$D_{2n} = \sum_{j=0}^{n} q^{jn} \begin{bmatrix} n+1 \\ j \end{bmatrix} D_{n-1-j} \quad \text{and} \quad D_n = 1 + \sum_{j=1}^{n} q^j D_{j-2}.$$

13. Let the "kth excess" of the partition $\lambda = (\lambda_1 \lambda_2 \cdots \lambda_r)$ denote $\lambda_1 - \lambda_{k+1}$. The generating function for partitions with j parts whose first excess is at most i is

$$\frac{(1-q^{i+1})q^j}{(1-q)(1-q^2)\cdots(1-q^j)}.$$

14. Example 13 may be extended to show that for any $k < j$, the generating function for partitions with j parts and with kth excess at most i is

$$\frac{(1-q^{i+1})(1-q^{i+2})\cdots(1-q^{i+k})q^j}{(1-q)(1-q^2)\cdots(1-q^j)}.$$

15. Theorem 3.1 is a corollary of Example 14.

16. The number of permutations of $\{1^{m_1} 2^{m_2} \cdots r^{m_r}\}$ with even greater index equals the number with odd greater index if and only if at least two m_i are odd.

17. We may define a several-variable analog of the Rogers–Szegö polynomials by

$$H_n(x_1,\ldots,x_s) = \sum_{j_1,\ldots,j_s\geqslant 0} \begin{bmatrix} n \\ j_1, j_2,\ldots, j_s, n-j_1-\cdots-j_s \end{bmatrix} x_1^{j_1}\cdots x_s^{j_s}.$$

The related generating function is

$$\sum_{n=0}^{\infty} \frac{H_n(x_1,\ldots,x_s)t^n}{(q)_n} = \frac{1}{(t)_\infty (tx_1)_\infty \cdots (tx_s)_\infty}.$$

18. A *protruded partition* of n is a decreasing sequence of positive integers $\lambda_1 \geqslant \lambda_2 \geqslant \cdots \geqslant \lambda_r > 0$, together with a sequence of nonnegative integers satisfying $0 \leqslant \mu_i \leqslant \lambda_i$ for $1 \leqslant i \leqslant r$, such that $\Sigma(\lambda_i + \mu_i) = n$. (The μ_i are the *protrusions* of the ordinary partition $(\lambda_1\lambda_2\cdots\lambda_r)$. The product

$$U_m = \prod_{j=1}^{m} (1 - q^j - q^{j+1} - \cdots - q^{2j})^{-1}$$

is the generating function for the protruded partitions with each $\lambda_i \leqslant m$.

19. From (3.3.13) (with $a = x^{-\frac{1}{2}}$, $b = x^{-\frac{1}{2}}$, $c = q^2/(1-q)$, t replaced by $xzq^2/(1-q)$) it follows when $x \to 0$ that

$$\sum_{m=0}^{\infty} U_m z^m = \left(\prod_{s=0}^{\infty} (1 - zq^s)^{-1}\right) \cdot \sum_{n=0}^{\infty} \frac{q^{n(n+1)}z^n}{(1-q)(1-q^2)\cdots(1-q^n)(1-q-q^2)\cdots(1-q-q^{n+1})}.$$

Notes

Gauss (1863, p. 16) introduced Gaussian polynomials, and he observed the facts in Theorem 3.2. Theorem 3.1 was perhaps first stated by Sylvester (1884–1886); however, it may be trivially deduced from (3.3.6) and (3.3.7) identities of Cauchy (see Cauchy's *Collected Works*, 1893, p. 46). Equation (3.3.8) is due to Gauss (1863) and was used by him in the treatment of Gaussian sums; (3.3.9) is also due to Gauss (1863). Equation (3.3.10) is actually a special case of Corollary 2.4 due to Heine (1847); actually it is the q-analog of the celebrated Chu–Vandermonde summation (R. Askey 1975a; see also 1975b, p. 60) and our proof parallels the standard elementary proof of that summation. As we remarked in the text, (3.3.11) is due to F. H. Jackson (1910); one of its most important applications was by G. N. Watson (1929), and it has also been applied in R. P. Stanley's (1972) recent work in combinatorics.

All the work in Section 3.4 can be found in important works by P. A. MacMahon (1913, 1914, 1915–1916, 1916) which foreshadowed much of the recent work in the combinatorial theory of permutations. It would lead us afield to delve deeply into recent work; however, we should mention the work of Foata (1965, 1968), Foata and Schützenberger (1970), and R. P. Stanley (1972). Interestingly enough, MacMahon's proof of Theorem 3.7 is of principal importance for Stanley's (1972) theory of (P, ω)-partitions (Lemma 6.1 and Theorem 6.2). Applications of (P, ω)-partitions to permutation problems are well described by Stanley (1972, Section 25).

The material in Section 3.5 comes from Andrews (1975).

Examples 1, 2. See Hermite (1891, pp. 155–156).

Examples 3–9. Carlitz (1956), Szegö (1926), Rogers (1893a, b).

Examples 10, 11. Andrews (1970).

Example 12. Andrews (1974).

Examples 13–15 are due to F. Franklin and appear in Sylvester (1882–1884).

Example 16. MacMahon (1913, 1915–1916).

Examples 18–19. Stanley (1972, Section 24). In Section 23 of Stanley's memoir, he considers stacks and *V*-partitions, and these objects also give rise to identities related to Corollary 2.3.

References

Andrews, G. E. (1970). "A polynomial identity which implies the Rogers–Ramanujan identities," *Scripta Math.* **28**, 297–305.

Andrews, G. E. (1974). "Combinatorial analysis and Fibonacci numbers," *Fibonacci Quart.* **12**, 141–146.

Andrews, G. E. (1975). "A theorem on reciprocal polynomials with applications to permutations and compositions," *Amer. Math. Monthly* **82**, 830–833.

Askey, R. A. (1975a). "A note on the history of series," Math. Res. Center Tech. Summary Rept. No. 1532, Madison, Wisconsin.

Askey, R. A. (1975b). *Orthogonal Polynomials and Special Functions*, No. 21. Regional Conf. Ser. Appl. Math., SIAM, Philadelphia.

Carlitz, L. (1956). "Some polynomials related to theta functions," *Ann. Math. Pura Appl.* (4) **41**, 359–373.

Cauchy, A. (1893). *Oeuvres, Ser.* 1, Vol. 8. Gauthier-Villars, Paris.

Foata, D. (1965). "Etude algébrique de certains problèmes d'Analyse Combinatoire et du Calcul des Probabilités," *Publ. Inst. Statist. Univ. Paris* **14**, 81–241.

Foata, D. (1968). "On the Netto inversion number of a sequence," *Proc. Amer. Math. Soc.* **19**, 236–240.

Foata, D., and Schutzenberger, M. P. (1970). *Théorie géométrique des polynômes Eulériens* (Lecture Notes in Math. No. 138). Springer, New York.

Gauss, C. F. (1863). *Werke*, Vol. 2. Königliche Gesellschaft der Wissenschaften, Göttingen.

Heine, E. (1847). "Untersuchungen über die Reihe ...," *J. Reine Angew. Math.* **34**, 285–328.

Hermite, C. (1891). *Oeuvres*, Vol. 2. Gauthier-Villars, Paris.

Jackson, F. H. (1910). "Transformations of *q*-series," *Messenger of Math.* **39**, 145–151.

MacMahon, P. A. (1913). "The indices of permutations and the derivation therefrom of functions of a single variable associated with the permutations of any assemblage of objects," *Amer. J. Math.* **35**, 281–322.

MacMahon, P. A. (1914). "The superior and inferior indices of permutations," *Trans. Cambridge Phil. Soc.* **29**, 55–60.

MacMahon, P. A. (1915–1916). *Combinatory Analysis*, 2 vols. Cambridge Univ. Press, London and New York (reprinted by Chelsea, New York, 1960).

MacMahon, P. A. (1916). "Two applications of general theorems in combinatory analysis: (1) to the theory of inversions of permutations; (2) to the ascertainment of the numbers of terms in the development of a determinant which has amongst its elements an arbitrary number of zeros," *Proc. London Math. Soc.* (2) **15**, 314–321.

Rogers, L. J. (1893a). "On a three-fold symmetry in the elements of Heine's series," *Proc. London Math. Soc.* **24**, 171–179.

Rogers, L. J. (1893b). "On the expansion of certain infinite products," *Proc. London Math. Soc.* **24**, 337–352.

Stanley, R. P. (1972). "Ordered structures and partitions," *Mem. Amer. Math. Soc.* **119**.

Sylvester, J. J. (1882–1884). "A constructive theory of partitions in three acts, an interact, and an exodion," *Amer. J. Math.* **5**, 251–330; **6**, 334–336 (or pp. 1–83 of the *Collected Papers of J. J. Sylvester*, Vol. 4, Cambridge Univ. Press, London and New York, 1912; reprinted by Chelsea, New York, 1974).

Szegö, G. (1926). "Ein Beitrag zur Theorie der Thetafunktionen," *S.B. Preuss. Akad. Wiss. Phys.-Math. Kl.* 242–252.

Watson, G. N. (1929). "A new proof of the Rogers–Ramanujan identities," *J. London Math. Soc.* **4**, 4–9.

CHAPTER 4

Compositions and Simon Newcomb's Problem

4.1 Introduction

The first three chapters treated elementary properties of partitions. *Compositions* are merely partitions in which the order of the summands is considered. For example, there are five partitions of 4: (4), (13), (2^2), $(1^2 2)$, (1^4); there are eight compositions of 4: (4), (13), (31), (22), (112), (121), (211), (1111). We shall find that *compositions of vectors* (or *compositions of multipartite numbers*) also have intrinsic interest and important applications, especially to permutation problems of the Simon Newcomb type.

4.2 Compositions of Numbers

DEFINITION 4.1. We let $c(m, n)$ denote the number of compositions of n with exactly m parts.

THEOREM 4.1

$$c(m, n) = \binom{n-1}{m-1} = \frac{(n-1)!}{(m-1)!(n-m)!}.$$

First proof of Theorem 4.1. The argument used to prove Theorem 1.1 may be easily adapted to show that

$$\begin{aligned}\sum_{n=0}^{\infty} c(m, n)q^n &= (q + q^2 + q^3 + q^4 + \cdots)^m \\ &= \frac{q^m}{(1-q)^m} \\ &= q^m \sum_{r=0}^{\infty} \binom{r+m-1}{r} q^r \qquad \text{(by the binomial series)} \\ &= \sum_{n=m}^{\infty} \binom{n-1}{n-m} q^n = \sum_{n=0}^{\infty} \binom{n-1}{m-1} q^n. \qquad (4.2.1)\end{aligned}$$

ENCYCLOPEDIA OF MATHEMATICS and Its Applications, Gian-Carlo Rota (ed.). 2, George E. Andrews, The Theory of Partitions

Comparing coefficients of q^n in the extremes of (4.2.1), we obtain the desired result. ∎

Second proof of Theorem 4.1. We introduce a graphical representation for the compositions of n. To the composition $(a_1 a_2 \cdots a_m)$ of n we associate m segments of the interval $[0, n]$; the first segment is of length a_1, the second of length a_2, and so on. Thus the composition (3 2 3 1 2) of 11 is represented as

0 1 2 3 4 5 6 7 8 9 10 11

We now observe that we may construct each of the $c(m, n)$ compositions of n with m parts by choosing $m - 1$ of the first $n - 1$ integers as end points for the m segments dividing $[0, n]$. Since these choices can be made in $\binom{n-1}{m-1}$ ways, we see that

$$c(m, n) = \binom{n-1}{m-1}. \quad ∎$$

DEFINITION 4.2. We let $c(N, M, n)$ denote the number of compositions of n with exactly M parts, each $\leqslant N$.

Clearly $c(N, M, n) = c(M, n)$ whenever $N \geqslant n$.

Interestingly enough, $c(N, M, n)$ possesses symmetry properties resembling those described in Theorem 3.10 for $p(N, M, n)$.

THEOREM 4.2. *For all* $N, M, n \geqslant 1$

$$c(N, M, n) = c(N, M, MN + M - n); \tag{4.2.2}$$

$$c(N, M, n) - c(N, M, n - 1) \geqslant 0 \quad \textit{for} \quad 0 < n \leqslant M(N + 1)/2. \tag{4.2.3}$$

Proof. Following the argument in the first proof of Theorem 3.1, we see directly that

$$\sum_{n \geqslant 0} c(N, M, n) q^n = (q + q^2 + \cdots + q^N)^M$$
$$= (0 + q + q^2 + \cdots + q^N + 0 \cdot q^{N+1})^M.$$

Thus $\sum_{n \geqslant 0} c(N, M, n) q^n$ is a product of M unimodal, reciprocal polynomials with nonnegative coefficients (the leading coefficient was not necessarily assumed to be nonzero); hence, by repeated application of Theorem 3.9, $\sum_{n \geqslant 0} c(N, M, n) q^n$ is a unimodal reciprocal polynomial with nonnegative coefficients. Equations (4.2.2) and (4.2.3) follow at once from this fact. ∎

The simplicity of Theorem 4.1 allows us to obtain sharp asymptotic estimates for certain partition functions by relating them to composition functions. This approach is due to H. Gupta (1942).

DEFINITION 4.3. Let $p_M(n)$ denote the number of partitions of n with exactly M parts.

Clearly $\sum_{M=0}^{n} p_M(n) = p(n)$, and $p(n, M, n) - p(n, M-1, n) = p_M(n)$.

THEOREM 4.3 (Erdös-Lehner). *As* $n \to \infty$

$$p_M(n) \sim \frac{1}{M!}\binom{n-1}{M-1}$$

provided that $M = o(n^{1/3})$.

Proof. To each partition of n with exactly M parts we may associate the set of compositions of n made up of the same parts. There are $M!$ such compositions if all the parts in the partition are distinct; otherwise there are less. Hence

$$M!p_M(n) \geqslant c(M, n)$$

or

$$p_M(n) \geqslant \frac{1}{M!}c(M, n) = \frac{1}{M!}\binom{n-1}{M-1}. \tag{4.2.4}$$

On the other hand, if $q_M(n)$ denotes the number of partitions of n with M parts all distinct, then

$$q_M(n + \tfrac{1}{2}M(M-1)) = p_M(n). \tag{4.2.5}$$

The truth of (4.2.5) may be seen by the correspondence

$$(A_1 + M - 1, A_2 + M - 2, A_3 + M - 3, \ldots, A_M) \leftrightarrow (A_1, A_2, \ldots, A_M)$$

where $A_1 \geqslant A_2 \geqslant \cdots \geqslant A_M$ and $\sum A_i = n$.

Now each partition enumerated by $q_M(n)$ corresponds to exactly $M!$ compositions under the association described in the first paragraph of this proof. Therefore

$$M!q_M(n) \leqslant c(M, n).$$

Hence

$$\begin{aligned} p_M(n) &= q_M(n + \tfrac{1}{2}M(M-1)) \\ &\leqslant \frac{1}{M!}c(M, n + \tfrac{1}{2}M(M-1)) \\ &= \frac{1}{M!}\binom{n + \tfrac{1}{2}M(M-1) - 1}{M-1}. \end{aligned} \tag{4.2.6}$$

Combining (4.2.4) and (4.2.6), we see that

$$1 \leqslant \frac{p_M(n)}{\frac{1}{M!}\binom{n-1}{M-1}} \leqslant \frac{(n + \frac{1}{2}M(M-1) - 1)!(n-M)!}{(n-1)!(n + \frac{1}{2}M(M-3))!}$$

$$= \frac{\Gamma(n - M + 1)\Gamma(n + \frac{1}{2}M(M-1))}{\Gamma(n)\Gamma(n + \frac{1}{2}M(M-3) + 1)}$$

$$\sim n^{-M+1}(n + \tfrac{1}{2}M(M-1))^{M-1} \qquad (\text{provided } M = o(n^{1/2}))$$

$$= \left(1 + \frac{\frac{1}{2}M(M-1)}{n}\right)^{M-1}$$

$$= \exp\left\{(M-1)\log\left[1 + \frac{M(M-1)}{2n}\right]\right\}$$

$$= \exp\left\{\frac{M(M-1)^2}{2n} + O\left(\frac{M^2(M-1)^3}{n^2}\right)\right\}$$

and this last expression approaches 1 as $n \to \infty$ provided $M = o(n^{1/3})$. ∎

It is reasonably clear that numerous asymptotic questions for partition functions may be attacked by devices similar to those utilized above.

4.3 Vector Compositions

The title of this section would have been "Compositions of Multipartite Numbers" if it had been chosen by P. A. MacMahon; however, the title "Vector Compositions" is more self-explanatory and reflects more recent usage. The vectors we shall consider will just be ordered r-tuples of nonnegative integers not all of which are zero.

DEFINITION 4.4. A *partition* of $(\alpha_1, \alpha_2, \ldots, \alpha_r)$ is a set of vectors $(\beta_1^{(i)}, \ldots, \beta_r^{(i)})$, $1 \leqslant i \leqslant s$ (order disregarded), such that $\sum_{i=1}^{s} (\beta_1^{(i)}, \ldots, \beta_r^{(i)}) = (\alpha_1, \alpha_2, \ldots, \alpha_r)$ (here as explained earlier all vectors have nonnegative integral coordinates not all zero). If the order of the parts is taken into account, we call $(\beta_1^{(1)}, \ldots, \beta_r^{(1)}), \ldots, (\beta_1^{(s)}, \ldots, \beta_r^{(s)})$ a *composition* of $(\alpha_1 \ldots, \alpha_s)$.

DEFINITION 4.5. We let $P_=(\alpha_1, \alpha_2, \ldots, \alpha_r; m)$ denote the number of partitions of $(\alpha_1, \alpha_2, \ldots, \alpha_r)$ with m parts, and we let $c(\alpha_1, \alpha_2, \ldots, \alpha_r; m)$ denote the number of compositions of $(\alpha_1, \alpha_2, \ldots, \alpha_r)$ with m parts.

Thus $P_=(2, 1, 1; 2) = 5$, since there are five partitions of $(2, 1, 1)$ into two parts: $(2, 1, 0)(0, 0, 1)$, $(2, 0, 1)(0, 1, 0)$, $(2, 0, 0)(0, 1, 1)$, $(1, 1, 0)(1, 0, 1)$, $(1, 1, 1)(1, 0, 0)$; $c(2, 1, 1; 2) = 10$ since each of the five partitions produces two compositions.

DEFINITION 4.6. We let $P(\alpha_1, \ldots, \alpha_r)$ (respectively $c(\alpha_1, \ldots, \alpha_r)$) denote the total number of partitions (respectively compositions) of $(\alpha_1, \alpha_2, \ldots, \alpha_r)$. For later convenience we define $c(0, 0, \ldots, 0) = \frac{1}{2}$.

Note that $\sum_{m \geq 1} P_=(\alpha_1, \ldots, \alpha_r; m) = P(\alpha_1, \ldots, \alpha_r)$, and $\sum_{m \geq 1} c(\alpha_1, \ldots, \alpha_r; m) = c(\alpha_1, \ldots, \alpha_r)$.

THEOREM 4.4

$$\sum_{\alpha_1, \ldots, \alpha_r \geq 0} c(\alpha_1, \ldots, \alpha_r) t_1^{\alpha_1} \cdots t_r^{\alpha_r} = \frac{1}{4(1 - t_1)(1 - t_2) \cdots (1 - t_r) - 2}. \quad (4.3.1)$$

Proof. We begin by observing that for $m \geq 1$,

$$\begin{aligned} &\sum_{\substack{\alpha_1, \ldots, \alpha_r \geq 0 \\ \text{not all zero}}} c(\alpha_1, \ldots, \alpha_r; m) t_1^{\alpha_1} \cdots t_r^{\alpha_r} \\ &= \Bigg(\sum_{\substack{\alpha_1, \ldots, \alpha_r \geq 0 \\ \text{not all zero}}} t_1^{\alpha_1} \cdots t_r^{\alpha_r} \Bigg)^m \\ &= \left(\frac{1}{(1 - t_1)(1 - t_2) \cdots (1 - t_r)} - 1 \right)^m. \end{aligned} \quad (4.3.2)$$

Hence

$$\begin{aligned} &\sum_{\alpha_1, \ldots, \alpha_r \geq 0} c(\alpha_1, \ldots, \alpha_r) t_1^{\alpha_1} \cdots t_r^{\alpha_r} \\ &= \tfrac{1}{2} + \sum_{m=1}^{\infty} \sum_{\substack{\alpha_1, \ldots, \alpha_r \geq 0 \\ \text{not all zero}}} c(\alpha_1, \ldots, \alpha_r; m) t_1^{\alpha_1} \cdots t_r^{\alpha_r} \\ &= \tfrac{1}{2} + \sum_{m=1}^{\infty} \left(\frac{1}{(1 - t_1)(1 - t_2) \cdots (1 - t_r)} - 1 \right)^m \\ &= \tfrac{1}{2} + \frac{\dfrac{1}{(1 - t_1)(1 - t_2) \cdots (1 - t_r)} - 1}{2 - \dfrac{1}{(1 - t_1)(1 - t_2) \cdots (1 - t_r)}} \\ &= \tfrac{1}{2} + \frac{1 - (1 - t_1)(1 - t_2) \cdots (1 - t_r)}{2(1 - t_1)(1 - t_2) \cdots (1 - t_r) - 1} \\ &= \frac{\tfrac{1}{2}}{2(1 - t_1)(1 - t_2) \cdots (1 - t_r) - 1}, \end{aligned}$$

as desired. ∎

Now Eq. (4.3.2) provides a means for obtaining a reasonably simple formula for $c(\alpha_1, \ldots, \alpha_r; m)$.

THEOREM 4.5. *For $m > 0$,*

$$c(\alpha_1,\ldots,\alpha_r;m)=\sum_{i=0}^{m}(-1)^i\binom{m}{i}\binom{\alpha_1+m-i-1}{\alpha_1}$$
$$\times\binom{\alpha_2+m-i-1}{\alpha_2}\cdots\binom{\alpha_r+m-i-1}{\alpha_r}. \quad (4.3.3)$$

Proof. By Eq. (4.3.2) (assuming $c(0,\ldots,0;m)=0$)

$$\sum_{\alpha_1,\ldots,\alpha_r\geq 0} c(\alpha_1,\ldots,\alpha_r;m)t_1^{\alpha_1}t_2^{\alpha_2}\cdots t_r^{\alpha_r}$$
$$=\left(\frac{1}{(1-t_1)(1-t_2)\cdots(1-t_r)}-1\right)^m$$
$$=\sum_{i=0}^{m}\binom{m}{i}(-1)^i(1-t_1)^{-m+i}(1-t_2)^{-m+i}\cdots(1-t_r)^{-m+i}$$
$$=\sum_{i=0}^{m}\binom{m}{i}(-1)^i\sum_{\alpha_1\geq 0}\binom{\alpha_1+m-i-1}{\alpha_1}t_1^{\alpha_1}\cdots\sum_{\alpha_r\geq 0}\binom{\alpha_r+m-i-1}{\alpha_r}t_r^{\alpha_r}$$
$$=\sum_{\alpha_1,\ldots,\alpha_r\geq 0}\left(\sum_{i=0}^{m}(-1)\binom{m}{i}\binom{\alpha_1+m-i-1}{\alpha_1}\cdots\right.$$
$$\left.\cdots\binom{\alpha_r+m-i-1}{\alpha_r}\right)t_1^{\alpha_1}\cdots t_r^{\alpha_r}. \quad (4.3.4)$$

Comparing coefficients of $t_1^{\alpha_1}t_2^{\alpha_2}\cdots t_r^{\alpha_r}$ in the extremes of (4.3.4), we obtain the desired result. ■

As an example of Theorem 4.5, we note that

$$c(2,1,1;2)=\binom{3}{2}\binom{2}{1}\binom{2}{1}-\binom{2}{1}\binom{2}{2}\binom{1}{1}\binom{1}{1}+\binom{1}{2}\binom{0}{1}\binom{0}{1}$$
$$=3\cdot 2\cdot 2-2\cdot 1\cdot 1\cdot 1+0\cdot 0\cdot 0=10.$$

4.4 Simon Newcomb's Problem

The problem we shall treat now is one that P. A. MacMahon solved neatly and completely in 1907. The problem and its history were succinctly described by MacMahon in *Combinatory Analysis*, Vol. 1 (1915, p. 187):

The problem was suggested to the late Professor Newcomb by a game of "patience" played with ordinary playing cards which he found to be a recreation in the few hours that he could spare from astronomical work. It may be stated as follows:

A pack of cards of any specification is taken—say there are m_1 cards marked 1, m_2 cards 2, m_3 cards 3, and so on—and being shuffled is dealt out on a table; so long as the cards that appear have numbers that are in ascending order of magnitude, equality of number counted as ascending order, they are placed together in one pack, but directly as the ascending order is broken a fresh pack is commenced and so on until all the cards have been dealt. The probability that there will result exactly m packs or at most m packs is required.

We may easily translate this problem into one involving permutations of a multiset:

DEFINITION 4.7. Let $N(m_1, m_2, \ldots, m_r; n)$ denote the number of permutations of $\{1^{m_2}2^{m_2}\cdots r^{m_r}\}$ with exactly n runs (a run in the permutation $\xi_1\xi_2\cdots\xi_{m_1+\cdots+m_r}$ is a maximal subsequence of the form $\xi_i \leqslant \xi_{i+1} \leqslant \xi_{i+2} \leqslant \cdots \leqslant \xi_{j-1} \leqslant \xi_j$).

We shall have the answer to Simon Newcomb's problem if we can find a closed expression for $N(m_1, m_2, \ldots, m_r; n)$.

LEMMA 4.6. *Each of the following relations implies the other*:

$$a_n = \sum_{j=0}^{n-1} \binom{r-n+j}{j} b_{n-j} \qquad \textit{for all} \quad n \geqslant 1; \tag{4.4.1}$$

$$b_n = \sum_{j=0}^{n-1} \binom{r-n+j}{j} (-1)^j a_{n-j} \qquad \textit{for all} \quad n \geqslant 1. \tag{4.4.2}$$

Proof. We begin by observing that we need only show that (4.4.2) always implies (4.4.1), since once this assertion is established the reverse implication follows by considering $b_n' = (-1)^n b_n$ and $a_n' = (-1)^n a_n$.

Now assuming (4.4.2), we see that

$$\sum_{j=0}^{n-1} \binom{r-n+j}{j} b_{n-j} = \sum_{j=0}^{n-1} \binom{r-n+j}{j} \sum_{k=0}^{n-j-1} \binom{r-n+j+k}{k} (-1)^k a_{n-j-k}$$

$$= \sum_{\substack{j+k\leqslant n-1\\ j,k\geqslant 0}} \binom{r-n+j}{j} \binom{r-n+j+k}{k} (-1)^k a_{n-j-k}$$

$$= \sum_{h=0}^{n-1} a_{n-h} \sum_{\substack{j+k=h\\ j,k\geqslant 0}} \binom{r-n+j}{j} \binom{r-n+h}{k} (-1)^k$$

$$= \sum_{h=0}^{n-1} a_{n-h}(-1)^h \sum_{j=0}^{n} \binom{n-r-1}{j} \binom{r-n+h}{h-j}$$

$$= a_n + \sum_{h=1}^{n-1} a_{n-h}(-1)^h \binom{h-1}{h}$$

(by the Chu–Vandermonde summation; i.e., (3.3.10) with $q = 1$)

$$= a_n,$$

as desired. ■

LEMMA 4.7

$$c(\alpha_1, \alpha_2, \ldots, \alpha_r; n) = \sum_{j=0}^{n-1} \binom{\alpha_1 + \alpha_2 + \cdots + \alpha_r - n + j}{j} N(\alpha_1, \ldots, \alpha_r; n - j). \tag{4.4.3}$$

Proof. The identity (4.4.3) is easily established once we provide appropriate alternative combinatorial interpretations for

$$c(\alpha_1, \alpha_2, \ldots, \alpha_r; n) \quad \text{and} \quad N(\alpha_1, \alpha_2, \ldots, \alpha_r; n).$$

First we see that $c(\alpha_1, \ldots, \alpha_r; n)$ enumerates the number of ways that α_1 balls labeled 1, α_2 balls labeled 2, ..., α_r balls labeled r may be distributed among n labeled boxes so that no box is empty. To see this we note that the composition

$$(\beta_1^{(1)}, \ldots, \beta_r^{(1)})(\beta_1^{(2)}, \ldots, \beta_r^{(2)}) \cdots (\beta_1^{(n)}, \ldots, \beta_1^{(n)})$$

of $(\alpha_1, \ldots, \alpha_r)$ corresponds to the distribution that places exactly $\beta_i^{(j)}$ balls labeled i in the jth box.

A very similar interpretation can be given to $N(\alpha_1, \ldots, \alpha_r; n)$. Indeed the jth run in a permutation may be viewed as specifying the elements to be placed in the jth box. It is important to note that the function $N(\alpha_1, \ldots, \alpha_r; n)$ does not enumerate all possible distributions (as does $c(\alpha_1, \ldots, \alpha_r; n)$), but only those for which the smallest element in each box is strictly smaller than the largest element in the preceding box. Therefore to each permutation enumerated by $N(\alpha_1, \ldots, \alpha_r; n - j)$ we may associate a unique set of distributions enumerated by $c(\alpha_1, \ldots, \alpha_r; n)$ as follows: First place a vertical line between each of the $n - j$ runs in the given permutation. There are now $\alpha_1 + \alpha_2 + \cdots + \alpha_r - (n - j - 1) - 1$ pairs $\xi_i \xi_{i+1}$ that have no vertical bar between them; place a vertical bar between j of these pairs and we have a distribution of α_1 ones, α_2 twos, and so on, into n labeled nonempty boxes. Since the final set of j vertical bars may be placed in

$$\binom{\alpha_1 + \alpha_2 + \cdots + \alpha_r - n + j}{j}$$

ways, we see that exactly this many of the distributions enumerated by $c(\alpha_1, \ldots, \alpha_r; n)$ correspond to a single permutation enumerated by $N(\alpha_1, \ldots, \alpha_r; n - j)$. Hence

$$c(\alpha_1, \ldots, \alpha_r; n) = \sum_{j=0}^{n} \binom{\alpha_1 + \cdots + \alpha_r - n + j}{j} N(\alpha_1, \ldots, \alpha_r; n - j)$$

as desired. ■

THEOREM 4.8. *For* $n > 0$,

$$N(\alpha_1, \ldots, \alpha_r; n) = \sum_{s=0}^{n-1} (-1)^s \binom{\alpha_1 + \cdots + \alpha_r + 1}{s} \binom{n + \alpha_1 - 1 - s}{\alpha_1} \binom{n + \alpha_2 - 1 - s}{\alpha_2} \cdots \binom{n + \alpha_r - 1 - s}{\alpha_r}. \tag{4.4.4}$$

Proof. By applying Lemma 4.6 to Eq. (4.4.3), we see that

$$\begin{aligned} N(\alpha_1, \ldots, \alpha_r; n) &= \sum_{j=0}^{n-1} (-1)^j \binom{\alpha_1 + \cdots + \alpha_r - n + j}{j} c(\alpha_1, \ldots, \alpha_r; n - j) \\ &= \sum_{j=0}^{n-1} (-1)^j \binom{\alpha_1 + \cdots + \alpha_r - n + j}{j} \sum_{i=0}^{n-j} (-1)^i \binom{n-j}{i} \\ &\quad \times \binom{\alpha_1 + n - j - i - 1}{\alpha_1} \cdots \binom{\alpha_r + n - j - i - 1}{\alpha_r} \qquad \text{(by Theorem 4.5)} \\ &= \sum_{s=0}^{n} (-1)^s \binom{\alpha_1 + n - s - 1}{\alpha_1} \binom{\alpha_2 + n - s - 1}{\alpha_2} \cdots \binom{\alpha_r + n - s - 1}{\alpha_r} \\ &\quad \times \sum_{\substack{i+j=s \\ i \geq 0, j \geq 0}} \binom{\alpha_1 + \cdots + \alpha_r - n + j}{j} \binom{n-j}{i}. \end{aligned} \tag{4.4.5}$$

Now

$$\begin{aligned} \sum_{\substack{i+j=s \\ i \geq 0, j \geq 0}} \binom{A - n + j}{j} \binom{n-j}{i} &= \sum_{j=0}^{s} (-1)^j \binom{n - A - 1}{j} \binom{n-j}{s-j} \\ &= (-1)^s \sum_{j=0}^{s} \binom{n - A - 1}{j} \binom{s - n - 1}{s - j} \end{aligned}$$

$$= (-1)^s \binom{s - A - 2}{s} \quad \text{(by (3.3.9) with } q = 1\text{)}$$

$$= \binom{A + 1}{s}. \tag{4.4.6}$$

Consequently, applying (4.4.6) to (4.4.5), we obtain (4.4.4). ■

Examples

1. Let F_n be the nth *Fibonacci number*: $F_0 = 0, F_1 = 1, F_n = F_{n-1} + F_{n-2}$ for $n > 1$. The number of compositions of n in which no 1's appear is F_{n-1}.

2. More generally, let ${}_kF_n$ be defined by ${}_kF_0 = \cdots = {}_kF_{k-2} = 0, {}_kF_{k-1} = 1$, and ${}_kF_n = {}_kF_{n-1} + {}_kF_{n-k}$. The number of compositions of n in which all parts are $\geqslant k$ is equal to ${}_kF_{n-1}$.

3. Example 2 implies that there are 2^{n-1} compositions of n.

4. Let $c_k(m, n)$ denote the number of compositions of n into exactly m parts, each $\geqslant k$. Then

$$c_k(m, n) = \binom{n - (k - 1)m - 1}{m - 1}.$$

5. Examples 2 and 4 imply that

$$\sum_{m \geqslant 1} \binom{n - (k - 1)m - 1}{m - 1} = {}_kF_{n-1}.$$

6. Let us say $(m_1, \ldots, m_r)$ is of *real dimension* h if $m_h \neq 0, m_{h+1} = 0, \ldots, m_r = 0$. Then the total number of all compositions of all vectors $(\beta_1, \ldots, \beta_r)$ with $0 \leqslant \beta_i \leqslant \alpha_i$ in which exactly h_i parts have real dimension i is

$$\binom{\alpha_1}{h_1}\binom{\alpha_2}{h_2}\cdots\binom{\alpha_r}{h_r}\binom{\alpha_1 + h_2 + \cdots + h_r}{\alpha_1}$$
$$\times \binom{\alpha_2 + h_3 + \cdots + h_r}{\alpha_2}\cdots\binom{\alpha_{r-1} + h_r}{\alpha_{r-1}}.$$

7. Example 6 implies that

$$c(\alpha_1, \ldots, \alpha_r) = \frac{1}{2} \sum_{\substack{h_1 \geq 0 \\ \vdots \\ h_r \geq 0}} \binom{\alpha_1}{h_1}\binom{\alpha_2}{h_2}\cdots\binom{\alpha_r}{h_r}\binom{\alpha_1 + h_2 + \cdots + h_r}{\alpha_1}$$
$$\times \binom{\alpha_2 + h_3 + \cdots + h_r}{\alpha_2}\cdots\binom{\alpha_{r-1} + h_r}{\alpha_{r-1}}.$$

8. The *Long operator* L_i is a linear operator on polynomials in x_i over the reals:

$$L_i x_i^k = \binom{\alpha_i}{\alpha_{i+1} + \cdots + \alpha_r - k}.$$

One can prove by induction on j that

$$L_i(x_i^k(1 + x_i)^j) = \binom{\alpha_i + j}{\alpha_{i+1} + \cdots + \alpha_r - k}.$$

9. If the polynomial

$$2^{\alpha_i - 1}\prod_{i=2}^{r}\{x_1x_2\cdots x_{i-1} + (1+x_1)(1+x_2)\cdots(1+x_{i-1})\}^{\alpha_i}$$

is fully expanded and then each x_i^k is replaced by

$$\binom{\alpha_i}{\alpha_{i+1} + \cdots + \alpha_r - k},$$

the result is $c(\alpha_1, \ldots, \alpha_r)$.

The foregoing assertion is merely that

$$c(\alpha_1, \ldots, \alpha_r) = L_1 L_2 \cdots L_{r-1} 2^{\alpha_1 - 1} \prod_{i=2}^{r} \{x_1 x_2 \cdots x_{i-1} + (1+x_1)(1+x_2)\cdots(1+x_{i-1})\}^{\alpha_i}.$$

We prove this through the expansion of each factor by the binomial theorem and then invoking Example 8 to reduce the identity to Example 7.

10. By expanding $(1 + x_i)^j$ through the binomial theorem, we may deduce the Chu–Vandermonde (Eq. (3.3.10) when $q = 1$) summation from Example 8.

$$\sum_{s=0}^{j}\binom{j}{s}\binom{\alpha_i}{\alpha_{i+1} + \cdots + \alpha_r - k - s} = \binom{\alpha_i + j}{\alpha_{i+1} + \cdots + \alpha_r - k}.$$

11. The case of $q = 1$ of Eq. (3.3.11) can be deduced easily through the use of Long operators. We write the desired result equivalently as

$$\sum_{r\geq 0}\binom{m}{r}\binom{n+\mu}{n-r}\binom{\mu+\nu+r}{m+n+\mu} = \binom{\mu+\nu}{\mu+m}\binom{\nu}{n}.$$

If we consider linear operators L_1 and L_2, defined by

$$L_1(x^r) = \binom{n}{n-r}, \qquad L_2(y^s) = \binom{\mu+\nu}{m+n+\mu-s},$$

then Example 8 implies that

$$L_1(x^r(1+x)^t) = \binom{n+t}{n-r} \quad \text{and} \quad L_2(y^s(1+y)^u) = \binom{\mu+\nu+u}{m+n+\mu-s}.$$

From here it follows that

$$\begin{aligned}
\sum_{r\geq 0}\binom{m}{r}\binom{n+\mu}{n-r}\binom{\mu+\nu+r}{m+n+\mu} &= L_1L_2((1+x)^\mu(1+x(1+y))^m) \\
&= L_1L_2((1+x)^{\mu+m}(1+xy(1+x)^{-1})^m) \\
&= \sum_{j\geq 0}\binom{m}{j}\binom{\mu+\nu}{m+n+\mu-j}\binom{n+\mu+m-j}{n-j} \\
&= \binom{\mu+\nu}{\mu+m}\sum_{j\geq 0}\binom{m}{j}\binom{\nu-m}{n-j} \\
&= \binom{\mu+\nu}{\mu+m}\binom{\nu}{n} \qquad \text{(by Example 10).}
\end{aligned}$$

Notes

The material in this chapter is primarily an extended and modernized version of portions of P. A. MacMahon's (1894, 1908) work on compositions (see also Section IV of MacMahon, 1915). Theorem 4.2 is due to Z. Star (1976; see also Andrews, 1975a), who also obtains nice asymptotic formulas for the $c(N, M, n)$. Theorem 4.3 is due to P. Erdös and J. Lehner (1941); the proof we present is by H. Gupta (1942).

Work on vector compositions and the Simon Newcomb problem has been extensive in recent years. The papers listed in the references by Carlitz, Dillon, Foata, Kreweras, Roselle and Schützenberger provide some of the primary references. We should also point out that the (P, ω)-partitions of R. P. Stanley are applicable to compositions and Simon Newcomb type problems (see Stanley, 1972, Section 25). Some of the papers reviewed in Section P80 of LeVeque (1964) concern compositions.

Example 1. Cayley (1876).

Examples 6–10. Long (1970), Andrews (1975b, 1976a, b).

References

Andrews, G. E. (1975a). "A theorem on reciprocal polynomials with applications to permutations and compositions," *Amer. Math. Monthly* **82**, 830–833.

Andrews, G. E. (1975b). "The theory of compositions, II: Simon Newcomb's problem," *Utilitas Math.* **7**, 33–54.

Andrews, G. E. (1976a). "The theory of compositions, I: The ordered factorizations of n and a conjecture of C. Long," *Canadian Math. Bull.* **18**, 479–484.

Andrews, G. E. (1976b). "The theory of compositions, III: The MacMahon formula and the Stanton–Cowan numbers," *Utilitas Math.* **9**, 283–290.

Carlitz, L. (1959). "Eulerian numbers and polynomials," *Math. Mag.* **33**, 247–260.

Carlitz, L. (1960). "Eulerian numbers and polynomials of higher order," *Duke Math. J.* **27**, 401–424.

Carlitz, L. (1964). "Extended Bernoulli and Eulerian numbers," *Duke Math. J.* **31**, 667–690.

Carlitz, L. (1972a). "Enumeration of sequences by rises and falls: A refinement of the Simon Newcomb problem," *Duke Math. J.* **39**, 267–280.

Carlitz, L. (1972b). "Sequences, paths, and ballot numbers," *Fibonacci Quart.* **10**, 531–549.

Carlitz, L. (1972c). "Eulerian numbers and operators," *The Theory of Arithmetic Functions* (A. A. Gioia and D. L. Goldsmith, eds.) (Lecture Notes in Math. No. 251). Springer, New York.

Carlitz, L. (1973a). "Enumeration of a special class of permutations by rises," *Publ. Elek. Fak. Univ. Beogradu* **451**, 189–196.

Carlitz, L. (1973b). "Enumeration of up-down permutations by number of rises," *Pacific J. Math.* **45**, 49–58.

Carlitz, L. (1973c). "Enumeration of up-down sequences," *Discrete Math.* **4**, 273–286.

Carlitz, L. (1973d). "Permutations with prescribed pattern," *Math. Nachr.* **58**, 31–53.

Carlitz, L. (1974a). "Permutations and sequences," *Advances in Math.* **14**, 92–120.

Carlitz, L. (1974b). "Up-down and down-up partitions," *Proc. Eulerian Series and Applications Conf.* Pennsylvania State Univ.

Carlitz, L., and Riordan, J. (1955). "The number of labeled two-terminal series-parallel networks," *Duke Math. J.* **23**, 435–446.

Carlitz, L., and Riordan, J. (1971). "Enumeration of some two-line arrays by extent," *J. Combinatorial Theory* **10**, 271–283.

Carlitz, L., and Scoville, R. (1972). "Up-down sequences," *Duke Math. J.* **39**, 583–598.

Carlitz, L., and Scoville, R. (1973). "Enumeration of rises and falls by position," *Discrete Math.* **5**, 45–59.

Carlitz, L., and Scoville, R. (1974). "Generalized Eulerian numbers: Combinatorial applications," *J. Reine Angew. Math.* **265**, 110–137.

Carlitz, L., and Scoville, R. (1975). "Enumeration of up-down permutations by upper records," *Arch. Math. (Basel)*

Carlitz, L., Roselle, D., and Scoville, R. (1966). "Permutations and sequences with repetitions by number of increases," *J. Combinatorial Theory* **1**, 350–374.

Cayley, A. (1876). "Theorems in trigonometry and on partitions," *Coll. Math. Papers of A. Cayley* **10**, 16.

Dillon, J. F., and Roselle, D. (1968). "Eulerian numbers of higher order," *Duke Math. J.* **35**, 247–256.

Dillon, J. F., and Roselle, D. (1969). "Simon Newcomb's problem," *SIAM J. Appl. Math.* **17**, 1086–1093.

Erdös, P., and Lehner, J. (1941). "The distribution of the number of summands in the partitions of a positive integer," *Duke Math. J.* **8**, 335–345.

Foata, D. (1965). "Etude algébrique de certains problèmes d'analyse combinatoire et du calcul des probabilités, " *Publ. Inst. Statist. Univ. Paris* **14**, 81–241.

Foata, D., and Riordan, J. (1974). "Mappings of acyclic and parking functions," *Aequationes Math.* **10**, 10–22.

Foata, D., and Schützenberger, M.-P. (1970). *Théorie géométrique des polynômes eulériens*, (Lecture Notes in Math. No. 138). Springer, New York.

Foata, D., and Schützenberger, M.-P. (1973). "Nombres d'Euler et permutations alternantes," *A Survey of Combinatorial Theory* (J. N. Srivastava et al., eds.), pp. 173–187. North-Holland, Amsterdam.

Foata, D., and Strehl, V. (1974). "Rearrangements of the symmetric group and enumerative properties of the tangent and secant numbers," *Math. Z.* **137**, 257–264.

Foulkes, H. O. (1975). "Enumeration of permutations with prescribed up–down and inversion sequence," *Discrete Math.* **9**, 365–374.

Fray, R. D., and Roselle, D. (1971). "Weighted lattice paths," *Pacific J. Math.* **37**, 85–96.

Gupta, H. (1942). "On an asymptotic formula in partitions," *Proc. Indian Acad. Sci.* **A16**, 101–102.

Gupta, H. (1955). "Partitions in general," *Res. Bull. Panjab Univ.* **67**, 31–38.

Gupta, H. (1970). "Partitions – a survey," *J. Res. Nat. Bur. Standards* **74B**, 1–29.

Kreweras, G. (1965). "Sur une classe de problèmes de dénombrement liés au treillis des partitions des entiers," *Cahiers du B.U.R.O.* No. 6.

Kreweras, G. (1966). "Dénombrements de chemins minimaux a sauts imposés," *C. R. Acad. Sci. Paris* **263**, 1–3.

Kreweras, G. (1967). "Traitement simultane du 'Problème de Young' et du 'Problème de Simon Newcomb'," *Cahiers du B.U.R.O.* No. 10.

Kreweras, G. (1969a). "Inversion des polynômes de Bell bidimensionnels et application au dénombrement des relations binaires connexes," *C. R. Acad. Sci. Paris* **268**, 577–579.

Kreweras, G. (1969b). "Dénombrement systématique de relations binaires externes," *Math. Sci. Humaines* **7**, 5–15.

Kreweras, G. (1970). "Sur les éventails de segments," *Cahiers du B.U.R.O.*, No. 15.

LeVeque, W. J. (1974). *Reviews in Number Theory*, Vol. 4. Amer. Math. Soc., Providence, R.I.

Long, C. (1970). "On a problem in partial difference equations," *Canadian Math. Bull.* **13**, 333–335.

MacMahon, P. A. (1894). "Memoir on the theory of the compositions of numbers," *Philos. Trans. Roy. Soc. London* **A184**, 835–901.

MacMahon, P. A. (1908). "Second memoir on the composition of numbers," *Philos. Trans. Roy. Soc. London* **A207**, 65–134.

MacMahon, P. A. (1915). *Combinatory Analysis*, Vol. 1. Cambridge Univ. Press, London and New York (reprinted by Chelsea, New York, 1960).

Roselle, D. (1969). "Permutations by number of rises and successions," *Proc. Amer. Math. Soc.* **19**, 8–16.

Stanley, R. P. (1972). "Ordered structures and partitions," *Mem. Amer. Math. Soc.* **119**.

Star, Z. (1976). "An asymptotic formula in the theory of compositions," *Aequationes Math.* **13**, 279–284.

CHAPTER 5

The Hardy-Ramanujan-Rademacher Expansion of p(n)

5.1 Introduction

Having seen many elementary properties of partitions and compositions and having seen an elementary asymptotic formula for $p_M(n)$ (Theorem 4.3), we now move to one of the crowning achievements in the theory of partitions: *the exact formula* for $p(n)$, an achievement undertaken and mostly completed by G. H. Hardy and S. Ramanujan and fully completed and perfected by H. Rademacher.

The story of the Hardy and Ramanujan collaboration on this formula is an amazing one, and is perhaps best told by J. E. Littlewood in his fascinating review of the *Collected Papers of Srinivasa Ramanujan* in the *Mathematical Gazette*, Vol. 14 (1929, pp. 427–428):

I must say something finally of the paper on partitions ... written jointly with Hardy. The number $p(n)$ of the partitions of n increases rapidly with n, thus:

$$p(200) = 3972999029388.$$

The authors show that $p(n)$ is the integer nearest

$$\frac{1}{2\sqrt{2}} \sum_{q=1}^{v} \sqrt{q}A_q(n)\psi_q(n), \tag{1}$$

where $A_q(n) = \sum \omega_{p,q} e^{-2n p\pi i/q}$, the sum being over p's prime to q and less than it, $\omega_{p,q}$ is a certain $24q$th root of unity, v is of the order of $\sqrt{n}$, and

$$\psi_q(n) = \frac{d}{dn}\left(\exp\left\{C\sqrt{\left(n - \frac{1}{24}\right)}\Big/ q\right\}\right), \qquad C = \pi\sqrt{\tfrac{2}{3}}.$$

We may take $v = 4$ when $n = 100$. For $n = 200$ we may take $v = 5$; five terms of the series (1) predict the correct value of $p(200)$. We may always

ENCYCLOPEDIA OF MATHEMATICS and Its Applications, Gian-Carlo Rota (ed.). 2, George E. Andrews, The Theory of Partitions

take $v = a\sqrt{n}$ (or rather its integral part), where a is any positive constant we please, provided n exceeds a value $n_0(a)$ depending only on a.

The reader does not need to be told that this is a very astonishing theorem, and he will readily believe that the methods by which it was established involve a new and important principle, which has been found very fruitful in other fields. The story of the theorem is a romantic one. (To do it justice I must infringe a little the rules about collaboration. I therefore add that Prof. Hardy confirms and permits my statements of bare fact.) One of Ramanujan's Indian conjectures was that the first term of (1) was a very good approximation to $p(n)$; this was established without great difficulty. At this stage the $n - (1/24)$ was represented by a plain n – the distinction is irrelevant. From this point the real attack begins. The next step in development, not a very great one, was to treat (1) as an "asymptotic" series, of which a fixed number of terms (e.g. $v = 4$) were to be taken, the error being of the order of the next term. But from now to the very end Ramanujan always insisted that much more was true than had been established: "there must be a formula with error $O(1)$." This was his most important contribution; it was both absolutely essential and most extraordinary. A severe numerical test was now made, which elicited the astonishing facts about $p(100)$ and $p(200)$. Then v was made a function of n; this *was* a very great step, and involved new and deep function-theory methods that Ramanujan obviously could not have discovered by himself. The complete theorem thus emerged. But the solution of the final difficulty was probably impossible without one more contribution from Ramanujan, this time a perfectly characteristic one. As if its analytical difficulties were not enough, the theorem was entrenched also behind almost impregnable defences of a purely formal kind. The form of the function $\psi_q(n)$ is a kind of indivisible unit; among many asymptotically equivalent forms it is essential to select exactly the right one. Unless this is done at the outset, and the $-1/24$ (to say nothing of the d/dn) is an extraordinary stroke of formal genius, the complete result can never come into the picture at all. There is, indeed, a touch of real mystery. If only we *knew* there was a formula with error $O(1)$, we might be forced, by slow stages, to the correct form of ψ_q. But why was Ramanujan so certain there *was* one? *Theoretical* insight, to be the explanation, had to be of an order hardly to be credited. Yet it is hard to see what numerical instances could have been available to suggest so strong a result. And unless the form of ψ_q was known already, *no* numerical evidence could suggest anything of the kind – there seems no escape, at least, from the conclusion that the discovery of the correct form was a single stroke of insight. We owe the theorem to a singularly happy collaboration of two men, of quite unlike gifts, in which each contributed the best, most characteristic, and most fortunate work that was in him. Ramanujan's genius did have this one opportunity worthy of it.

The formula that we shall prove is in the final form obtained by Rademacher:

THEOREM 5.1

$$p(n) = \frac{1}{\pi\sqrt{2}} \sum_{k=1}^{\infty} A_k(n) k^{\frac{1}{2}} \left[\frac{d}{dx} \frac{\sinh((\pi/k)(\frac{2}{3}(x - 1/24))^{\frac{1}{2}})}{(x - 1/24)^{\frac{1}{2}}} \right]_{x=n}, \qquad (5.1.1)$$

where

$$A_k(n) = \sum_{\substack{h \bmod k \\ (h,k)=1}} \omega_{h,k} e^{-2\pi inh/k}$$

with $\omega_{h,k}$ *a certain 24kth root of unity defined in Section 5.2.*

This unbelievable identity wherein the left-hand side is the humble arithmetic function $p(n)$ and the right-hand side is an infinite series involving π, square roots, complex roots of unity, and derivatives of hyperbolic functions provides not only a theoretical formula for $p(n)$ but also a formula which admits relatively rapid computation.

For example, $p(200) = 3972999029388$, while if we compute the first eight terms of the series in (1) we find that the result is

$$\begin{array}{r} +\ 3{,}972{,}998{,}993{,}185.896 \\ +\ 36{,}282.978 \\ -\ 87.555 \\ +\ 5.147 \\ +\ 1.424 \\ +\ 0.071 \\ +\ 0.000 \\ +\ 0.043 \\ \hline 3{,}972{,}999{,}029{,}388.004 \end{array}$$

which is the correct value of $p(n)$ within 0.004. It is a simple matter to determine explicitly the error when the infinite series is truncated, so we can determine with certainty values of $p(n)$ directly from (5.1.1). For small values of n we can, of course, obtain $p(n)$ via the recurrence given in Corollary 1.8.

Furthermore it is a simple matter to show that each term of the infinite series in (5.1.1) is $O(\exp[\pi(2n)^{1/2}/k\sqrt{3}])$. Consequently the term $k = 1$ provides an asymptotic formula for $p(n)$, and if we note that $A_1(n) = 1$, we see without much trouble that as $n \to \infty$

$$p(n) \sim \frac{1}{4n\sqrt{3}} \exp\left[\pi\left(\frac{2n}{3}\right)^{\frac{1}{2}}\right]. \tag{5.1.2}$$

The truth of (5.1.1) is intimately tied to the fact that

$$\eta(\tau) = \exp\left(\frac{\pi i\tau}{12}\right)\prod_{m=1}^{\infty}[1 - \exp(2\pi im\tau)] = \frac{\exp(\pi i\tau/12)}{\sum_{n=0}^{\infty} p(n)\exp(2\pi in\tau)}$$

is actually a modular form. In order to prove (5.1.1) we must make use of very fundamental properties of $\eta(\tau)$, in particular its behavior under transformations of the modular group: $\tau \to (a\tau + b)/(c\tau + d)$, $ad - bc = 1$. A totally complete proof of (5.1.1) would now require at least two further chapters for this book; one in which the modular group and its basic properties were fully developed (see Knopp, 1970, Chapter 1, or LeVeque, 1956, Chapter 1), and a second in which the fundamental transformation formulas for $\eta(\tau)$ are derived (see Knopp, 1970, Chapter 3). Such a project would take us well afield from partitions, and so at the beginning of Section 5.2 we shall state the necessary transformation formula.

5.2 The Formula for $p(n)$

Instead of using the transformation for $\eta(\tau)$ explicitly, we shall utilize an equivalent result that will be most suited to our purposes. Namely, we let

$$P(q) = \sum_{n \geq 0} p(n)q^n = \prod_{n=1}^{\infty} (1 - q^n)^{-1}, \tag{5.2.1}$$

then

$$P(\exp(2\pi i(h + iz)/k)) = \omega_{h,k} z^{\frac{1}{2}} \exp[\pi(z^{-1} - z)/12k] P(\exp[2\pi i(h' + iz^{-1})/k]); \tag{5.2.2}$$

here $\operatorname{Re} z > 0$, the principal branch of $z^{\frac{1}{2}}$ is selected, h' is a solution of the congruence

$$hh' \equiv -1 \pmod{k}, \tag{5.2.3}$$

and $\omega_{h,k}$ is a 24kth root of unity given by

$$\omega_{h,k} = \begin{cases} \left(\dfrac{-k}{h}\right) \exp(-\pi i(\frac{1}{4}(2 - hk - h) + \frac{1}{12}(k - k^{-1})(2h - h' + h^2h'))) & \text{if } h \text{ odd}, \\ \left(\dfrac{-h}{k}\right) \exp(-\pi i(\frac{1}{4}(k - 1) + \frac{1}{12}(k - k^{-1})(2h - h' + h^2h'))) & \text{if } k \text{ odd}, \end{cases} \tag{5.2.4}$$

with (a/b) the Legendre–Jacobi symbol. See Knopp (1970, Chapter 3) or Rademacher (1973, Chapter 9) for a proof of (5.2.2); there is a proof of (5.2.2) due to B. Berndt that is sketched in Examples 6–17 at the end of this chapter.

In passing we mention the elegant representation of Rademacher for $\omega_{h,k}$:

$$\omega_{h,k} = \exp(\pi i s(h, k)) \tag{5.2.5}$$

where $s(h, k)$ is the Dedekind sum:

$$s(h, k) = \sum_{\mu=1}^{k-1} \left(\frac{\mu}{k} - \left[\frac{\mu}{k}\right] - \frac{1}{2}\right)\left(\frac{h\mu}{k} - \left[\frac{h\mu}{k}\right] - \frac{1}{2}\right). \tag{5.2.6}$$

Now we come to the truly *remarkable* approach to the evaluation of $p(n)$ that was developed by Hardy and Ramanujan. Clearly Cauchy's integral theorem implies that

$$p(n) = \frac{1}{2\pi i}\int_C \frac{P(x)}{x^{n+1}}\,dx \tag{5.2.7}$$

where, say, C is a circle centered on the origin and inside the unit circle: $|x| = 1$. How can we evaluate this integral? By examining the generating function $P(x) = \prod_{n=1}^{\infty}(1 - x^n)^{-1}$, we see that each partial product $\prod_{n=1}^{N}(1 - x^n)^{-1}$ has a pole of order N at $x = 1$, a pole of order $[N/2]$ at $x = -1$, poles of order $[N/3]$ at $x = \exp(2\pi i/3)$ and $\exp(4\pi i/3)$, and so on. Furthermore, we note that (5.2.2) gives us extremely good information on the behavior of $P(x)$ near $\exp(2\pi ih/k)$, namely, as $z \to 0$ ($\mathrm{Re}\, z > 0$):

$$P\left[\exp\left(\frac{2\pi ih}{k} - \frac{2\pi z}{k}\right)\right] \sim \omega_{h,k} z^{\frac{1}{2}} \exp\left[\frac{\pi(z - z^{-1})}{12k}\right]. \tag{5.2.8}$$

Clearly then we should divide the circle of integration into segments so that our knowledge from (5.2.8) may be most powerfully applied, depending on which "rational point" $\exp(2\pi ih/k)$ we are near (if α is rational, we shall call $\exp(2\pi i\alpha)$ a rational point of the unit circle). Unfortunately, the rational points are dense on the unit circle and so the preceding prescription is not exactly helpful. To gain some idea of how to make this approach effective, we recall that the points $\exp(2\pi ih/k)$ where k is small seem to be the most important of the singularities; hence instead of worrying about the dense set of all rational points, we restrict our attention to the discrete set of those rational points $\exp(2\pi ih/k)$ with $0 < k \leqslant N$, a fixed positive integer. Thus we are interested in F_N: the set of proper Farey fractions of order N. For example, when $N = 5$,

$$F_5 = \{0, \tfrac{1}{5}, \tfrac{1}{4}, \tfrac{1}{3}, \tfrac{2}{5}, \tfrac{1}{2}, \tfrac{3}{5}, \tfrac{2}{3}, \tfrac{3}{4}, \tfrac{4}{5}, 1\}.$$

Seeing the Farey fractions arise in this natural manner, we surmise that we should divide our circle C of radius ρ into segments so that each segment is in some sense "centered" on the successive rational points $\rho e^{2\pi ih/k}$ where $h/k \in F_N$. We must now rely on the elementary properties of F_N to suggest the optimal choice of intervals. As is well known (and easily proved), if h/k and h_1/k_1 are successive terms in F_N, then the rational number with least denominator lying strictly between h/k and h_1/k_1 is $(h + h_1)/(k + k_1)$ and this number is called the mediant of h/k and h_1/k_1. Thus using the mediants as the end points for our intervals seems a natural dissection of C. Now if h_0/k_0, h/k, h_1/k_1 are three successive terms in F_N, let us write

$$\theta'_{0,1} = \frac{1}{N+1};$$

$$\theta'_{h,k} = \frac{h}{k} - \frac{h_0 + h}{k_0 + h} \qquad \text{for} \quad h > 0,$$

$$\theta''_{h,k} = \frac{h_1 + h}{k_1 + h} - \frac{h}{k}. \tag{5.2.9}$$

Then clearly

$$p(n) = \frac{1}{2\pi i}\int_C \frac{P(x)}{x^{n+1}}\,dx$$

$$= \rho^{-n}\int_0^1 P[\rho\exp(2\pi i\phi)]\exp(-2\pi i n\phi)\,d\phi$$

$$= \rho^{-n}\sum_{\substack{k=1\\(h,k)=1\\0\leqslant h<k}}^{N}\int_{-\theta'_{h,k}}^{\theta''_{h,k}} P\left[\rho\exp\left(\frac{2\pi i h}{k} + 2\pi i\phi\right)\right]\exp\left(-\frac{2\pi i n h}{k} - 2\pi i n\phi\right)d\phi. \tag{5.2.10}$$

All that awaits now before the application of (5.2.2) is an appropriate choice of ρ; we could leave the matter indefinite and choose ρ as well as possible when compelled to do so. However, for simplicity we shall make the "right" choice immediately:

$$\rho = \exp\left(-\frac{2\pi}{N^2}\right), \tag{5.2.11}$$

and shall content ourselves with subsequent observations on the necessity of this choice. Hence

$$p(n) = \exp\left(\frac{2\pi n}{N^2}\right)\sum_{\substack{k=1\\(h,k)=1\\0\leqslant h<k}}^{N}\exp\left(-\frac{2\pi i h n}{k}\right)$$

$$\times\int_{-\theta'_{h,k}}^{\theta''_{h,k}} P\left\{\exp\left[\frac{2\pi i h}{k} - \frac{2\pi}{k}\left(\frac{k}{N^2} - ik\phi\right)\right]\right\}\exp(-2\pi i n\phi)\,d\phi; \tag{5.2.12}$$

thus to apply (5.2.2) we must define

$$z = k(N^{-2} - i\phi). \tag{5.2.13}$$

Consequently, applying (5.2.2) to the integrand in (5.2.12), we find that

$$p(n) = \exp\left(\frac{2\pi n}{N^2}\right) \sum_{\substack{k=1 \\ (h,k)=1 \\ 0 \leqslant h < k}}^{N} \exp\left(-\frac{2\pi i h n}{k}\right) \omega_{h,k}$$

$$\times \int_{-\theta'_{h,k}}^{\theta''_{h,k}} z^{\frac{1}{2}} \exp\left[\pi \frac{(z^{-1} - z)}{12k}\right] P\left\{\exp\left[2\pi i \frac{(h' + iz^{-1})}{k}\right]\right\} \exp(-2\pi i n\phi)\, d\phi. \tag{5.2.14}$$

Now as $z \to 0$ with Re $z > 0$, we see that $\exp[2\pi i(h' + iz^{-1})/k] \to 0$ rapidly. Therefore the obvious way to evaluate (5.2.14) is to replace in the integrand $P(x)$ by $1 + (P(x) - 1)$. Hence

$$p(n) = \exp\left(\frac{2\pi n}{N^2}\right) \sum_{\substack{k=1 \\ (h,k)=1 \\ 0 \leqslant h < k}}^{N} \exp\left(-\frac{2\pi i h n}{k}\right) \omega_{h,k} \int_{-\theta'_{h,k}}^{\theta''_{h,k}} z^{\frac{1}{2}} \exp\left[\pi \frac{(z^{-1} - z)}{12k} - 2\pi i n\phi\right] d\phi$$

$$+ \exp\left(\frac{2\pi n}{N^2}\right) \sum_{\substack{k=1 \\ (h,k)=1 \\ 0 \leqslant h < k}}^{N} \exp\left(-\frac{2\pi i h n}{k}\right) \omega_{hk} \int_{-\theta'_{h,k}}^{\theta''_{h,k}} z^{\frac{1}{2}} \exp\left[\pi \frac{(z^{-1} - z)}{12k}\right]$$

$$\times \left(P\left\{\exp\left[2\pi i \frac{(h' + iz^{-1})}{k}\right]\right\} - 1\right) \exp(-2\pi i n\phi)\, d\phi = \Sigma_1 + \Sigma_2. \tag{5.2.15}$$

It is our natural expectation that Σ_1 will contribute our principal estimate for $p(n)$ and that the contribution of Σ_2 will be negligible. We undertake our long analysis of (5.2.15) by proving that Σ_2 is indeed negligible: First (recalling that $z = kN^{-2} - ik\phi$),

$$z^{\frac{1}{2}} \exp\left[\pi \frac{(z^{-1} - z)}{12k}\right] \left(P\left\{\exp\left[2\pi i \frac{(h' + iz^{-1})}{k}\right]\right\} - 1\right)$$

$$\leqslant |z|^{\frac{1}{2}} \exp\left(-\frac{\pi}{12N^2}\right) \sum_{m=1}^{\infty} p(m) \exp\left[-2\pi \operatorname{Re}(z^{-1}) \frac{(m - 1/24)}{k}\right]. \tag{5.2.16}$$

Now

$$\frac{1}{z} = \frac{1}{kN^{-2} - ik\phi} = \frac{N^{-2} + i\phi}{k(N^{-4} + \phi^2)}.$$

From (5.2.9) it is immediate that each of $\theta'_{h,k}$ and $\theta''_{h,k}$ satisfies $1/2kN \leqslant \theta_{h,k} < 1/kN$, and since $-\theta'_{h,k} \leqslant \phi \leqslant \theta''_{h,k}$, we see that

$$\frac{1}{k}\operatorname{Re}(z^{-1}) = \frac{N^{-2}}{k^2(N^{-4} + \phi^2)} > \frac{N^{-2}}{k^2N^{-4} + N^{-2}} = \frac{1}{1 + k^2N^{-2}} \geqslant \tfrac{1}{2}. \quad (5.2.17)$$

Also

$$|z|^{\frac{1}{2}} = (k^2N^{-4} + k^2\phi^2)^{\frac{1}{4}} < (k^2N^{-4} + N^{-2})^{\frac{1}{4}} \leqslant 2^{\frac{1}{4}}N^{-\frac{1}{2}}. \quad (5.2.18)$$

Hence by (5.2.16), (5.2.17), and (5.2.18) we have the following estimate for Σ_2:

$$|\Sigma_2| \leqslant \exp\left(\frac{2\pi n}{N^2}\right) \sum_{\substack{k=1 \\ (h,k)=1 \\ 0 \leqslant h < k}}^{N} 2^{\frac{1}{4}}N^{-\frac{1}{2}} \exp\left(-\frac{\pi}{12N^2}\right)$$

$$\times \sum_{m=1}^{\infty} p(m) \exp\left[-\pi\left(m - \frac{1}{24}\right)\right] \int_{-\theta'_{h,k}}^{\theta''_{h,k}} d\phi$$

$$\leqslant \exp\left(\frac{2\pi n}{N^2} - \frac{\pi}{12N^2}\right) 2^{\frac{1}{4}}N^{-\frac{1}{2}} \sum_{m=1}^{\infty} p(m) \exp\left[-\pi\left(m - \frac{1}{24}\right)\right] \sum_{\substack{k=1 \\ (h,k)=1 \\ 0 \leqslant h < k}}^{N} \int_{-\theta'_{h,k}}^{\theta''_{h,k}} d\phi$$

$$= \exp\left(\frac{2\pi n}{N^2} - \frac{\pi}{12N^2}\right) 2^{\frac{1}{4}}N^{-\frac{1}{2}} \sum_{m=1}^{\infty} p(m) \exp\left[-\pi\left(m - \frac{1}{24}\right)\right]$$

$$\leqslant CN^{-\frac{1}{2}} \exp\left(\frac{2\pi n}{N^2}\right). \quad (5.2.19)$$

This estimate suffices for our purposes, since $N^{-\frac{1}{2}}\exp(2\pi n N^{-2}) \to 0$ as $N \to \infty$ for fixed n.

We now handle Σ_1, the main term in Eq. (5.2.15). Here we shall show that the integral is the most significant portion of a Hankel-type loop integral. Once we establish this fact (Eq. (5.2.26)), it will then be a simple matter to complete our evaluation of $p(n)$.

In the integral for Σ_1 (see (5.2.15)) we set $\omega = N^{-2} - i\phi$, and thus the integral becomes

$$I_{h,k} \equiv \int_{N^{-2}+i\theta'_{h,k}}^{N^{-2}-i\theta''_{h,k}} (k\omega)^{\frac{1}{2}} \exp\left[\frac{\pi}{12k}\left(\frac{1}{k\omega} - k\omega\right) + 2\pi n(\omega - N^{-2})\right] i\, d\omega$$

$$= \exp(-2\pi n N^{-2})\, k^{\frac{1}{2}} i^{-1} \int_{N^{-2}-i\theta''_{h,k}}^{N^{-2}+i\theta'_{h,k}} \omega^{\frac{1}{2}} \exp\left[\frac{\pi}{12k^2\omega}\right] \exp\left[2\pi\left(n - \frac{1}{24}\right)\omega\right] d\omega$$

$$= \exp(-2\pi n N^{2})\, k^{\frac{1}{2}} i^{-1} \int_{N^{-2}+i\theta'_{h,k}}^{N^{-2}-i\theta''_{h,k}} g(\omega)\, d\omega, \tag{5.2.20}$$

where

$$g(\omega) = \omega^{\frac{1}{2}} \exp\left[2\pi\left(n - \frac{1}{24}\right)\omega + \frac{\pi}{12k^2\omega}\right].$$

The integrand of this last integral is single valued and analytic in the complex ω-plane cut along the entire negative real axis. Hence we may write (invoking Cauchy's theorem):

$$\exp(2\pi n N^{-2}) I_{h,k} = \frac{k^{\frac{1}{2}}}{i}\left(\int_{-\infty}^{(0+)} - \int_{-\infty}^{-\varepsilon} - \int_{-\varepsilon}^{-\varepsilon - i\theta''_{h,k}} - \int_{-\varepsilon - i\theta''_{h,k}}^{N^{-2}-i\theta''_{h,k}} \right.$$

$$\left. - \int_{N^{-2}+i\theta'_{h,k}}^{-\varepsilon+i\theta'_{h,k}} - \int_{-\varepsilon+i\theta'_{h,k}}^{-\varepsilon} - \int_{-\varepsilon}^{-\infty}\right) g(\omega)\, d\omega \tag{5.2.21}$$

where the integral $\int_{-\infty}^{(0+)}$ is the loop integral along the contour $\mathscr{L}$ in Fig. 5.1. We assume $0 < \varepsilon < N^{-2}$ and we shall mostly be interested in what happens as $\varepsilon \to 0$. For brevity, we rewrite (5.2.21) as

$$\exp(2\pi n N^{-2}) I_{h,k} = k^{\frac{1}{2}} i^{-1}\{L_k - I_1 - I_2 - I_3 - I_4 - I_5 - I_6\} \tag{5.2.22}$$

and our next task is to show that each of the four integrals I_2, I_3, I_4, and I_5 is negligible:

$$|I_2| \leqslant \int_0^{-\theta''_{h,k}} (\varepsilon^2 + v^2)^{\frac{1}{4}} \exp\left[\frac{\pi}{12k^2} \operatorname{Re}\left(\frac{1}{-\varepsilon + iv}\right)\right] \exp\left[-2\pi\left(n - \frac{1}{24}\right)\varepsilon\right] |dv|.$$

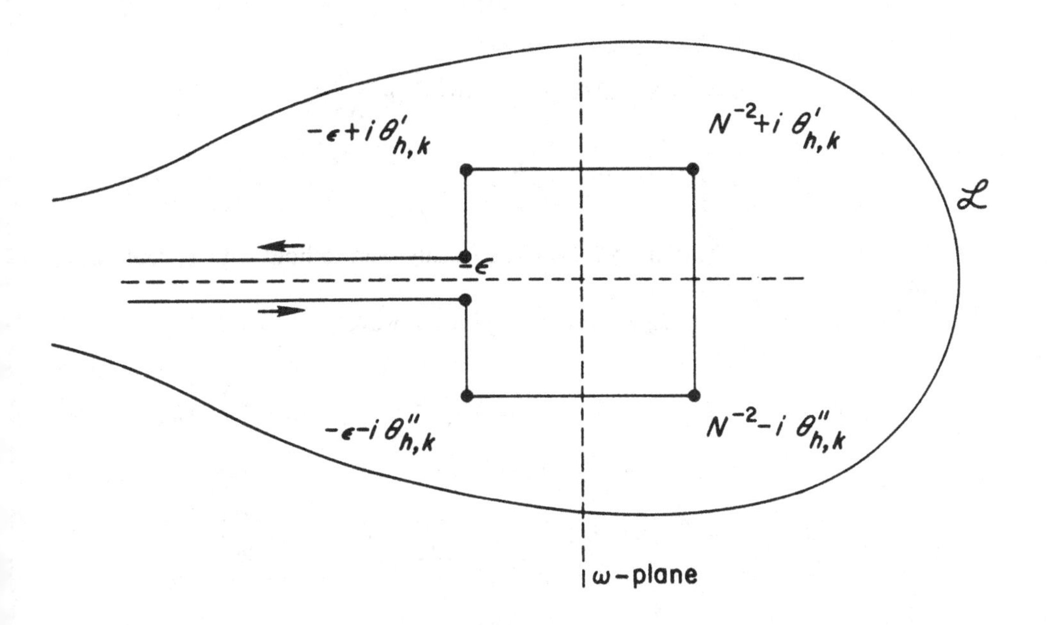

Figure 5.1

But $\mathrm{Re}[(-\varepsilon + iv)^{-1}] = -\varepsilon/(\varepsilon^2 + v^2) < 0$, and so

$$|I_2| < (\varepsilon^2 + \theta''^2_{hk})^{\frac{1}{4}}\theta''_{h,k} < \left(\varepsilon^2 + \frac{1}{k^2N^2}\right)^{\frac{1}{4}} \frac{1}{kN} \to k^{-3/2}N^{-3/2} \qquad \text{as} \quad \varepsilon \to 0. \tag{5.2.23}$$

Integral I_5 is treated in exactly the same way and (5.2.23) is also valid with I_2 replaced by I_5. As for I_3,

$$|I_3| \leqslant \int_{-\varepsilon}^{N^{-2}} (u^2 + \theta''^2_{h,k})^{\frac{1}{4}} \exp\left[\frac{\pi}{12k^2}\mathrm{Re}\left(\frac{1}{u - i\theta''_{hk}}\right)\right] \exp\left[2\pi\left(n - \frac{1}{24}\right)u\right] du.$$

Now

$$\frac{1}{k^2}\mathrm{Re}\,[(u - i\theta''_{h,k})^{-1}] = \frac{u}{k^2(u^2 + \theta''^2_{hk})} \leqslant \frac{N^{-2}}{k^2\theta''^2_{h,k}} \leqslant 4.$$

Hence

$$|I_3| < (N^{-4} + \theta''^2_{h,k})^{\frac{1}{4}} \exp\left(\frac{\pi}{3}\right) \exp\left(\frac{2\pi n}{N^2}\right)(\varepsilon + N^{-2})$$

$$< (N^{-4} + k^{-2}N^{-2})^{1/4} \exp\left(\frac{\pi}{3} + \frac{2\pi n}{N^2}\right)(\varepsilon + N^{-2})$$

$$\leqslant (\varepsilon + N^{-2})2^{1/4}k^{-1/2}N^{-1/2} \exp\left(\frac{\pi}{3} + \frac{2\pi n}{N^2}\right)$$

$$\to 2^{1/4}k^{-1/2}N^{-5/2} \exp\left(\frac{\pi}{3} + \frac{2\pi n}{N^2}\right) \quad \text{as} \quad \varepsilon \to 0 \tag{5.2.24}$$

and the final inequality in (5.2.24) is also easily seen to hold with I_3 replaced by I_5.

The integrals I_1 and I_6 are not negligible; however,

$$I_1 + I_6 = \int_{-\infty}^{-\varepsilon} \sqrt{|u|} \exp\left(-\frac{\pi i}{2}\right) \exp\left[\frac{\pi}{12k^2 u} + 2\pi\left(n - \frac{1}{24}\right)u\right] du$$

$$+ \int_{-\varepsilon}^{-\infty} \sqrt{|u|} \exp\left(\frac{\pi i}{2}\right) \exp\left[\frac{\pi}{12k^2 u} + 2\pi\left(n - \frac{1}{24}\right)u\right] du$$

$$= -2i \int_{\varepsilon}^{\infty} t^{1/2} \exp\left[-2\pi\left(n - \frac{1}{24}\right)t - \frac{\pi}{12k^2 t}\right] dt$$

$$= -2iH_k. \tag{5.2.25}$$

Hence, letting $\varepsilon \to 0$ in (5.2.22), we may simplify that equation to

$$\exp(2\pi n N^{-2})I_{hk} = \frac{k^{1/2}}{i} L_k + 2k^{1/2}H_k + O(k^{-1}N^{-3/2}) + O\left[\exp\left(\frac{2\pi n}{N^2}\right)N^{-5/2}\right], \tag{5.2.26}$$

where the constants implied in the O terms are absolute. Consequently, writing

$$\psi_k(n) = \frac{k^{1/2}}{i} L_k + 2k^{1/2}H_k, \tag{5.2.27}$$

we see from (5.2.15), (5.2.19), and (5.2.26) that

$$p(n) = \sum_{\substack{k=1 \\ (h,k)=1 \\ 0 \leqslant h < k}}^{N} \left\{\omega_{h,k} \exp\left(-\frac{2\pi i h n}{k}\right)\psi_k(n)\right\} + O\left[\exp\left(\frac{2\pi n}{N^2}\right)N^{-1/2}\right]$$

$$+ O\left(\sum_{\substack{k=1 \\ (h,k)=1 \\ 0 \leqslant h < k}}^{N} k^{-1}N^{-3/2}\right) + O\left[\sum_{\substack{k=1 \\ (h,k)=1 \\ 0 \leqslant h < k}}^{N} \exp\left(\frac{2\pi n}{N^2}\right)N^{-5/2}\right]$$

$$= \sum_{k=1}^{N} A_k(n)\psi_k(n) + O\left[N^{-\frac{1}{2}} \exp\left(\frac{2\pi n}{N^2}\right)\right] + O(N^{-\frac{1}{2}}). \tag{5.2.28}$$

Now $A_k(n)$ is precisely as it appears in (5.1.1) and the error terms here are all $\to 0$ as $N \to \infty$. Hence the final problem for us is to show that

$$\psi_k(n) = \frac{k^{\frac{1}{2}}}{\pi\sqrt{2}}\left[\frac{d}{dx}\frac{\sinh(\pi/k(\frac{2}{3}(x-1/24))^{\frac{1}{2}})}{(x-1/24)^{\frac{1}{2}}}\right]_{x=n}. \tag{5.2.29}$$

We may do this fairly easily, assuming certain classical results about Hankel's integral. First

$$\frac{1}{i}L_k = \frac{1}{i}\int_{-\infty}^{(0+)} \omega^{\frac{1}{2}} \exp\left[\frac{\pi}{12k^2\omega} + 2\pi\left(n - \frac{1}{24}\right)\omega\right] d\omega$$

$$= \frac{1}{i}\int_{-\infty}^{(0+)} \omega^{\frac{1}{2}} \exp\left[2\pi\left(n - \frac{1}{24}\right)\omega\right] \sum_{s=0}^{\infty} \frac{(\pi/12k^2\omega)^s}{s!} d\omega$$

$$= \frac{1}{i}\sum_{s=0}^{\infty} \frac{(\pi/12k^2)^s}{s!} \int_{-\infty}^{(0+)} \omega^{\frac{1}{2}-s} \exp\left[2\pi\left(n - \frac{1}{24}\right)\omega\right] d\omega$$

$$= 2\pi \sum_{s=0}^{\infty} \frac{(\pi/12k^2)^s}{s!}\left[2\pi\left(n - \frac{1}{24}\right)\right]^{s-3/2} \frac{1}{2\pi i}\int_{-\infty}^{(0+)} e^z z^{-s+\frac{1}{2}}\, dz$$

$$= (2\pi)^{-\frac{1}{2}}\left(n - \frac{1}{24}\right)^{-3/2} \sum_{s=0}^{\infty} \frac{[\pi^2(n-1/24)/6k^2]^s}{s!\,\Gamma(s-1/2)}$$

where the final equation follows from Hankel's loop integral formula for the reciprocal of the gamma function:

$$\frac{1}{\Gamma(s-\frac{1}{2})} = \frac{1}{2\pi i}\int_{-\infty}^{(0+)} e^z z^{-s+\frac{1}{2}}\, dz.$$

Now

$$\Gamma(s-\tfrac{1}{2}) = (s-3/2)(s-5/2)\cdots\tfrac{1}{2}\Gamma(\tfrac{1}{2})$$

$$= 2^{-s+1}\pi^{\frac{1}{2}}(2s-3)(2s-5)\cdots 3\cdot 1.$$

Hence

$$\sum_{s=0}^{\infty} \frac{(\frac{1}{4}Y^2)^s}{s!\Gamma(s-\frac{1}{2})} = \frac{1}{2\sqrt{\pi}}\left[-1 + Y^2\left(\frac{1}{2!} + \frac{3Y^2}{4!} + \frac{5Y^4}{6!} + \cdots\right)\right]$$

$$= \frac{1}{2\sqrt{\pi}}\left[-1 + Y^2 \frac{d}{dY}\sum_{n=0}^{\infty} \frac{Y^{2n+1}}{(2n+2)!}\right]$$

$$= \frac{1}{2\sqrt{\pi}}\left[-1 + Y^2 \frac{d}{dY}\left(\frac{\cosh Y - 1}{Y}\right)\right];$$

and so with $Y = (\pi/k)(2/3(x - 1/24))^{\frac{1}{2}}$, we see that

$$\frac{1}{i} L_k = (2\pi)^{-\frac{1}{2}}\left(n - \frac{1}{24}\right)^{-3/2} \frac{1}{2\sqrt{\pi}}\left[-1 + Y^2 \frac{d}{dY}\left(\frac{\cosh Y - 1}{Y}\right)\right]_{x=n}$$

$$= 2^{-3/2}\pi^{-1}\left(n - \frac{1}{24}\right)^{-3/2}\left[Y^2 \frac{d}{dY}\left(\frac{\cosh Y}{Y}\right)\right]_{x=n}$$

$$= 2^{-3/2}\pi^{-1}\left(n - \frac{1}{24}\right)^{-3/2}\left\{Y^2\left[\frac{d}{dx}\left(\frac{\cosh Y}{Y}\right)\Big/\frac{dY}{dx}\right]\right\}_{x=n}$$

$$= 2^{-3/2}\pi^{-1}\left(n - \frac{1}{24}\right)^{-3/2}\left[\frac{3k^2Y^3}{\pi^2}\frac{d}{dx}\left(\frac{\cosh Y}{Y}\right)\right]_{x=n}$$

$$= 3^{-\frac{1}{2}}k^{-1}\left\{\frac{d}{dx}\frac{\cosh((\pi/k)[\frac{2}{3}(x - 1/24)]^{\frac{1}{2}})}{(\pi/k)[\frac{2}{3}(x - 1/24)]^{\frac{1}{2}}}\right\}_{x=n}$$

$$= \frac{1}{\pi\sqrt{2}}\left\{\frac{d}{dx}\frac{\cosh((\pi/k)[\frac{2}{3}(x - 1/24)]^{\frac{1}{2}})}{(x - 1/24)^{\frac{1}{2}}}\right\}_{x=n}. \tag{5.2.30}$$

Finally we treat the H_k in (5.2.27) in a similar manner. We begin with a classical evaluation of a definite integral:

$$\int_0^{\infty} \exp(-c^2t - a^2t^{-1})t^{-\frac{1}{2}}\,dt = 2\int_0^{\infty} \exp(-c^2u^2 - a^2u^{-2})\,du = \frac{\sqrt{\pi}}{c}\exp(-2ac).$$

Hence

$$\int_0^{\infty} \exp(-c^2t - a^2t^{-1})t^{\frac{1}{2}}\,dt = -\frac{\sqrt{\pi}}{2c}\frac{d}{dc}\left[\frac{\exp(-2ac)}{c}\right], \tag{5.2.31}$$

and applying (5.2.31) to H_k we see that

$$H_k = -\frac{1}{4\sqrt{\pi}(n - 1/24)}\left(\frac{d}{dc}\frac{\exp[-\pi(2/3)^{\frac{1}{2}}c/k]}{(2\pi)^{\frac{1}{2}}c}\right)_{c=(n-1/24)^{\frac{1}{2}}}$$

$$= -\frac{1}{2\pi\sqrt{2}}\left\{\frac{d}{dx}\,\frac{\exp[-\pi(2/3)^{\frac{1}{2}}(x-1/24)^{\frac{1}{2}}/k]}{(x-1/24)^{\frac{1}{2}}}\right\}_{x=n}. \qquad (5.2.32)$$

Thus we obtain, from (5.2.27), (5.2.30), and (5.2.32), that

$$\begin{aligned}\psi_k(n) &= \frac{k^{\frac{1}{2}}}{\pi\sqrt{2}}\left\{\frac{d}{dx}\,\frac{\cosh((\pi/k)[\frac{2}{3}(x-1/24)]^{\frac{1}{2}})}{(x-1/24)^{\frac{1}{2}}}\right\}_{x=n} \\ &\quad - \frac{k^{\frac{1}{2}}}{\pi\sqrt{2}}\left(\frac{d}{dx}\,\frac{\exp\{-(\pi/k)[\frac{2}{3}(x-1/24)]^{\frac{1}{2}}\}}{(x-1/24)^{\frac{1}{2}}}\right)_{x=n} \\ &= \frac{k^{\frac{1}{2}}}{\pi\sqrt{2}}\left\{\frac{d}{dx}\,\frac{\sinh((\pi/k)[\frac{2}{3}(x-1/24)]^{\frac{1}{2}})}{(x-1/24)^{\frac{1}{2}}}\right\}_{x=n}, \qquad (5.2.33)\end{aligned}$$

and since (5.2.33) is just (5.2.29), we see that Theorem 5.1 is established. ■

Examples

1. The number of partitions of n into at most two parts ($= p(\text{“}\{1, 2\}\text{”}, n)$ by Theorem 1.4) may be shown to be $[n/2] + 1$ by examination of the decomposition of the generating function

$$\frac{1}{(1-q)(1-q^2)} = \frac{1/2}{(1-q)^2} + \frac{1/2}{(1-q^2)}.$$

2. The number of partitions of n into at most three parts ($= p(\text{“}\{1, 2, 3\}\text{”}, n)$ by Theorem 1.4) may be shown to be the nearest integer to $(1/12)(n + 3)^2$ by examination of the following decomposition of the generating function:

$$\frac{1}{(1-q)(1-q^2)(1-q^3)} = \frac{1/6}{(1-q)^3} + \frac{1/4}{(1-q)^2} + \frac{1/4}{(1-q^2)} + \frac{1/3}{(1-q^3)}.$$

We remark that Examples 1 and 2 are special cases of Cayley's general decomposition:

$$\frac{1}{(1-q)(1-q^2)\cdots(1-q^i)} = \sum \frac{1}{1^{p_1}2^{p_2}3^{p_3}\cdots i^{p_i}p_1!p_2!\cdots p_i!} \times \frac{1}{(1-q)^{p_1}(1-q^2)^{p_2}\cdots(1-q^i)^{p_i}}$$

where the summation runs over all partitions $(1^{p_1}2^{p_2}2^{p_3}\cdots i^{p_i})$ of i. We shall give this formula in Example 1 of Chapter 12 as a special case of results on Bell polynomials.

3*. There are many modular functions besides $\eta(\tau)$ that have transformation formulas like (5.2.2). Indeed it has been shown by Hagis and Subramanyasastri that if H is the set of integers $\equiv \pm a_1, \pm a_2, \ldots, \pm a_r \pmod{k}$, then there exists an asymptotic series expansion for $p(\text{“}H\text{”}, n)$ of the same form as that given in Theorem 5.1. In fact, the same assertion is true of $p(\text{"}H\text{"}(\leqslant t), n)$ as well. As an example of these results, we present the series for $p(\mathscr{D}, n) = p(\mathscr{O}, n)$ (Corollary 1.2), which was found through different techniques by Hua, Iseki, and Hagis:

$$p(\mathscr{D}, n) = \pi \sum_{\substack{k=1 \\ k \text{ odd}}}^{\infty} \left[\sum_{\substack{h \bmod k \\ (h,k)=1}} \chi(h, \kappa) \exp\left(\frac{-2\pi inh}{k}\right) \right] \frac{1}{k(24n+1)^{\frac{1}{2}}} I_1 \left[\frac{\pi}{12k} (48n+2)^{\frac{1}{2}} \right];$$

here

$$I_1(z) = \sum_{n=0}^{\infty} \frac{(z/2)^{2n+1}}{n!(n+1)!},$$

and $\chi(h, k)$ is the root of unity arising in the modular transformation formula for $\sum_{n \geqslant 0} p(\mathscr{D}, n) q^n$ precisely the way $\omega_{h,k}$ arises from (5.2.2).

4*. Apart from the modular functions, there exists a class of functions called mock-theta functions (originally studied by S. Ramanujan and G. N. Watson) for which reasonable asymptotic series for the coefficients may be derived. For example, let

$$f(q) = 1 + \sum_{n=1}^{\infty} \frac{q^{n^2}}{(1+q)^2(1+q^2)^2 \cdots (1+q^n)^2}$$

$$= \sum_{n=0}^{\infty} a_n q^n.$$

Then it can be shown that

$$a_n = \tfrac{1}{2} \sum_{k=1}^{\sqrt{n}} (-1)^{[(k+1)/2]} A_{2k}[n - \tfrac{1}{4}(1 + (-1)^k)k][k(n - 1/24)]^{-\frac{1}{2}}$$

$$\times \exp\left[\frac{\pi}{k\sqrt{6}} (n - 1/24)^{\frac{1}{2}} \right] + O(n^{\varepsilon})$$

where $A_k(n)$ is the same as the exponential sum appearing after (5.1.1).

Dragonette has pointed out the possibility that the error involved in the expression for a_n may generally be less than $\frac{1}{2}$ in absolute value. In particular, the series at $n = 100$ is -18520.206 while $a_{100} = -18520$, at $n = 200$ the series yields -2660007.847 while $a_{200} = -2660008$.

5. The proof of the result in Example 4 relies heavily on the fact that

$$(q)_{\infty} f(q) = 1 + 4 \sum_{n=1}^{\infty} \frac{(-1)^n q^{\frac{1}{2}n(3n+1)}}{1+q^n}.$$

This formula is due to Watson and is derived by techniques similar to those used in Section 7.2.

Examples 6–17 sketch a proof of the modular transformation formula (5.2.2) for $\eta(z) = e^{\pi iz/12}/P(e^{2\pi iz})$. The following functions all play a role in this development:

$$G(z, s) = \sum_{\substack{m,n=-\infty \\ (m,n)\neq(0,0)}}^{\infty} (mz + n)^{-s} \quad \text{for } -\pi \leqslant \arg s < \pi, \quad \operatorname{Im} z > 0, \quad \operatorname{Re} s > 2;$$

$$Vz = (az + b)/(cz + d)$$

where a, b, c, d are integers with $c > 0$, and $ad - bc = 1$;

$$g(z, s) = \sum_{\substack{m \leqslant 0 \\ dm - cn > 0}} (mz + n)^{-s}$$

(the conditions for $G(z, s)$ also apply to $g(z, s)$);

$$h(z, s) = \sum_{\substack{m > 0 \\ n > dm/c}} (mz + n)^{-s}$$

(with again the same conditions);

$$\Gamma(s) = \int_0^{\infty} u^{s-1} e^{-u}\, du, \qquad \operatorname{Re} s > 0$$

(this is Euler's integral representation of the gamma function);

$$\zeta(s) = \sum_{n=1}^{\infty} n^{-s}, \qquad \operatorname{Re} s > 1$$

(the Riemann zeta function);

$$A(z, s) = \sum_{m,n \geqslant 1} n^{s-1} e^{2\pi i m n z}, \qquad \operatorname{Im} z > 0;$$

$$H(z, s) = (1 + e^{\pi i s}) A(z, s);$$

$$L(z, s) = \sum_{j=1}^{c} \int_{\mathscr{C}} \frac{u^{s-1} e^{-(cz+d)ju/c} e^{\{jd/c\}u}}{(1 - e^{-(cz+d)u})(e^u - 1)}\, du$$

where $\{x\} = x - [x]$, with $[x]$ the largest integer not exceeding x, and where $\mathscr{C}$ is the following loop in the u plane oriented counterclockwise:

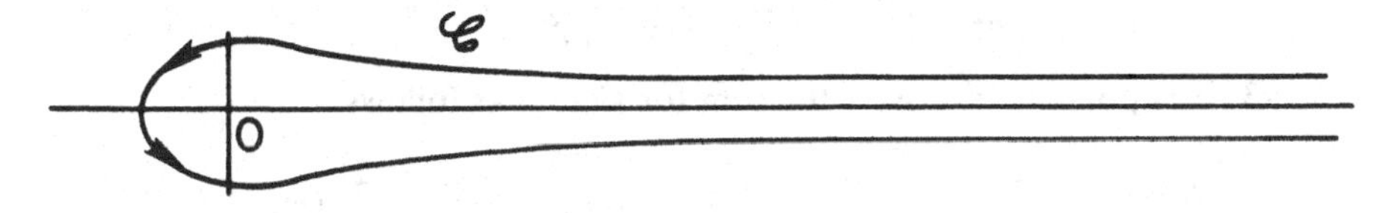

6. It is easy to show that $A(z, 0) = (\pi i z/12) - \log \eta(z)$; we need only use the fact that $\log(1 - w)^{-1} = \sum_{n=1}^{\infty} w^n/n$.

7. $$(cz + d)^{-s}G(Vz, s) = e^{-2\pi is} \sum_{\substack{m \leqslant 0 \\ \text{and } dm - cn > 0}} (mz + n)^{-s} + \sum_{\substack{m > 0 \\ \text{or } dm - cn \leqslant 0}} (mz + n)^{-s}$$
$$= G(z, s) + (e^{-2\pi is} - 1)g(z, s).$$

8. $g(z, s) = e^{\pi is}\zeta(s) + e^{\pi is}h(z, s)$.

9. For Re $z > -d/c$, Im $z > 0$,

$$\Gamma(s)h(z, s) = \sum_{\substack{m > 0 \\ n > dm/c}} \int_0^\infty u^{s-1} \exp(-mzu - nu)\, du$$

$$= \sum_{m', n' = 0}^{\infty} \int_0^\infty u^{s-1} \exp(-(m' + 1)zu - (n' + 1 + [(m'd + d)/c])u) du$$

where $m' = m - 1$ and $n' = n - [md/c] - 1$, with $[x]$ the greatest integer function.

10. If in the second sum in Example 9 we replace n' by n and set $m' = pc + j - 1$, $0 \leqslant p < \infty$, $1 \leqslant j \leqslant c$, then it follows that

$$\Gamma(s)h(z, s) = \sum_{j=1}^{c} \int_0^\infty u^{s-1} \exp\left\{-jzu - \left(1 + \left[\frac{jd}{c}\right]\right)u\right\}$$

$$\times \sum_{p, n \geqslant 0} \exp(-p(cz + d)u - nu)\, du$$

$$= \sum_{j=1}^{c} \int_0^\infty \frac{u^{s-1} \exp(-(cz + d)ju/c + \{jd/c\}u)}{\{1 - \exp(-(cz + d)u)\}(\exp(u) - 1)}\, du$$

$$= (1 - \exp(2\pi is))^{-1} \sum_{j=1}^{c} \int_{\mathscr{C}} \frac{u^{s-1} \exp(-(cz + d)ju/c + \{jd/c\}u)}{(1 - \exp(-(cz + d)u))(\exp(u) - 1)}\, du.$$

11. Applying Example 10 to Example 8 and hence to Example 7, we find that for Re $s > 2$, Re $z > -d/c$, Im $z > 0$,

$$(cz + d)^{-s}\Gamma(s)G(Vz, s) = \Gamma(s)G(z, s) - 2i \sin \pi s\Gamma(s)\zeta(s) + e^{-\pi is}L(z, s).$$

12*. (The Lipschitz Summation Formula.) For Im $z > 0$ and Re $s > 1$,

$$\sum_{n=-\infty}^{\infty} (n + z)^{-s} = \frac{(-2\pi i)^s}{\Gamma(s)} \sum_{n+\alpha > 0} n^{s-1} e^{2\pi inz}.$$

13. It is possible to dissect the sum for $G(z, s)$ as follows.

$$G(z, s) = \sum_{\substack{n=-\infty \\ n \neq 0}}^{\infty} n^{-s} + \sum_{m<0} \sum_{n=-\infty}^{\infty} (mz + n)^{-s} + \sum_{m>0} \sum_{n=-\infty}^{\infty} (mz + n)^{-s}$$
$$= (1 + e^{\pi is})\zeta(s) + S_2 + S_3$$

where by Example 12

$$S_2 = e^{\pi is}(-2\pi i)^s A(z, s)/\Gamma(s), \qquad S_3 = (-2\pi i)^s A(z, s)/\Gamma(s).$$

14. The result in Example 13 may be rewritten as

$$G(z, s) = (1 + e^{\pi is})\zeta(s) + (1 + e^{\pi is})(-2\pi i)^s A(z, s)/\Gamma(s),$$

which gives the analytic continuation of $G(z, s)$ to the entire complex s plane.

15. From Example 14 and the definition of $H(z, s)$, it is possible to rewrite Example 11 in terms of $H(z, s)$:

$$(cz + d)^{-s} H(Vz, s) = H(z, s) - e^{\pi is}(2\pi i)^{-s}(cz + d)^{-s}\Gamma(s)(1 + e^{\pi is})\zeta(s) + (2\pi i)^{-s}\Gamma(s)(1 + e^{\pi is})\zeta(s) + (2\pi i)^{-s} L(z, s).$$

16*. The calculus of residues may be utilized to show that

$$L(z, 0) = 2\pi i\left(\frac{-1}{12c(cz + d)} - \frac{(cz + d)}{12c} + s(d, c) - \tfrac{1}{4}\right),$$

where $s(d, c)$ is given in Eq. (5.2.6).

17. Setting $s = 0$ in Example 15 and applying Example 16, we may deduce that

$$\log \eta(Vz) = \log \eta(z) - \frac{\pi i}{4} + \tfrac{1}{2}\log(cz + d) - \pi i s(d, c) + \frac{\pi i(a + d)}{12c},$$

a result equivalent to (5.2.2).

Notes

The origin of this chapter is the epoch-making paper by Hardy and Ramanujan (1918). Our presentation follows closely Rademacher's (1937) original exposition of this theorem. Subsequently, Rademacher (1943, 1973) refined the circle method using what are called *Ford circles*. Although this procedure is elegant and has been useful in later developments of modular function theory, we have chosen Rademacher's original approach because it is easily adapted to problems concerning nonmodular generating functions (see Chapter 6). The books by Ayoub (1963), Hardy (1940), Knopp (1970), Lehner (1964), and Rademacher (1973) present excellent accounts of the role of modular functions in partition theory. Certain important contributions to partition asymptotics by E. Grosswald (1958, 1960) are related to both this chapter and Chapter 6. Reviews of the recent work extending the Hardy–Ramanujan–Rademacher method are found in Sections P68 and P72 of LeVeque (1974).

Examples 1, 2. Cayley (1898), MacMahon (1916), Arkin (1970).

Example 3. Hagis (1962, 1963, 1964a, b, 1965a, b, c, 1966) treats many cases when k is prime; Subramanyasastri (1972) treats arbitrary k. See also Iseki (1959, 1960, 1961).

Example 4. Watson (1936), Dragonette (1952), Andrews (1966).

Example 5. Watson (1929, 1936).

The proof of (5.2.2) presented in Examples 6–17 is due to B. Berndt (private communication) and is presented in much greater generality in Berndt (1973,

1975). Berndt has shown this method to be applicable to numerous modular functions and to be simpler conceptually and computationally than alternative methods; for these reasons this approach should have wide ramifications in partition theory.

References

Andrews, G. E. (1966). "On the theorems of Watson and Dragonette for Ramanujan's mock theta functions," *Amer. J. Math.* **88**, 454–490.

Arkin, J. (1970). "Researches on partitions," *Duke Math. J.* **38**, 403–409.

Ayoub, R. (1963). An Introduction to the Analytic Theory of Numbers, American Mathematical Soc., Providence.

Berndt, B. (1973). "Generalized Dedekind eta-functions and generalized Dedekind sums," *Trans. Amer. Math. Soc.* **178**, 495–508.

Berndt, B. (1975). "Generalized Eisenstein series and modified Dedekind sums," *J. Reine Angew. Math.* **272**, 182–193.

Cayley, A. (1898). "Researches in the partition of numbers," *Collected Math. Papers* **2**, 235–249, 506–512.

Dragonette, L. A. (1952). "Some asymptotic formulae for the mock theta series of Ramanujan," *Trans. Amer. Math. Soc.* **72**, 474–500.

Grosswald, E. (1958). "Some theorems concerning partitions," *Trans. Amer. Math. Soc.* **89**. 113–128.

Grosswald, E. (1960). "Partitions into prime powers," *Michigan Math. J.* **7**, 97–122.

Hagis, P. (1962). "A problem on partitions with a prime modulus $p \geqslant 3$," *Trans. Amer. Math. Soc.* **102**, 30–62.

Hagis, P. (1963). "Partitions into odd summands," *Amer. J. Math.* **85**, 213–222.

Hagis, P. (1964a). "On a class of partitions with distinct summands," *Trans. Amer. Math. Soc.* **112**, 401–415.

Hagis, P. (1964b). "Partitions into odd and unequal parts," *Amer. J. Math.* **86**, 317–324.

Hagis, P. (1965a). "A correction of some theorems on partitions," *Trans. Amer. Math. Soc.* **118**, 550.

Hagis, P. (1965b). "Partitions with odd summands – some comments and corrections," *Amer. J. Math.* **87**, 218–220.

Hagis, P. (1965c). "On the partitions of an integer into distinct odd summands," *Amer. J. Math.* **87**, 867–873.

Hagis, P. (1966). "Some theorems concerning partitions into odd summands," *Amer. J. Math.* **88**, 664–681.

Hardy, G. H. (1940). *Ramanujan*. Cambridge Univ. Press, London and New York (reprinted by Chelsea, New York).

Hardy, G. H., and Ramanujan, S. (1918). "Asymptotic formulae in combinatory analysis," *Proc. London Math. Soc.* (2) **17**, 75–115. (Also, *Collected Papers of S. Ramanujan*, pp. 276–309. Cambridge Univ. Press, London and New York, 1927; reprinted by Chelsea, New York, 1962.)

Hua, L. K. (1942). "On the number of partitions of a number into unequal parts," *Trans. Amer. Math. Soc.* **51**, 194–201.

Iseki, S. (1959). "A partition function with some congruence condition," *Amer. J. Math.* **81**, 939–961.

Iseki, S. (1960). "On some partition functions," *J. Math. Soc. Japan* **12**, 81–88.

Iseki, S. (1961). "Partitions in certain arithmetic progressions," *Amer. J. Math.* **83**, 243–264.

Knopp, M. I. (1970). *Modular Functions in Analytic Number Theory*. Markham, Chicago.

Lehmer, D. H. (1939). "On the remainders and convergence of the series for the partition function," *Trans. Amer. Math. Soc.* **46**, 362–373.

Lehner, J. (1964). *Discontinuous Groups and Automorphic Functions*. Mathematical Surveys, No. VIII, Amer. Math. Soc., Providence.

LeVeque, W. J. (1956). *Topics in Number Theory*, Vol. 2. Addison-Wesley, Reading, Mass.

LeVeque, W. J. (1974). *Reviews in Number Theory*, Vol. 4. Amer. Math. Soc., Providence, R.I.

MacMahon, P. A. (1916). *Combinatory Analysis*, Vol. 2. Cambridge Univ. Press, London and New York (reprinted by Chelsea, New York, 1960).

Rademacher, H. (1937). "On the partition function $p(n)$," *Proc. London Math. Soc.* (2) **43**, 241–254.

Rademacher, H. (1943). "On the expansion of the partition function in a series," *Ann. of Math.* **44**, 416–422.

Rademacher, H. (1958). "On the Selberg formula for $A_k(n)$," *J. Indian Math. Soc.* (*N.S.*) **21**, 41–55.

Rademacher, H. (1973). *Topics in Analytic Number Theory*. Springer, Berlin.

Subramanyasastri, V. V. (1972). "Partitions with congruence conditions," *J. Indian Math. Soc.* **36**, 177–194.

Watson, G. N. (1936). "The final problem: an account of the mock-theta functions," *J. London Math. Soc.* **11**, 55–80.

Whiteman, A. L. (1947). "A sum connected with the partition function," *Bull. Amer. Math. Soc.* **53**, 598–603.

Whiteman, A. L. (1956). "A sum connected with the series for the partition function," *Pacific J. Math.* **6**, 159–176.

CHAPTER 6

The Asymptotics of Infinite Product Generating Functions

6.1 Introduction

As we saw in Chapter 1, many partition functions have infinite products as the associated generating function. Numerous researchers have considered the asymptotics of such partition functions. We shall present the general theorem of G. Meinardus, which gives an asymptotic formula that includes numerous partition functions. His method is also amenable to generalization and we shall discuss some possibilities for generalization at the end of this chapter.

The infinite product we shall consider is

$$f(\tau) = \prod_{n=1}^{\infty} (1 - q^n)^{-a_n}$$

$$= 1 + \sum_{n=1}^{\infty} r(n) q^n, \tag{6.1.1}$$

where $q = e^{-\tau}$ and $\operatorname{Re} \tau > 0$ (or equivalently $|q| < 1$). We restrict the a_n to be nonnegative real numbers. We also consider an auxiliary Dirichlet series:

$$D(s) = \sum_{n=1}^{\infty} \frac{a_n}{n^s} \qquad (s = \sigma + it), \tag{6.1.2}$$

which is assumed to converge for $\sigma > \alpha$, a positive real number. Furthermore, we assume that $D(s)$ possesses an analytic continuation in the region $\sigma \geqslant -C_0$ $(0 < C_0 < 1)$ and that in this region $D(s)$ is analytic except for a pole of order 1 at $s = \alpha$ with residue A. Finally, we assume that

$$D(s) = O(|t|^{C_1}) \tag{6.1.3}$$

uniformly in $\sigma \geqslant -C_0$ as $|t| \to \infty$, where C_1 is a fixed positive real number.

ENCYCLOPEDIA OF MATHEMATICS and Its Applications, Gian-Carlo Rota (ed.). 2, George E. Andrews. The Theory of Partitions

We must finally make use of the function

$$g(\tau) = \sum_{n=1}^{\infty} a_n q^n, \qquad q = e^{-\tau}. \tag{6.1.4}$$

Now if $\tau = y + 2\pi i x$ (x, y real), we shall also assume that for $|\arg \tau| > \pi/4$, $|x| \leqslant \frac{1}{2}$,

$$\mathrm{Re}(g(\tau)) - g(y) \leqslant -C_2 y^{-\varepsilon} \tag{6.1.5}$$

for sufficiently small y, where ε is an arbitrary fixed positive number and C_2 is a suitably chosen positive real number depending on ε.

With the definitions above in mind, we state Meinardus's main theorem:

THEOREM 6.2. *As* $n \to \infty$,

$$r(n) = Cn^{\kappa} \exp\left\{n^{\alpha/(\alpha+1)}\left(1 + \frac{1}{\alpha}\right)[A\Gamma(\alpha + 1)\zeta(\alpha + 1)]^{1/(\alpha+1)}\right\}(1 + O(n^{-\kappa_1})) \tag{6.1.6}$$

where $\zeta(s) = \sum_{m=1}^{\infty} m^{-s}$ *is the Riemann zeta function, and*

$$C = e^{D'(0)}[2\pi(1 + \alpha)]^{-\frac{1}{2}}[A\Gamma(\alpha + 1)\zeta(\alpha + 1)]^{(1-2D(0))/(2+2\alpha)}, \tag{6.1.7}$$

$$\kappa = \frac{D(0) - 1 - \frac{1}{2}\alpha}{1 + \alpha}, \tag{6.1.8}$$

$$\kappa_1 = \frac{\alpha}{\alpha + 1} \min\left(\frac{C_0}{\alpha} - \frac{\delta}{4}, \frac{1}{2} - \delta\right), \tag{6.1.9}$$

δ *an arbitrary real number.*

The proof of this theorem relies on application of the saddle point method. In the examples at the end of the chapter, we shall present numerous applications of Theorem 6.2.

6.2 Proof of Theorem 6.2

In applying the saddle point method, we must have information on the behavior of $f(\tau)$ in the half-plane $\mathrm{Re}(\tau) > 0$ especially, near $\tau = 0$; this is supplied by the following:

LEMMA 6.1. *Under the assumptions on* $f(\tau)$, $D(s)$, *and* $g(\tau)$ *presented in Section* 6.1 (*with* $\tau = y + 2\pi i x$),

$$f(\tau) = \exp[A\Gamma(\alpha)\zeta(\alpha + 1)\tau^{-\alpha} - D(0) \log \tau + D'(0) + O(y^{C_0})] \tag{6.2.1}$$

uniformly in x as $y \to 0$, provided $|\arg \tau| \leqslant \pi/4$, $|x| \leqslant \frac{1}{2}$; there exists a positive ε_1 such that

$$f(y + 2\pi ix) = O(\exp[A\Gamma(\alpha)\zeta(\alpha + 1)y^{-\alpha} - C_3 y^{-\varepsilon_1}]) \qquad (6.2.2)$$

uniformly in x with $y^\beta \leqslant |x| \leqslant \frac{1}{2}$, as $y \to 0$, where

$$\beta = 1 + \frac{\alpha}{2}\left(1 - \frac{\delta}{2}\right) \quad \text{with} \quad 0 < \delta < \tfrac{2}{3}, \qquad (6.2.3)$$

and C_3 a fixed real number.

Proof. We note that

$$\log f(\tau) = -\sum_{\nu=1}^{\infty} a_\nu \log(1 - e^{-\nu\tau})$$

$$= \sum_{k=1}^{\infty} \frac{1}{k} \sum_{\nu=1}^{\infty} a_\nu e^{-\nu k\tau}. \qquad (6.2.4)$$

Now recall that $e^{-\tau}$ is the Mellin transform of $\Gamma(s)$; that is,

$$e^{-\tau} = \frac{1}{2\pi i} \int_{\sigma_0 - i\infty}^{\sigma_0 + i\infty} \tau^{-s}\Gamma(s)\, ds \qquad (\text{for } \operatorname{Re}\tau > 0,\ \sigma_0 > 0). \qquad (6.2.5)$$

Applying (6.2.5) to the exponential function in (6.2.4), we see that

$$\log f(\tau) = \frac{1}{2\pi i} \int_{1+\alpha - i\infty}^{1+\alpha + i\infty} \tau^{-s}\Gamma(s)\zeta(s+1)D(s)\, ds; \qquad (6.2.6)$$

the interchange of summation and integration necessary to obtain (6.2.6) is permissible due to the absolute convergence of the resulting integrated series.

Our object now is to shift the line of integration from $\operatorname{Re}\tau = 1 + \alpha$ to $\operatorname{Re}\tau = -C_0$. We first note that the integrand in (6.2.6) has a first-order pole at $s = \alpha$ and a second-order pole at $s = 0$; the residue at $s = \alpha$ is clearly $\tau^{-\alpha}\Gamma(\alpha)\zeta(\alpha + 1)A$. Around $s = 0$,

$$\tau^{-s}\Gamma(s)\zeta(s+1)D(s)$$

$$= (1 - s\log\tau + \cdots)\left(\frac{1}{s} - \gamma + \cdots\right)\left(\frac{1}{s} + \gamma + \cdots\right)(D(0) + D'(0)s + \cdots)$$

$$= \frac{1}{s^2} + (D'(0) - D(0)\log\tau)\frac{1}{s} + \cdots;$$

hence, the residue of the integrand in (6.2.6) at $s = 0$ is $D'(0) - D(0)\log\tau$. Hence, we see that

$$\log f(\tau) = A\Gamma(\alpha)\zeta(\alpha+1)\tau^{-\alpha} - D(0)\log\tau + D'(0) + \frac{1}{2\pi i}\int_{-C_0-i\infty}^{-C_0+i\infty} \tau^{-s}\Gamma(s)\zeta(s+1)D(s)\,ds; \tag{6.2.7}$$

the shift of the line of integration made in passing from (6.2.6) to (6.2.7) is permissible since for $|\arg\tau| \leqslant \pi/4$, we see that

$$|\tau^{-s}| = |\tau|^{-\sigma}\exp(t\arg\tau) \leqslant |\tau|^{-\sigma}\exp(\pi|t|/4)$$

and for $\sigma \geqslant C_0$,

$$D(s) = O(|t|^{C_1})$$

(by assumption), while classical results on the ζ and Γ functions assert that

$$\zeta(s+1) = O(|t|^{C_4}),$$

and

$$\Gamma(s) = O\left(\exp\left(-\frac{\pi}{2}|t|\right)|t|^{C_5}\right)$$

as $t \to \infty$.

Finally we observe that

$$\left|\frac{1}{2\pi i}\int_{-C_0-i\infty}^{-C_0+i\infty} \tau^{-s}\Gamma(s)\zeta(s+1)D(s)\,ds\right|$$

$$= O\left(|\tau|^{C_0}\int_{-\infty}^{\infty}\exp\left(-\frac{\pi}{4}|t|\right)|t|^{C_1+C_4+C_5}\,dt\right)$$

$$= O(|\tau|^{C_0}) = O(y^{C_0}), \tag{6.2.8}$$

(since $|\arg\tau| \leqslant \pi/4$ implies $|2\pi x| \leqslant y$ and thus $|\tau| \leqslant \sqrt{2}\,y$). Equation (6.2.1) now follows from (6.2.7) once we estimate the integral in (6.2.7) by means of (6.2.8).

We now turn to Eq. (6.2.2); here we consider two cases: (1) $y^\beta \leqslant |x| \leqslant y/2\pi$ and (2) $y/2\pi \leqslant |x| \leqslant \frac{1}{2}$. (One or both of these cases may be vacuous, depending on y, but since we are interested only in y approaching 0, we may always assume y sufficiently small to make both nonvacuous.)

In case 1, we see that

$$\tan|\arg \tau| = \frac{2\pi|x|}{y} \leqslant 1 \qquad \text{or} \qquad |\arg \tau| \leqslant \frac{\pi}{4},$$

as assumed above. Hence we may estimate (6.2.7) in the way done before, and we determine that

$$|\log f(y + 2\pi i x)| \leqslant A\Gamma(\alpha)\zeta(\alpha + 1)|\tau|^{-\alpha} + C_6|\log y|. \tag{6.2.9}$$

The extra terms we might expect on the right-hand side of (6.2.9) have all been accounted for by $C_6|\log y|$, since the other terms are of lower order of magnitude (remember y is approaching 0, so that $|\log y|$ dominates all nonnegative powers of y). Recalling that $|\tau|^2 = y^2 + 4\pi^2x^2$, we see from (6.2.9) that

$$\begin{aligned}
&\log|f(y + 2\pi i x)| \\
&\quad \leqslant A\Gamma(a)\zeta(\alpha + 1)y^{-\alpha} + A\Gamma(\alpha)\zeta(\alpha + 1)y^{-\alpha}[(1 + 4\pi^2x^2y^{-2})^{-\alpha/2} - 1] \\
&\qquad + C_6|\log y| \\
&\quad \leqslant A\Gamma(\alpha)\zeta(\alpha + 1)y^{-\alpha} - C_7y^{-\alpha}|x|^2y^{-2} \leqslant A\Gamma(\alpha)\zeta(\alpha + 1)y^{-\alpha} - C_7y^{-\alpha + 2(\beta - 1)}.
\end{aligned} \tag{6.2.10}$$

The next to last inequality follows from the observation that as $Y \to 0^+$, $(1 + AY)^{-\alpha/2} - 1 \sim -\alpha AY/2$ and the $|\log y|$ is dominated by any negative power of y as y approaches 0.

Now by (6.2.3), we see that $-\alpha + 2(\beta - 1) = -\alpha\delta/2 \leqslant -\varepsilon_1$, say. Hence, in case 1,

$$\log|f(y + 2\pi i x)| \leqslant A\Gamma(\alpha)\zeta(\alpha + 1)y^{-\alpha} - C_3y^{-\varepsilon_1}, \tag{6.2.11}$$

which is equivalent to (6.2.2).

In case 2 (i.e., $y/2\pi \leqslant |x| \leqslant \frac{1}{2}$), we see from (6.1.4) and (6.2.4) that

$$\begin{aligned}
\log|f(y + 2\pi i x)| - \operatorname{Re}(g(\tau)) &= \sum_{k=2}^{\infty} \frac{1}{k} \sum_{\nu=1}^{\infty} a_\nu e^{-\nu k y}\cos(2\pi k\nu x) \\
&\leqslant \sum_{k=2}^{\infty} \frac{1}{k} \sum_{\nu=1}^{\infty} a_\nu e^{-\nu k y} = \log f(y) - g(y),
\end{aligned} \tag{6.2.12}$$

since the a_ν are all nonnegative. Now we have the hypotheses necessary to apply (6.1.5). Hence in case 2

$$\begin{aligned}\log|f(y + 2\pi ix)| &\leqslant \log f(y) + \mathrm{Re}(g(\tau)) - g(y)\\ &\leqslant A\Gamma(\alpha)\zeta(\alpha + 1)y^{-\alpha} - C_8 y^{-\varepsilon}\\ &\leqslant A\Gamma(\alpha)\zeta(\alpha + 1)y^{-\alpha} - C_3 y^{-\varepsilon_1}.\end{aligned} \tag{6.2.13}$$

Hence our lemma is completely proved. ■

We are now prepared to obtain the asymptotic formula for $r(n)$.

Proof of Theorem 6.2. We recall by the Cauchy integral theorem that

$$r(n) = \frac{1}{2\pi i}\int_{\tau_0}^{\tau_0+2\pi i} f(\tau)e^{n\tau}\,d\tau$$

$$= \int_{-\frac{1}{2}}^{\frac{1}{2}} f(y + 2\pi ix)e^{ny+2\pi inx}\,dx. \tag{6.2.14}$$

We wish to apply the saddle point method to the evaluation of this integral. Since the maximum absolute value of the integrand occurs for $x = 0$, and since for $x = 0$, Lemma 6.1 implies that the integrand is well approximated by

$$\exp[A\Gamma(\alpha)\zeta(\alpha + 1)y^{-\alpha} + ny],$$

the saddle point method suggests that we should minimize this expression; that is, we should require that y be chosen so that

$$\frac{d}{dy}\{\exp[A\Gamma(\alpha)\zeta(\alpha + 1)y^{-\alpha} + ny]\} = 0.$$

Hence we take

$$y = n^{-1/(\alpha+1)}(A\Gamma(\alpha + 1)\zeta(\alpha + 1))^{1/(\alpha+1)}. \tag{6.2.15}$$

For notational simplicity we define

$$m = ny = n^{\alpha/(\alpha+1)}[A\Gamma(\alpha + 1)\zeta(\alpha + 1)]^{1/(\alpha+1)}. \tag{6.2.16}$$

From (6.2.14), we see that

$$r(n) = e^m \int_{-y^\beta}^{y^\beta} f(y + 2\pi ix)e^{2\pi inx}\,dx + e^m R_1, \tag{6.2.17}$$

where

$$R_1 = \int_{-\frac{1}{2}}^{-y^\beta} + \int_{y^\beta}^{\frac{1}{2}} f(y + 2\pi i x)e^{2\pi i n x}\, dx \tag{6.2.18}$$

and where β is defined by (6.2.3). The interval of integration is being split into two parts in a manner related to the Hardy–Ramanujan Farey dissection applied to the partition generating function in Chapter 5. To actually understand how (6.2.3) arises we may note that we have already used explicitly in the proof of Lemma 6.1 that $-\alpha + 2(\beta - 1)$ should be a small negative number.

By Lemma 6.1,

$$R_1 = O\left(\exp\left[\left(\frac{m}{n}\right)^{-\alpha} A\Gamma(\alpha)\zeta(\alpha+1) - C_3\left(\frac{m}{n}\right)^{-\varepsilon_1}\right]\right) \tag{6.2.19}$$

as $n \to \infty$ (i.e., $y = m/n \to 0$). Hence

$$\exp(m)R_1 = O\left(\exp\left[\left(1 + \frac{1}{\alpha}\right)m - C_9 m^{\varepsilon_2}\right]\right) \tag{6.2.20}$$

as $n \to \infty$.

Equation (6.2.20) provides an adequate estimate for the second term in (6.2.17). We must now treat the main integral. Let us choose $n \geqslant n_2 \geqslant n_1$, where n_2 is sufficiently large that $2\pi(m/n)^{\beta-1} \leqslant 1$ (this is quite permissible, since $\beta > 1$ and $m/n \to 0$ as $n \to \infty$). Then by Lemma 6.1

$$r(n) = \exp\left[\left(1 + \frac{1}{\alpha}\right)m + D'(0)\right] \int_{-(m/n)^\beta}^{(m/n)^\beta} \exp(\phi_1(x))\, dx + \exp(m)R_1, \tag{6.2.21}$$

where

$$\phi_1(x) = \frac{m}{\alpha}\left[\left(1 + \frac{2\pi i x n}{m}\right)^{-\alpha} - 1\right] + 2\pi i n x - D(0)\log\left(\frac{m}{n} + 2\pi i x\right) + O(m^{-C_0/\alpha}); \tag{6.2.22}$$

note that the choice of n_1 guarantees that throughout the interval of integration $|x| \leqslant \frac{1}{2}$, and the choice of n_2 guarantees that $\arg\tau \leqslant \pi/4$ as well.

We now make the change of variable $2\pi x = (m/n)\omega$, and we obtain

$$r(n) = \exp\left[\left(1 + \frac{1}{\alpha}\right)m - (D(0)-1)\log\frac{m}{n} + D'(0) - \log 2\pi\right]\cdot I + \exp(m)\, R_1, \tag{6.2.23}$$

where

$$I = \int_{-C_{10}m^{(1-\beta)/\alpha}}^{C_{10}m^{(1-\beta)/\alpha}} \exp(\phi_2(\omega))\, d\omega, \tag{6.2.24}$$

and

$$\phi_2(\omega) = m\left\{\frac{1}{\alpha(1+i\omega)^\alpha} - \frac{1}{\alpha} + i\omega\right\} - D(0)\log(1+i\omega) + O(m^{-C_0/\alpha}) \tag{6.2.25}$$

as $m \to \infty$.

Our problem is now reduced to obtaining an asymptotic expression for I. First

$$I = \int_{-C_{10}m^{(1-\beta)/\alpha}}^{C_{10}m^{(1-\beta)/\alpha}} \exp\left[-\frac{m(\alpha+1)}{2}\omega^2\right] d\omega + R_2, \tag{6.2.26}$$

where

$$R_2 = \int_{-C_{10}m^{(1-\beta)/\alpha}}^{C_{10}m^{(1-\beta)/\alpha}} \exp\left[-m\frac{(\alpha+1)}{2}\omega^2\right][\exp(\phi_3(\omega)) - 1]\, d\omega, \tag{6.2.27}$$

with

$$\phi_3(\omega) = m\left[\frac{1}{\alpha(1+i\omega)^\alpha} - \frac{1}{\alpha} + i\omega + \frac{\alpha+1}{2}\omega^2\right] - D(0)\log(1+i\omega) + O(m^{-C_0/\alpha}) \tag{6.2.28}$$

as $m \to \infty$. For $n \geqslant n_3 \geqslant n_2$, choose n_3 sufficiently large that $|\omega| < 1$ throughout the interval of integration. Hence

$$\begin{aligned} m&\left[\frac{1}{\alpha(1+i\omega)^\alpha} - \frac{1}{\alpha} + i\omega + \frac{\alpha+1}{2}\omega^2\right] \\ &= m\left[\frac{1}{\alpha}\sum_{j\geqslant 0}\binom{\alpha+j-1}{j}(-1)^j i^j \omega^j - \frac{1}{\alpha} + i\omega + \frac{\alpha+1}{2}\omega^2\right] \\ &= \frac{m}{\alpha}\sum_{j\geqslant 3}\binom{\alpha+j-1}{j}(-1)^j i^j\omega^j = O(m|\omega|^3) = O(m^{(\alpha+3(1-\beta))/\alpha}) \end{aligned} \tag{6.2.29}$$

while

$$\log(1 + i\omega) = -\sum_{j \geqslant 1} \frac{(-1)^j i^j \omega^j}{j} = O(|\omega|) = O(m^{(1-\beta)/\alpha}). \tag{6.2.30}$$

Therefore, as $m \to \infty$

$$\exp(\phi_3(\omega)) - 1 = O(|\phi_3(\omega)|)$$
$$= O(m^{[\alpha + 3(1-\beta)]/\alpha} + m^{(1-\beta)/\alpha} + m^{-C_0/\alpha}) = O(m^{-\mu_1}), \tag{6.2.31}$$

where

$$\mu_1 = \min\left(\frac{C_0}{\alpha}, \frac{1}{2} - \frac{3\delta}{4}\right), \tag{6.2.32}$$

by (6.2.3).

Since the length of the interval of integration in (6.2.27) is $O(m^{(1-\beta)/\alpha})$, we see that as $m \to \infty$

$$R_2 = O(m^{-\mu_2}) \tag{6.2.33}$$

where

$$\mu_2 = \min\left(\frac{C_0}{\alpha} + \frac{1}{2} - \frac{\delta}{4}, 1 - \delta\right). \tag{6.2.34}$$

As for the integral in (6.2.26), we see that

$$\int_{-C_{10}m^{(1-\beta)/\alpha}}^{C_{10}m^{(1-\beta)/\alpha}} \exp\left[-\frac{m(\alpha+1)}{2}\omega^2\right] d\omega = [\tfrac{1}{2}m(\alpha+1)]^{-\frac{1}{2}} \int_{-C_{11}m^{\delta/4}}^{C_{11}m^{\delta/4}} \exp(-z^2)\, dz$$
$$= \left[\frac{2\pi}{m(\alpha+1)}\right]^{\frac{1}{2}} + O(m^{-\frac{1}{2}} \exp(-C_{12}m^{\delta/4})). \tag{6.2.35}$$

Hence by (6.2.26), (6.2.33), and (6.2.35), we obtain, as $m \to \infty$,

$$I = \left[\frac{2\pi}{m(\alpha+1)}\right]^{\frac{1}{2}} (1 + O(m^{-\mu_3})) \tag{6.2.36}$$

where

$$\mu_3 = \min\left(\frac{C_0}{\alpha} - \frac{\delta}{4}, \frac{1}{2} - \delta\right). \tag{6.2.37}$$

Finally by (6.2.17), (6.2.20), (6.2.33), and (6.2.36), we see that

$$r(n) = \exp\left[\left(1 + \frac{1}{\alpha}\right)m - (D(0) - 1)\log\frac{m}{n} + D'(0)\right](2\pi m(\alpha+1))^{-\frac{1}{2}}$$
$$\times (1 + O(m^{-\mu_3})) \tag{6.2.38}$$

as $m \to \infty$. Using (6.2.16) to replace m by a function of n, we obtain exactly (6.1.16). Thus Theorem 6.2 is proved. ■

6.3 Applications of Theorem 6.2

Theorem 6.2 applies to numerous partition functions considered in Chapters 1 and 2. We shall see in Chapter 11, Example 6, that it also produces asymptotic formulas for many plane partition functions.

THEOREM 6.3. (Cf. Eq. (5.1.2).)

$$p(n) \sim \frac{1}{4n\sqrt{3}} \exp(\pi(\tfrac{2}{3})^{\frac{1}{2}} n^{\frac{1}{2}}).$$

Proof. In Theorem 6.2, set $a_n = 1$ for all n; then $r(n) = p(n)$; $D(s) = \zeta(s)$ (and $\zeta'(0) = -\tfrac{1}{2}\log(2\pi)$, $\zeta(0) = -\tfrac{1}{2}$); and $g(\tau) = (1 - e^{-\tau})^{-1}$. It is an easy matter to check the various conditions listed in (6.1.3)–(6.1.5), and the result above follows from Theorem 6.2 once the indicated substitutions are made. ■

THEOREM 6.4. *Let $H_{k,a}$ denote all positive integers congruent to a modulo k. Then for $1 \leqslant a \leqslant k$,*

$$p(\text{“}H_{k,a}\text{”}, n) \sim Cn^{\kappa} \exp\left(\pi\left(\frac{2n}{3k}\right)^{\frac{1}{2}}\right)$$

where

$$C = \Gamma\left(\frac{a}{k}\right)\pi^{(a/k)-1}2^{-(3/2)-(a/2k)}3^{-(a/2k)}k^{-\frac{1}{2}+(a/2k)}$$

and

$$\kappa = -\tfrac{1}{2}\left(1 + \frac{a}{k}\right).$$

This result follows in exactly the way Theorem 6.3 was proved. Now $D(s) = k^{-s}\zeta(s, a/k)$, where $\zeta(s, h) = \sum_{n \geqslant 0} (n + h)^{-s}$ is the Hurwitz zeta function, and $g(s) = e^{-a\tau}(1 - e^{-k\tau})^{-1}$. ■

Examples

1. Let $H_{k,\pm a}$ denote the set of positive integers congruent to a or $-a$ modulo k. Then for k prime and $1 \leqslant a < k/2$,

$$p(\text{“}H_{k,\pm a}\text{”}, n) \sim \frac{\csc(\pi a/k)}{4\pi 3^{\frac{1}{4}} k^{\frac{1}{4}}} n^{-3/4} \exp\left(2\pi\left(\frac{n}{3k}\right)^{\frac{1}{2}}\right).$$

2. In order to compute explicitly the $r(n)$ in Theorem 6.2, we use the well-known logarithmic derivative:

$$\frac{\sum_{n=1}^{\infty} nr(n)q^n}{\sum_{n=0}^{\infty} r(n)q^n} = q\frac{d}{dq}\log \prod_{n=1}^{\infty}(1-q^n)^{-a_n}$$

$$= \sum_{N=1}^{\infty}\left(\sum_{d|N} da_d\right)q^N.$$

From this it follows that

$$nr(n) = \sum_{j=1}^{n} r(n-j)D_j$$

where

$$D_j = \sum_{d|j} da_d.$$

3. It can be shown by partial summation that the condition $\sum_{1\leqslant n\leqslant t} a_n = At^\alpha/\alpha + O(t^{\alpha-\varepsilon})$ implies $D(s)$ is continuable for $\operatorname{Re} s > \alpha - \varepsilon$ with a first order pole of residue A at $s = \alpha$.

4. If $\sum_{1\leqslant n\leqslant t} a_n \leqslant At^\alpha/\alpha$ for all $t \geqslant 0$, then for $\lambda > 0$

$$\sum_{m=1}^{\infty} ma_m e^{-\lambda m} \leqslant A\lambda^{-\alpha-1}\Gamma(\alpha+1).$$

5. If it is true that the a_n are nonnegative and that $\sum_{1\leqslant n\leqslant t} a_n \leqslant \frac{At^\alpha}{\alpha}$ for all $t \geqslant 0$, then we can prove by essentially elementary means that

$$r(n) < \exp\left(n^{\alpha/(\alpha+1)}\left(1+\frac{1}{\alpha}\right)(A\Gamma(\alpha+1)\zeta(\alpha+1))^{1/(\alpha+1)}\right) = \exp(cn^{\alpha/(\alpha+1)}).$$

To see this we note from Example 2 that

$$nr(n) = \sum_{\substack{k=1\\ k\upsilon<n}}^{n}\sum_{\upsilon=1}^{n} a_\upsilon \upsilon r(n-k\upsilon).$$

The desired inequality is trivial for $n = 1$, and we may use the recurrence to prove the full inequality by induction. We sketch the procedure:

$$nr(n) < \sum_{\substack{k=1\\ k\upsilon<n}}^{n}\sum_{\upsilon=1}^{n} a_\upsilon \upsilon \exp[c(n-k\upsilon)^{\alpha/(\alpha+1)}]$$

$$< \sum_{k=1}^{\infty}\sum_{\upsilon=1}^{\infty} a_\upsilon \upsilon \exp\left[cn^{\alpha/(\alpha+1)} - \frac{c\alpha k\upsilon}{(\alpha+1)}n^{-1/(\alpha+1)}\right]$$

$$= \exp(cn^{\alpha/(\alpha+1)})A \sum_{k=1}^{\infty} \lambda^{-\alpha-1}\Gamma(\alpha+1)$$

(by Example 4 with $\lambda = c\alpha k n^{-(\alpha+1)^{-1}}(\alpha+1)^{-1}$)

$$= \exp(cn^{\alpha/(\alpha+1)})A\Gamma(\alpha+1)c^{-\alpha-1}\alpha^{-\alpha-1}n(\alpha+1)^{\alpha+1}\zeta(\alpha+1)$$

$$= n\exp(cn^{\alpha/(\alpha+1)})c^{-\alpha-1}[A\alpha^{-\alpha-1}(\alpha+1)^{\alpha+1}\Gamma(\alpha+1)\zeta(\alpha+1)]$$

$$= n\exp(cn^{\alpha/(\alpha+1)}) \quad \left(\text{since} \quad c = \left(1+\frac{1}{\alpha}\right)(A\Gamma(\alpha+1)\zeta(\alpha+1))^{1/(\alpha+1)}\right).$$

Unfortunately not much is known about problems when a series rather than a product is involved. The following examples treat some of the very intricate researches of Meinardus on partitions with difference d between parts.

6. If $\Delta_d(n)$ denotes the number of partitions of n with difference at least d between parts, then

$$F_d(q) = \sum_{n\geqslant 0} \Delta_d(n)q^n = \sum_{n=0}^{\infty} \frac{q^{\frac{1}{2}dn(n-1)+n}}{(q)_n}.$$

7*. Using Eqs. (2.2.5) and (2.2.10), we can prove that

$$F_d(q) = \operatorname*{Res}_{\omega=0}\left\{\frac{1}{\omega}(\omega q)_\infty^{-1}(q^d;q^d)_\infty(-\omega^{-1};q^d)_\infty(-\omega q^d;q^d)_\infty\right\}.$$

8*. Example 7 provides a formula for $F_d(q)$ that can yield the behavior of $F_d(q)$ for $|q|$ near 1. Using the circle method, we can then show that

$$\Delta_d(n) \sim C_d n^{-3/4}\exp(2(A_d n)^{1/2})$$

where

$$C_d = \frac{1}{2\sqrt{\pi}} A_d^{\frac{1}{4}}[(\alpha_d)^{d-1}(d(\alpha_d)^{d-1}+1)]^{-1/2},$$

$$A_d = \frac{d}{2}\log^2\alpha_d + \sum_{r=1}^{\infty}\frac{(\alpha_d)^{rd}}{r^2},$$

and α_d is the positive real root of $x^d + x - 1 = 0$.

9. It is possible to prove that $A_2 = \pi^2/15$ by comparing the result in Example 1 with $a = 1$, $k = 5$. (Corollary 8.6 must also be invoked to obtain that $\Delta_2(n) = r(n)$).

10*. If H is the union of a finite number of arithmetic progressions of the form $\{a_i + nM\}_{n=0}^{\infty}$ $(1 \leqslant i \leqslant r,\ 1 \leqslant a_i \leqslant M)$, then $p(\text{“}H\text{”}(\leqslant 1), n)$, the number of partitions of n into distinct parts taken from H, is monotonically increasing for n sufficiently large if and only if g.c.d.$(a_1, a_2, \ldots, a_r, M) = 1$.

11*. The partition function $p(\text{"}H\text{"}(\leqslant 1), n)$ defined in Example 10 satisfies the following asymptotic relation

$$p(\text{"}H\text{"}(\leqslant 1), n) = 2^{((r-3)/2-(a_1+a_2+\cdots+a_r)/M)}3^{-\frac{1}{4}}n^{-3/4}e^{\pi\sqrt{rn/3M}}\{1 + O(n^{-\frac{1}{4}+\delta})\},$$

provided g.c.d.$(a_1, a_2, \ldots, a_r, M) = 1$.

12*. The number $p_{n,k}$ $(= p(n, k, n) - p(n, k-1, n))$ of partitions of n into exactly k parts is unimodal in k if n is sufficiently large (i.e. there exists $k_1 = k_1(n)$ such that $p_{n,k} > p_{n,k+1}$ for $k > k_1$, and $p_{n,k} < p_{n,k+1}$ for $k < k_1$).

13*. Almost all partitions π_0 of n (i.e. with the exception of $o(p(n))$) have $\#(\pi_0)$ summands where

$$\left| \#(\pi_0) - \frac{\sqrt{6n}}{2\pi} \log n \right| < \sqrt{n}\, w(n)$$

provided only that $w(n)$ is nondecreasing and tends to infinity with n.

14*. If $0 < \alpha < \frac{1}{2}$, then almost all partitions of n (i.e. with the exception of $o(p(n))$) contain

$$(1 + o(1))\alpha \frac{\sqrt{6n}}{\pi} \log n$$

summands which are $\leqslant n^{\alpha}(\sqrt{6}/\pi)$.

15*. Let $k \geqslant 2$, $\delta > 0$ be fixed. For each partition π_i of n, we denote by S_i the number of different summands of π_i. Then the set of integers that are summands common to each of the partitions $\pi_1, \pi_2, \ldots, \pi_k$ of n is at least as large as $((1/k) - \delta) \max_{1 \leqslant j \leqslant k} S_j$ for almost all k-tuples $(\pi_1, \pi_2, \ldots, \pi_k)$ of partitions of n (i.e. with the exception of $o(p(n)^k)$ k-tuples of partitions of n).

16*. If the k-tuples of partitions in Example 14 are restricted to those with distinct parts then an analogous result holds with $((1/k) - \delta)$ replaced by $[(1/(k2^k \log 2)) - \delta]$, and $p(n)$ replaced by $p(\mathscr{D}, n)$.

Notes

We have chosen to present the work of Meinardus. However, there have been numerous significant papers on the subject of this chapter; in particular, we cite those of Ingham (1941), Brigham (1950), Auluck and Haselgrove (1952), Haselgrove and Temperley (1954), Kohlbecker (1958), Roth and Szekeres (1954), Szekeres (1951, 1953), and Wright (1931, 1933, 1934). Recently B. Richmond (1972, 1975a, b) has extended the important work of Roth and Szekeres (1954); and W. Schwarz (1967, 1968, 1969) has also made important contributions to this area. Reviews of recent work occur in Section P72 of LeVeque (1974).

Example 1. Livingood (1945).

Example 2. The case in which $a_d = 1$ was treated by Vahlen (1893), although very similar results were found by Euler; $a_d = d$ was done by MacMahon (1923).

Examples 3–5 present a modification of work done by Erdös (1942) on $p(n)$.

Examples 6–8 appear in Meinardus (1954b).

Example 9. See Watson (1937) for a history of such problems.

Example 10. Bateman and Erdös (1956).

Example 11. Richmond (1972).

Example 12. Szekeres (1953).

Example 13. Erdös and Lehner (1941).

Example 14. Turán (1975).

Example 15. Turán (1973, 1975).

Example 16. Turán (1974, 1975).

In Examples 13–16 some of the interesting recent theorems on the statistical aspects of partitions are listed. For an extensive account of these and related work on the statistical aspects of group theory we refer the reader to Erdös and Lehner (1941), Erdös and Turán (1965, 1967a, b, c, 1968, 1971a, b), Lehmer (1972a, b, c), Richmond (1975c), Turán (1973, 1974, 1975).

References

Auluck, F. C., and Haselgrove, C. B. (1952). "On Ingham's Tanberian theorem for partitions," *Proc. Cambridge Phil. Soc.* **48**, 566–570.

Bateman, P., and Erdös, P. (1956). "Monotonicity of partition functions," *Mathematika* **3**, 1–14.

Brigham, N. A. (1950). "A general asymptotic formula for partition functions," *Proc. Amer. Math. Soc.* **1**, 182–191.

Erdös, P. (1942). "On an elementary proof of some asymptotic formulas in the theory of partitions," *Ann. of Math.* **43**, 437–450.

Erdös, P., and Lehner, J. (1941). "The distribution of the number of summands in the partitions of a positive integer," *Duke Math. J.* **8**, 335–345.

Erdös, P., and Turán, P. (1965). "On some problems of a statistical group theory, I," *Z. Wahrscheinlichkeitstheorie und Verw. Gebiete* **4**, 175–186.

Erdös, P., and Turán, P. (1967a). "On some problems of a statistical group theory, II," *Acta Math. Acad. Sci. Hungar.* **18**, 151–163.

Erdös, P., and Turán, P. (1967b). "On some problems of a statistical group theory, III," *Acta Math. Acad. Sci. Hungar.* **18**, 171–173.

Erdös, P., and Turán, P. (1967c). "Certain problems of statistical group theory," *Magyar Tud. Akad. Mat. Fiz. Oszt. Közl.* **17**, 51–57.

Erdös, P., and Turán, P. (1968). "On some problems of a statistical group theory, IV," *Acta Math. Acad. Sci. Hungar.* **19**, 413–435.

Erdös, P., and Turán, P. (1971a). "On some general problems in the theory of partitions, I," *Acta Arith.* **18**, 53–62.

Erdös, P., and Turán, P. (1971b). "On some problems of a statistical group theory, V," *Period. Math. Hungar.* **1**, No. 1, 5–13.

Hardy, G. H., and Ramanujan, S. (1918). "Asymptotic formulae in combinatory analysis," *Proc. London Math. Soc.* (2) **17**, 75–115. (Also in *Collected Papers of S. Ramanujan*, pp. 276–309, Cambridge Univ. Press, Cambridge, 1927; reprinted by Chelsea, New York, 1962.)

Haselgrove, C. B., and Temperley, H. N. V. (1954). "Asymptotic formulae in the theory of partitions," *Proc. Cambridge Phil. Soc.* **50**, 225–241.

Ingham, A. E. (1941). "A Tauberian theorem for partitions," *Ann. of Math.* (2) **42**, 1075–1090.

Kohlbecker, E. E. (1958). "Weak asymptotic properties of partitions," *Trans. Amer. Math. Soc.* **88**, 346–365.

Lehmer, D. H. (1972a). "On reciprocally weighted partitions," *Acta Arith.* **21**, 379–388.

Lehmer, D. H. (1972b). "Calculating moments of partitions," *Proceedings of the Second Manitoba Conference on Numerical Mathematics*, Oct. 5–7, 1972.

Lehmer, D. H. (1972c). "Some structural aspects of the partitions of an integer," *Proceedings of the 1972 Number Theory Conference*, University of Colorado, Boulder, pp. 122–127.

LeVeque, W. J. (1974). *Reviews in Number Theory*, Vol. 4, Amer. Math. Soc., Providence, R.I.

Livingood, J. (1945). "A partition function with the prime modulus $p > 3$," *Amer. J. Math.* **67**, 194–208.

MacMahon, P. A. (1923). "The connexion between the sum of the squares of the divisors and the number of partitions of a given number," *Messenger of Math.* **52**, 113–116.

Meinardus, G. (1954a). "Asymptotische Aussagen über Partitionen," *Math. Z.* **59**, 388–398.

Meinardus, G. (1954b). "Über Partitionen mit Differenzenbedingungen," *Math. Z.* **61**, 289–302.

Richmond, B. (1972). "On a conjecture of Andrews," *Utilitas Math.* **2**, 3–8.

Richmond, B. (1975a). "Asymptotic relations for partitions," *J. Number Theory* **7**, 389–405.

Richmond, B. (1975b). "A general asymptotic result for partitions," *Canadian J. Math.* **27**, 1083–1091.

Richmond, B. (1975c). "The moments of partitions, I," *Acta Arith.* **26**, 411–425.

Roth, K. F., and Szekeres, G. (1954). "Some asymptotic formulae in the theory of partitions," *Quart. J. Math. Oxford Ser.* (2) **5**, 241–259.

Schwarz, W. (1967). "C. Mahler's Partitionsproblem," *J. Reine Angew. Math.* **229**, 182–188.

Schwarz, W. (1968). "Schwache asymptotische Eigenschaften von Partitionen," *J. Reine Angew. Math.* **232**, 1–16.

Schwarz, W. (1969). "Asymptotische Formeln für Partitionen," *J. Reine Angew. Math.* **234**, 174–178.

Szekeres, G. (1951). "An asymptotic formula in the theory of partitions, I," *Quart. J. Math. Oxford Ser.* (2) **2**, 85–108.

Szekeres, G. (1953). "An asymptotic formula in the theory of partitions, II," *Quart. J. Math. Oxford Ser.* (2), **4**, 96–111.

Turán, P. (1973). "Combinatorics, partitions, group theory," *Proceedings of the International Colloquium on Combinatorial Theories*, Rome, September 3–15, 1973.

Turán, P. (1974). "On a property of partitions," *J. Number Theory* **6**, 405–411.

Turán, P. (1975). "On some phenomena in the theory of partitions," *Journees Arithmetiques de Bordeaux*, Societe Math. de France, Asterisque 24–25, pp. 311–319.

Vahlen, K. T. (1893). "Beiträge zu einer additiven Zahlentheorie," *Crelle J.* **112**, 1–36.

Watson, G. N. (1937). "A note on Spence's logarithmic transcendant," *Quart. J. Math. Oxford Ser.* (2) **8**, 39–42.

Wright, E. M. (1931). "Asymptotic partition formulae, I," *Quart J. Math. Oxford Ser.* (2) **2**, 177–189.

Wright, E. M. (1933). "Asymptotic partition formulae, II," *Proc. London Math. Soc.* (2) **36**, 117–141.

Wright, E. M. (1934). "Asymptotic partition formulae, III," *Acta Math.* **63**, 143–191.

CHAPTER 7

Identities of the Rogers-Ramanujan Type

7.1 Introduction

The Rogers–Ramanujan identities provide one of the most fascinating chapters in the history of partitions. The story begins with the discovery of the Indian genius S. Ramanujan by G. H. Hardy. In his first letter to Hardy (dated January 16, 1913) Ramanujan stated several marvelous theorems on continued fractions. Two are of special interest to us now:

$$\cfrac{1}{1+\cfrac{e^{-2\pi}}{1+\cfrac{e^{-4\pi}}{1+\cfrac{e^{-6\pi}}{\ddots}}}} = \left(\sqrt{\frac{5+\sqrt{5}}{2}} - \frac{\sqrt{5}+1}{2}\right)e^{2\pi/5}$$

$$1-\cfrac{e^{-\pi}}{1+\cfrac{e^{-2\pi}}{1-\cfrac{e^{-3\pi}}{\ddots}}} = \left(\sqrt{\frac{5-\sqrt{5}}{2}} - \frac{\sqrt{5}-1}{2}\right)e^{\pi/5}.$$

Of these (and a related formula), Hardy says, in the article "The Indian Mathematician Ramanujan" (*Amer. Math. Monthly* **44** (1937), p. 144), "[These formulas] defeated me completely. I had never seen anything in the least like them before. A single look at them is enough to show that they could only be written down by a mathematician of the highest class. They must be true because, if they were not true, no one would have had the imagination to invent them."

Before we conclude our history, it is useful to see how these two continued fractions are related to mathematical constructs of the type previously considered. Let us consider $F(x)$ to be a function analytic in x at 0, $F(0) = 1$,

---

ENCYCLOPEDIA OF MATHEMATICS and Its Applications, Gian-Carlo Rota (ed.). 2, George E. Andrews, The Theory of Partitions

and such that

$$F(x) = F(xq) + xqF(xq^2), \tag{7.1.1}$$

a linear second-order q-difference equation. Then if $c(x, q) = F(x)/F(xq)$, we have that

$$\begin{aligned} c(x, q) &= 1 + \frac{xq}{c(xq, q)} \\ &= 1 + \frac{xq}{1 + (xq^2/c(xq^2, q))} \cdots \end{aligned} \tag{7.1.2}$$

and so Ramanujan's continued fraction theorems are precisely evaluations of $c(1, e^{-2\pi})$ and $c(1, -e^{-\pi})$.

If now we write $F(x) = \sum_{n \geq 0} A_n(q)x^n$, and we substitute this series into (7.1.1), we find, by comparing coefficients of x^n, that

$$A_n(q) = q^n A_n(q) + q^{2n-1}A_{n-1}(q). \tag{7.1.3}$$

Hence

$$\begin{aligned} A_n(q) &= \frac{q^{2n-1}}{1 - q^n} A_{n-1}(q) = \frac{q^{(2n-1)+(2n-3)}}{(1 - q^n)(1 - q^{n-1})} A_{n-2}(q) \\ &= \cdots = \frac{q^{1+3+\cdots+(2n-1)}}{(q)_n} A_0(q) = \frac{q^{n^2}}{(q)_n}. \end{aligned} \tag{7.1.4}$$

Hence

$$F(x) = \sum_{n=0}^{\infty} \frac{x^n q^{n^2}}{(q)_n}. \tag{7.1.5}$$

Surprisingly, $F(1)$ and $F(q)$ may be evaluated in terms of infinite products (as we shall prove later in this chapter), namely,

$$\begin{aligned} F(1) &= 1 + \frac{q}{1 - q} + \frac{q^4}{(1 - q)(1 - q^2)} + \frac{q^9}{(1 - q)(1 - q^2)(1 - q^3)} + \cdots \\ &= \prod_{n=0}^{\infty} (1 - q^{5n+1})^{-1}(1 - q^{5n+4})^{-1} \end{aligned} \tag{7.1.6}$$

and

$$\begin{aligned} F(q) &= 1 + \frac{q^2}{1 - q} + \frac{q^6}{(1 - q)(1 - q^2)} + \frac{q^{12}}{(1 - q)(1 - q^2)(1 - q^3)} + \cdots \\ &= \prod_{n=0}^{\infty} (1 - q^{5n+2})^{-1}(1 - q^{5n+3})^{-1}. \end{aligned} \tag{7.1.7}$$

From (7.1.6) and (7.1.7), we directly obtain an infinite product representation of $c(1, q) = F(1)/F(q)$, and we may then deduce Ramanujan's continued fraction theorems by an appeal to the theory of elliptic theta functions.

Hardy has asserted that "it would be difficult to find more beautiful formulae than the 'Rogers–Ramanujan' identities," and he succinctly fills in the remainder of their early history in the following passage, taken from *Ramanujan* (Cambridge Univ. Press, 1940, p. 91):

They were found first in 1894 by Rogers, a mathematician of great talent but comparatively little reputation, now remembered mainly from Ramanujan's rediscovery of his work. Rogers was a fine analyst, whose gifts were, on a smaller scale, not unlike Ramanujan's; but no one paid much attention to anything he did, and the particular paper in which he proved the formulae was quite neglected.

Ramanujan rediscovered the formulae sometime before 1913. He had then no proof (and knew that he had none), and none of the mathematicians to whom I communicated the formulae could find one. They are therefore stated without proof in the second volume of MacMahon's *Combinatory Analysis.*

The mystery was solved, trebly, in 1917. In that year Ramanujan, looking through old volumes of the *Proceedings of the London Mathematical Society*, came accidentally across Rogers's paper. I can remember very well his surprise, and the admiration which he expressed for Rogers's work. A correspondence followed in the course of which Rogers was led to a considerable simplification of his original proof. About the same time I. Schur, who was then cut off from England by the war, rediscovered the identities again. Schur published two proofs, one of which is "combinatorial" and quite unlike any other proof known. There are now seven published proofs, the four referred to already, the two much simpler proofs found later by Rogers and Ramanujan and published in the *Papers*, and a much later proof by Watson based on quite different ideas. None of these proofs can be called both "simple" and "straightforward," since the simplest are essentially verifications; and no doubt it would be unreasonable to expect a really easy proof.

Hardy's comments about the nonexistence of a really easy proof of the Rogers–Ramanujan identities are still true today. The approach that we shall use here is related to one of the "verification" proofs referred to by Hardy. On the one hand, this may seem inappropriate for this particular book; however, we may defend our choice in observing that our approach has been by far the most fruitful in extending our knowledge of Rogers–Ramanujan type identities.

We shall, in fact, show that other partition identities may be treated in the same way, and as an example, we shall prove not only the Rogers–Ramanujan identities, but also B. Gordon's beautiful generalization of them, as well as other identities due independently to H. Göllnitz and B. Gordon.

In every case, we find that theorems on partitions correspond to generating function identities. The partition-theoretic interpretations of (7.1.6) and (7.1.7) are given in Section 7.3. If the reader wishes he may proceed by accepting Lemmas 7.2–7.4 and then going directly to the partition theorems in Section 7.3.

7.2 The Generating Functions

We shall consider, for $|x| < |q|^{-1}$, $|q| < 1$,

$$H_{k,i}(a;x;q) = \sum_{n=0}^{\infty} \frac{x^{kn}q^{kn^2+n-in}a^n(1 - x^iq^{2ni})(axq^{n+1})_\infty(a^{-1})_n}{(q)_n(xq^n)_\infty}, \quad (7.2.1)$$

$$J_{k,i}(a;x;q) = H_{k,i}(a;xq;q) - xqaH_{k,i-1}(a;xq;q). \quad (7.2.2)$$

We note that any value for a is admissible, even $a = 0$, since $a^n(a^{-1})_n = (a-1)(a-q)\cdots(a-q^{n-1})$ is merely a polynomial in a whose value at 0 is $(-1)^nq^{n(n-1)/2}$.

Very little can be said in way of motivation for this section. Actually extensive work in the theory of basic hypergeometric series and partition identities shows that "well-poised" basic hypergeometric series provide the generating functions for numerous families of partition identities. The series in (7.2.1) (once the infinite products have been factored out: "$\sum \alpha_n(Aq^n)_\infty^{\pm 1} = (A)_\infty^{\pm 1} \sum \alpha_n(A)_n^{\mp 1}$") is an example of a well-poised series.

LEMMA 7.1

$$H_{k,i}(a;x;q) - H_{k,i-1}(a;x;q) = x^{i-1}J_{k,k-i+1}(a;x;q). \quad (7.2.3)$$

Proof. Noting that

$$q^{-in}(1 - x^iq^{2ni}) - q^{-(i-1)n}(1 - x^{i-1}q^{2n(i-1)})$$
$$= q^{-in}(1-q^n) + x^{i-1}q^{n(i-1)}(1 - xq^n),$$

we see that

$$H_{k,i}(a;x;q) - H_{k,i-1}(a;x;q)$$
$$= \sum_{n=0}^{\infty} \frac{x^{kn}q^{kn^2+n}a^n(axq^{n+1})_\infty(a^{-1})_n}{(q)_n(xq^n)_\infty} q^{-in}(1-q^n)$$
$$+ \sum_{n=0}^{\infty} \frac{x^{kn}q^{kn^2+n}a^n(axq^{n+1})_\infty(a^{-1})_n x^{i-1}q^{n(i-1)}(1-xq^n)}{(q)_n(xq^n)_\infty}$$

$$= \sum_{n=1}^{\infty} \frac{x^{kn} q^{kn^2+n} a^n (axq^{n+1})_\infty (a^{-1})_n q^{-in}}{(q)_{n-1}(xq^n)_\infty}$$

$$+ \sum_{n=0}^{\infty} \frac{x^{kn} q^{kn^2+n} a^n (axq^{n+1})_\infty (a^{-1})_n x^{i-1} q^{n(i-1)}}{(q)_n (xq^{n+1})_\infty}$$

$$= \sum_{n=0}^{\infty} \frac{x^{kn+k} q^{kn^2+n+2kn+k+1} a^{n+1} (axq^{n+2})_\infty (a^{-1})_{n+1} q^{-in-i}}{(q)_n (xq^{n+1})_\infty}$$

$$+ \sum_{n=0}^{\infty} \frac{x^{kn} q^{kn^2+n} a^n (axq^{n+1})_\infty (a^{-1})_n x^{i-1} q^{n(i-1)}}{(q)_n (xq^{n+1})_\infty}$$

$$= x^{i-1} \sum_{n=0}^{\infty} \frac{x^{kn} q^{kn^2+in} a^n (axq^{n+2})_\infty (a^{-1})_n}{(q)_n (xq^{n+1})_\infty} \left\{ (1 - axq^{n+1}) \right.$$

$$\left. + ax^{k-i+1} q^{2n(k-i)+k-i+1+n} \left(1 - \frac{q^n}{a}\right) \right\}$$

$$= x^{i-1} \sum_{n=0}^{\infty} \frac{x^{kn} q^{kn^2+in} a^n (axq^{n+2})(a^{-1})_n}{(q)_n (xq^{n+1})_\infty} [1 - (xq)^{k-i+1} q^{n[2(k-i+1)-1]}]$$

$$- x^{i-1} \sum_{n=0}^{\infty} \frac{x^{kn} q^{kn^2+in} a^n (axq^{n+2})(a^{-1})_n}{(q)_n (xq^{n+1})_\infty} \{axq^{n+1}[1-(xq)^{k-i} q^{n[2(k-i)-1]}]\}$$

$$= x^{i-1}[H_{k,k-i+1}(a; xq; q) - axqH_{k,k-i}(a; xq; q)]$$

$$= x^{i-1} J_{k,k-i+1}(a; x; q). \qquad \blacksquare$$

LEMMA 7.2

$$J_{k,i}(a; x; q) - J_{k,i-1}(a; x; q) = (xq)^{i-1}(J_{k,k-i+1}(a; xq; q) - aJ_{k,k-i+2}(a; xq; q)). \qquad (7.2.4)$$

Proof.

$$J_{k,i}(a; x; q) - J_{k,i-1}(a; x; q)$$

$$= (H_{k,i}(a; xq; q) - H_{k,i-1}(a; xq; q))$$

$$- axq(H_{k,i-1}(a; xq; q) - H_{k,i-2}(a; xq; q))$$

$$= (xq)^{i-1} J_{k,k-i+1}(a; xq; q) - a(xq)^{i-1} J_{k-i+2}(a; xq; q). \qquad \blacksquare$$

In treating problems, there are numerous instances in which we require an infinite product representation of a generating function. Such representations are given in the following lemmas.

LEMMA 7.3. *For* $1 \leqslant i \leqslant k$, $|q| < 1$,

$$J_{k,i}(0; 1; q) = \prod_{\substack{n=1 \\ n \not\equiv 0, \pm i (\mathrm{mod}\ 2k+1)}}^{\infty} (1 - q^n)^{-1}. \tag{7.2.5}$$

Proof. By (7.2.2)

$$\begin{aligned} J_{k,i}(0; 1; q) &= H_{k,i}(0; q; q) \\ &= (q)_\infty^{-1} \sum_{n=0}^{\infty} q^{kn^2 + (k-i+1)n}(-1)^n q^{n(n-1)/2}(1 - q^{(2n+1)i}) \\ &= (q)_\infty^{-1} \sum_{n=0}^{\infty} (-1)^n q^{(2k+1)n(n+1)/2 - in}(1 - q^{(2n+1)i}) \\ &= \prod_{\substack{n=1 \\ n \not\equiv 0, \pm i (\mathrm{mod}\ 2k+1)}}^{\infty} (1 - q^n)^{-1} \qquad \text{(by Corollary 2.9).} \qquad \blacksquare \end{aligned}$$

LEMMA 7.4. *For* $1 \leqslant i \leqslant k$, $|q| < 1$,

$$J_{k,i}(-q^{-1}; 1; q^2) = \prod_{\substack{n=1 \\ n \not\equiv 2 (\mathrm{mod} 4) \\ n \not\equiv 0, \pm(2i-1)(\mathrm{mod}\ 4k)}}^{\infty} (1 - q^n)^{-1}. \tag{7.2.6}$$

Proof.

$$\begin{aligned} J_{k,i}(-q^{-1}, 1; q^2) &= H_{k,i}(-q^{-1}; q^2; q^2) + qH_{k,i-1}(-q^{-1}; q^2; q^2) \\ &= \frac{(-q; q^2)_\infty}{(q^2; q^2)_\infty} \left\{ \sum_{n=0}^{\infty} (-1)^n q^{2kn + 2kn^2 - (2i-1)n} \right. \\ &\quad \times \frac{(1 - q^{2i+4ni})}{(1 + q^{2n+1})} + q \sum_{n=0}^{\infty} (-1)^n q^{2kn + 2kn^2 - (2i-3)n} \\ &\quad \left. \times \frac{(1 - q^{2i-2+4ni-4n})}{(1 + q^{2n+1})} \right\} \\ &= \frac{(-q; q^2)_\infty}{(q^2; q^2)_\infty} \sum_{n=0}^{\infty} (-1)^n q^{2kn + 2kn^2 + n} \\ &\quad \times \frac{(q^{-2in} - q^{2i+2ni} + q^{1-2in+2n} - q^{-1+2i+2ni-2n})}{(1 + q^{2n+1})} \\ &= \frac{(-q; q^2)_\infty}{(q^2; q^2)_\infty} \sum_{n=0}^{\infty} (-1)^n q^{2kn^2 + (2k+1-2i)n}(1 - q^{(2n+1)(2i-1)}) \end{aligned}$$

$$= \frac{(-q;q^2)_\infty}{(q^2;q^2)_\infty} \sum_{n=-\infty}^{\infty} (-1)^n q^{2kn^2+(2k-2i+1)n}$$

$$= \frac{(-q;q^2)_\infty}{(q^2;q^2)_\infty} \prod_{n=0}^{\infty} (1-q^{4kn+4k})(1-q^{4kn+2i-1})$$

$$\times (1-q^{4kn+4k-2i+1})$$

$$= \frac{1}{(q;q^2)_\infty (q^4;q^4)_\infty} \prod_{n=0}^{\infty} (1-q^{4k(n+1)})(1-q^{4kn+2i-1})$$

$$\times (1-q^{4k(n+1)-2i+1})$$

$$= \prod_{\substack{n=1 \\ n \not\equiv 2 \pmod 4 \\ n \not\equiv 0, \pm(2i-1) \pmod{4k}}}^{\infty} (1-q^n)^{-1}. \qquad \blacksquare$$

7.3 The Rogers-Ramanujan Identities and Gordon's Generalization

In this section we shall utilize the analytic work in Section 7.2 to prove the following theorem, which is due to B. Gordon.

THEOREM 7.5. *Let $B_{k,i}(n)$ denote the number of partitions of n of the form $(b_1 b_2 \cdots b_s)$, where $b_j - b_{j+k-1} \geqslant 2$, and at most $i - 1$ of the b_j equal 1. Let $A_{k,i}(n)$ denote the number of partitions of n into parts $\not\equiv 0, \pm i (\bmod 2k+1)$. Then $A_{k,i}(n) = B_{k,i}(n)$ for all n.*

Before we prove Theorem 7.5, it is appropriate to give center stage to its two most celebrated corollaries, the Rogers–Ramanujan identities (stated in terms of partitions).

COROLLARY 7.6 (The first Rogers–Ramanujan identity). *The partitions of an integer n in which the difference between any two parts is at least 2 are equinumerous with the partitions of n into parts $\equiv 1$ or 4 (modulo 5).*

Proof. Take $k = i = 2$ in Theorem 7.5. ■

COROLLARY 7.7 (The second Rogers–Ramanujan identity). *The partitions of an integer n in which each part exceeds 1 and the difference between any two parts is at least 2 are equinumerous with the partitions of n into parts $\equiv 2$ or 3 (modulo 5).*

Proof. Take $k = i + 1 = 2$ in Theorem 7.5. ■

Proof of Theorem 7.5. Let $b_{k,i}(m, n)$ denote the number of partitions $(b_1 b_2 \cdots b_m)$ of n with exactly m parts such that $b_j \geqslant b_{j+1}$, $b_j - b_{j+k-1} \geqslant 2$, and at most $i - 1$ of the b_j equal 1. Then for $1 \leqslant i \leqslant k$

$$b_{k,i}(m, n) = \begin{cases} 1 & \text{if } m = n = 0 \\ 0 & \text{if } m \leqslant 0 \text{ or } n \leqslant 0 \text{ but } (m, n) \neq (0, 0); \end{cases} \tag{7.3.1}$$

$$b_{k,0}(m, n) = 0; \tag{7.3.2}$$

for $1 \leqslant i \leqslant k$

$$b_{k,i}(m, n) - b_{k,i-1}(m, n) = b_{k,k-i+1}(m - i + 1, n - m). \tag{7.3.3}$$

Equations (7.3.1) and (7.3.2) are obvious once we recall that the only partition that is either of a nonpositive number or has a nonpositive number of parts is the empty partition of 0.

Equation (7.3.3) requires careful attention: $b_{k,i}(m, n) - b_{k,i-1}(m, n)$ enumerates the number of partitions among those enumerated by $b_{k,i}(m, n)$ that have exactly $i - 1$ appearances of 1. Let us transform this set of partitions by deleting the $i - 1$ ones, and then subtracting 1 from each of the remaining parts. The resulting partitions $(b_1' \cdots b_{m-i+1}')$ have $m - i + 1$ parts; they partition $n - m$, and the parts satisfy $b_j' - b_{j+k-1}' \geqslant 2$. Since originally 1 appeared $i - 1$ times and the total number appearances of ones and twos could not exceed $k - 1$ (due to the difference condition), we see that originally 2 appeared at most $(k - i + 1) - 1$ times, and thus after the transformation 1 appears at most $k - i + 1$ times. The transformation described above establishes a one-to-one correspondence between the partitions enumerated by $b_{k,i}(m, n) - b_{k,i-1}(m, n)$ and those enumerated by $b_{k,k-i+1}(m - i + 1, n - m)$. Hence (7.3.3) is established.

We now make a simple yet essential observation: the $b_{k,i}(m, n)$ $(0 \leqslant i \leqslant k)$ are *uniquely* determined by (7.3.1), (7.3.2), and (7.3.3). To see this, proceed by a double mathematical induction first on n and then on i. Equation (7.3.1) takes care of $n \leqslant 0$, $m \leqslant 0$, $i > 0$. Equation (7.3.2) handles all n when $i = 0$. Equation (7.3.3) represents $b_{k,i}(m, n)$ as a two-term sum in which the first term has a lower i index and the second a lower n index (since we can assume $m > 0$).

Now let us consider

$$J_{k,i}(0; x; q) = \sum_{m=0}^{\infty} \sum_{n=0}^{\infty} c_{k,i}(m, n) x^m q^n.$$

From the fact that for $1 \leqslant i \leqslant k$

$$J_{k,i}(0; 0; q) = J_{k,i}(0; x; 0) = 1,$$

we see that for $1 \leqslant i \leqslant k$

$$c_{k,i}(m, n) = \begin{cases} 1 & \text{if} \quad m = n = 0 \\ 0 & \text{if} \quad m \leqslant 0 \quad \text{or} \quad n \leqslant 0 \quad \text{but} \quad (m, n) \neq (0, 0). \end{cases} \tag{7.3.4}$$

From the fact that

$$J_{k,0}(0; X; q) = H_{k,0}(0; xq; q) = 0,$$

we see that

$$c_{k,0}(m, n) = 0. \tag{7.3.5}$$

Finally, by comparing coefficients of $x^m q^n$ on both sides of (7.2.4) with $a = 0$, we see that

$$c_{k,i}(m, n) - c_{k,i-1}(m, n) = c_{k,k-i+1}(m - i + 1, n - m). \tag{7.3.6}$$

So we see that the $c_{k,i}(m, n)$ also satisfy the system of equations (7.3.1)–(7.3.3) that uniquely defines the $b_{k,i}(m, n)$. Therefore, $b_{k,i}(m, n) = c_{k,i}(m, n)$ for all m and n with $0 \leqslant i \leqslant k$.

Hence, since $\sum_{m \geqslant 0} b_{k,i}(m, n) = B_{k,i}(n)$, we see that

$$\begin{aligned} \sum_{n \geqslant 0} B_{k,i}(n) q^n &= \sum_{m \geqslant 0} \sum_{n \geqslant 0} b_{k,i}(m, n) q^n \\ &= J_{k,i}(0; 1; q) \\ &= \prod_{\substack{n=1 \\ n \not\equiv 0, \pm i (\mathrm{mod}\, 2k+1)}}^{\infty} (1 - q^n)^{-1} \qquad (\text{by } (7.2.5)) \\ &= \sum_{n \geqslant 0} A_{k,i}(n) q^n \qquad (\text{by Theorem 1.1, Eq. } (1.2.3)). \end{aligned}$$

Comparing coefficients of q^n in the extremes of the string of equations above, we see that $A_{k,i}(n) = B_{k,i}(n)$. ∎

Theorem 7.5 has an analytic counterpart, which we shall prove before proceeding.

Theorem 7.8. *For $1 \leqslant i \leqslant k$, $k \geqslant 2$, $|q| < 1$*

$$\sum_{n_1, n_2, \ldots, n_{k-1} \geqslant 0} \frac{q^{N_1^2 + N_2^2 + \cdots + N_{k-1}^2 + N_i + N_{i+1} + \cdots + N_{k-1}}}{(q)_{n_1} (q)_{n_2} \cdots (q)_{n_{k-1}}} = \prod_{\substack{n=1 \\ n \not\equiv 0, \pm i (\mathrm{mod}\, 2k+1)}}^{\infty} (1 - q^n)^{-1} \tag{7.3.7}$$

where $N_j = n_j + n_{j+1} + \cdots + n_{k-1}$.

Proof. We shall prove that

$$J_{k,i}(0;x;q)=\sum_{n_1,n_2,\ldots,n_{k-1}\geqslant 0}\frac{x^{N_1+N_2+\cdots+N_{k-1}}q^{N_1^2+N_2^2+\cdots+N_{k-1}^2+N_i+N_{i+1}+\cdots+N_{k-1}}}{(q)_{n_1}(q)_{n_2}\cdots(q)_{n_{k-1}}}. \tag{7.3.8}$$

We obtain Eq. (7.3.7) from (7.3.8) by setting $x = 1$ and invoking (7.2.5).

Equation (7.3.8) itself follows from

$$J_{k,i}(0;x;q)=\sum_{n\geqslant 0}\frac{x^{(k-1)n}q^{(k-1)n^2+(k-i)n}}{(q)_n}J_{k-1,i}(0;xq^{2n};q), \tag{7.3.9}$$

which may be seen immediately by induction on k once we observe that $J_{k,k+1}(0;x;q) = J_{k,k}(0;x;q)$ (set $i = k + 1$ in (7.2.4) and recall that $J_{k,0}(0;xq;q) = H_{k,0}(0,xq^2;q) = 0$) and that $J_{1,1}(0;x;q) = 1$ (since by (7.2.4) $J_{1,1}(0;x;q) = J_{1,1}(0;xq;q) = J_{1,1}(0,xq^2;q) = \cdots = J_{1,1}(0;xq^n;q) \to J_{1,1}(0;0;q) = 1$).

To prove (7.3.9), we define

$$R_{k,i}(x;q)=\sum_{n\geqslant 0}\frac{x^{(k-1)n}q^{(k-1)n^2+(k-i)n}}{(q)_n}J_{k-1,i}(0;xq^{2n};q).$$

Then for $1 \leqslant i \leqslant k$,

$$R_{k,i}(0;q) = R_{k,i}(x;0) = 1 \tag{7.3.10}$$

and

$$R_{k,0}(x;q) = 0. \tag{7.3.11}$$

Finally

$$\begin{aligned}
&R_{k,i}(x;q) - R_{k,i-1}(x;q)\\
&\quad=\sum_{n\geqslant 0}\frac{x^{(k-1)n}q^{(k-1)n^2+(k-i)n}}{(q)_n}(J_{k-1,i}(0;xq^{2n};q) - q^nJ_{k-1,i-1}(0;xq^{2n};q))\\
&\quad=\sum_{n\geqslant 0}\frac{x^{(k-1)n}q^{(k-1)n^2+(k-i)n}}{(q)_n}(J_{k-1,i-1}(0;xq^{2n};q)\\
&\qquad+(xq^{2n+1})^{i-1}J_{k-1,k-i}(0;xq^{2n+1};q) - q^nJ_{k-1,i-1}(0;xq^{2n};q))\\
&\quad=\sum_{n\geqslant 1}\frac{x^{(k-1)n}q^{(k-1)n^2+(k-i)n}}{(q)_n}(1-q^n)J_{k-1,i-1}(0;xq^{2n};q)\\
&\qquad+(xq)^{i-1}\sum_{n\geqslant 0}\frac{x^{(k-1)n}q^{(k-1)n^2+(k+i-2)n}}{(q)_n}J_{k-1,k-i}(0;xq^{2n+1};q)
\end{aligned}$$

$$= x^{k-1}q^{2k-i-1}\sum_{n\geq 0}\frac{x^{(k-1)n}q^{(k-1)n^2+(3k-i-2)n}}{(q)_n}J_{k-1,i-1}(0;xq^{2n+2};q)$$

$$+ (xq)^{i-1}\sum_{n\geq 0}\frac{x^{(k-1)n}q^{(k-1)n^2+(k+i-2)n}}{(q)_n}(J_{k-1,k-i+1}(0;xq^{2n+1};q)$$

$$- (xq^{2n+2})^{k-i}J_{k-1,i-1}(0;xq^{2n+2};q))$$

$$= x^{k-1}q^{2k-i-1}\sum_{n\geq 0}\frac{x^{(k-1)n}q^{(k-1)n^2+(3k-i-2)n}}{(q)_n}J_{k-1,i-1}(0;xq^{2n+2};q)$$

$$+(xq)^{i-1}\sum_{n\geq 0}\frac{(xq)^{(k-1)n}q^{(k-1)n^2+(k-(k-i+1))n}}{(q)_n}J_{k-1,k-i+1}(0;xq^{2n+1};q)$$

$$- x^{k-1}q^{2k-i-1}\sum_{n\geq 0}\frac{x^{(k-1)n}q^{(k-1)n^2+(3k-i-2)n}}{(q)_n}J_{k-1,i-1}(0;xq^{2n+2};q)$$

$$= (xq)^{i-1}R_{k,k-i+1}(xq,q). \tag{7.3.12}$$

Recalling that the coefficients in the expansion of $J_{k,i}(0;x;q)$ were uniquely determined by (7.3.4), (7.3.5), and (7.3.6), we conclude that since $R_{k,i}(x;q)$ satisfies (7.3.10), (7.3.11), and (7.3.12), and thus its coefficients must satisfy (7.3.4), (7.3.5), and (7.3.6), therefore $R_{k,i}(x;q) = J_{k,i}(0;x;q)$ for $0 \leqslant i \leqslant k$. Thus we have (7.3.9) and with it Theorem 7.8. ∎

COROLLARY 7.9 (Eq. (7.1.6)).

$$1 + \frac{q}{1-q} + \frac{q^4}{(1-q)(1-q^2)} + \frac{q^9}{(1-q)(1-q^2)(1-q^3)} + \cdots$$

$$= \prod_{n=0}^{\infty}(1-q^{5n+1})^{-1}(1-q^{5n+4})^{-1}.$$

Proof. Set $k = i = 2$ in Theorem 7.8. ∎

COROLLARY 7.10 (Eq. (7.1.7)).

$$1 + \frac{q^2}{1-q} + \frac{q^6}{(1-q)(1-q^2)} + \frac{q^{12}}{(1-q)(1-q^2)(1-q^3)} + \cdots$$

$$= \prod_{n=0}^{\infty}(1-q^{5n+2})^{-1}(1-q^{5n+3})^{-1}.$$

Proof. Set $k = 2$, $i = 1$ in Theorem 7.8. ∎

7.4 The Göllnitz-Gordon Identities and Their Generalization

To appreciate the general area of partition identities such as the Rogers-Ramanujan identities, we turn to results discovered independently in the 1960s by H. Göllnitz and B. Gordon. Actually we shall prove a general theorem that reduces to the Göllnitz–Gordon identities in the special cases in which $k = i = 2$, $k = i + 1 = 2$.

THEOREM 7.11. *Let i and k be integers with $0 < i \leqslant k$. Let $C_{k,i}(n)$ denote the number of partitions of n into parts $\not\equiv 2 \pmod 4$ and $\not\equiv 0, \pm(2i-1) \pmod{4k}$. Let $D_{k,i}(n)$ denote the number of partitions $(b_1 b_2 \cdots b_s)$ of n in which no odd part is repeated, $b_j \geqslant b_{j+1}$, $b_j - b_{j+k-1} \geqslant 2$ if b_j odd, $b_j - b_{j+k-1} > 2$ if b_j even, and at most $i - 1$ parts are $\leqslant 2$. Then $C_{k,i}(n) = D_{k,i}(n)$.*

Remark. The proof of this theorem is very similar to that of Theorem 7.5. The role of $J_{k,i}(0; x; q)$ in Theorem 7.5 is now played by $J_{k,i}(q^{-1}; x; q^2)$. We shall, therefore, proceed somewhat rapidly through those portions of this theorem that are familiar ground.

Proof. Let $d_{k,i}(m, n)$ denote the number of partitions enumerated by $D_{k,i}(n)$ that have exactly m parts. As in Theorem 7.5, we see that for $1 \leqslant i \leqslant k$

$$d_{k,i}(m, n) = \begin{cases} 1 & \text{if} \quad m = n = 0 \\ 0 & \text{if} \quad m \leqslant 0 \quad \text{or} \quad n \leqslant 0 \quad \text{but} \quad (m, n) = (0, 0). \end{cases} \tag{7.4.1}$$

Next we assert that

$$d_{k,1}(m, n) = d_{k,k}(m, n - 2m), \tag{7.4.2}$$

and for $1 < i \leqslant k$

$$d_{k,i}(m, n) - d_{k,i-1}(m, n) = d_{k,k-i+1}(m - i + 1, n - 2m) + d_{k,k-i+2}(m - i + 1, n - 2m + 1). \tag{7.4.3}$$

The proof of (7.4.2) and (7.4.3) is exactly like the proof of (7.3.3). For example, $d_{k,i}(m, n) - d_{k,i-1}(m, n)$ enumerates the number of partitions of the type enumerated by $d_{k,i}(m, n)$ with the added condition that the total number of parts each $\leqslant 2$ is exactly $i - 1$. Such partitions may be split into two disjoint classes: (1) those that have 1 as a part; (2) those that do not. We transform the partitions in class 1 by deleting the 1 and the $(i-2)$2's, and subtracting 2 from each of the remaining parts. This transformation establishes a one-to-one correspondence between the partitions in class 1 and those enumerated by $d_{k,k-i+2}(m - i + 1, n - 2m + 1)$. We transform the partitions in class 2 by deleting the $i - 1$ 2's and subtracting 2 from each of the remaining parts. In this way, a one-to-one correspondence is established between the partitions in class 2 and those enumerated by $d_{k,k-i+1}(m - i + 1, n - 2m)$. This establishes (7.4.3), and (7.4.2) is treated similarly. Furthermore (just as the $b_{k,i}(m, n)$ were uniquely determined by (7.3.1)–(7.3.3)), we may easily prove that the $d_{k,i}(m, n)$ are uniquely determined by (7.4.1), (7.4.2), and (7.4.3).

Now we consider

$$J_{k,i}(-q^{-1}; x; q^2) = \sum_{m=0}^{\infty} \sum_{n=0}^{\infty} e_{k,i}(m, n)x^m q^n.$$

That Eq. (7.4.1) is also true for the $e_{k,i}(m, n)$ follows from

$$J_{k,i}(-q^{-1}, 0; q^2) = J_{k,i}(-q^{-1}, x; q^2)|_{q=0} = 1,$$

while the truth of (7.4.2) for the $e_{k,i}(m, n)$ follows from

$$\begin{aligned} J_{k,1}(-q^{-1}, x; q^2) &= H_{k,1}(-q^{-1}; xq; q^2) \qquad \text{(by (7.2.2))} \\ &= J_{k,k}(-q^{-1}; xq; q^2) \qquad \text{(by (7.2.3))}. \end{aligned}$$

Finally, we may deduce (7.4.3) for the $e_{k,i}(m, n)$ from the fact that

$$\begin{aligned} &J_{k,i}(-q^{-1}, x; q^2) - J_{k,i-1}(-q^{-1}, x; q^2) \\ &\quad = (xq^2)^{i-1}(J_{k,k-i+1}(-q^{-1}; xq^2; q^2) + q^{-1}J_{k,k-i+2}(-q^{-1}; xq^2; q^2)) \\ &\quad = x^{i-1}q^{2i-2}J_{k,k-i+1}(-q^{-1}; xq^2; q^2) \\ &\qquad + x^{i-1}q^{2i-3}J_{k,k-i+2}(-q^{-1}; xq^2; q^2). \end{aligned}$$

Hence, $d_{k,i}(m, n) = e_{k,i}(m, n)$ for $1 \leqslant i \leqslant k$ and all m and n.

Consequently, for $1 \leqslant i \leqslant n$

$$\begin{aligned} \sum_{n=0}^{\infty} D_{k,i}(n)q^n &= \sum_{m=0}^{\infty} \sum_{n=0}^{\infty} d_{k,i}(m, n)q^n \\ &= J_{k,i}(-q^{-1}, 1; q^2) \\ &= \prod_{\substack{n=1 \\ n \not\equiv 2 \pmod 4 \\ n \not\equiv 0, \pm(2i-1) \pmod{4k}}}^{\infty} (1 - q^n)^{-1} \qquad \text{(by (7.2.6))} \\ &= \sum_{n=0}^{\infty} C_{k,i}(n)q^n \qquad \text{(by Theorem 1.1, Eq. (1.2.3))}. \end{aligned}$$

Comparing coefficients of q^n in the extremes of the string of equations above, we see that $C_{k,i}(n) = D_{k,i}(n)$ for $1 \leqslant i \leqslant k$ and all n. ■

We conclude this section by remarking that a result analogous to Theorem 7.8 exists for $J_{k,i}(-q^{-1}, 1; q^2)$. For example, with $N_j = n_j + \cdots + n_{k-1}$

$$\sum_{n_1,\ldots,n_{k-1}\geqslant 0} \frac{(-q;q^2)_{N_1} q^{N_1^2+2(N_2^2+\cdots+N_{k-1}^2)}}{(q^2;q^2)_{n_1}(q^2;q^2)_{n_2}\cdots(q^2;q^2)_{n_{k-1}}}$$

$$= J_{k,k}(-q^{-1};1;q^2) = \prod_{\substack{n=1\\ n\not\equiv 2 \pmod 4\\ n\not\equiv 0,\pm(2k-1) \pmod{4k}}}^{\infty} (1-q^n)^{-1}. \qquad (7.4.4)$$

Examples

1. Let $s_j(m, n)$ denote the number of partitions $(\lambda_1\lambda_2\cdots\lambda_m)$ of n such that for $1 \leqslant i < m$, $\lambda_i - \lambda_{i+1} \geqslant 3$ with strict inequality if $3|\lambda_i$ and $\lambda_m \geqslant j$. We can prove (similar to the proof of Theorem 7.5) that if $f_j(x) = \sum_{m,n\geqslant 0} s_j(m, n) \times x^m q^n$, then

$$f_1(x) - f_2(x) = xqf_1(xq^3),$$

$$f_2(x) - f_3(x) = xq^2 f_2(xq^3),$$

$$f_3(x) - f_4(x) = xq^3 f_4(xq^3),$$

$$f_4(x) = f_1(xq^3).$$

2. From Example 1 it follows that

$$f_1(x) = (1 + xq + xq^2)f_1(xq^3) + xq^3(1 - xq^3)f_1(xq^6).$$

3. From Example 2, we deduce that

$$f_1(x) = (x;q^3)_\infty \sum_{n\geqslant 0} \frac{(-q;q^3)_n(-q^2;q^3)_n x^n}{(q^3;q^3)_n}.$$

4. From Example 3 we may deduce Schur's partition theorem: The number of partitions of n into parts congruent to 1 or 5 modulo 6 equals the number of partitions of n in which the difference between all parts is at least 3 and between multiples of 3 is at least 6.

5. The procedure described in Examples 1–4 may be extended to prove a family of identities like Schur's theorem: Let $\mathbf{j}$ denote the least positive residue of j modulo 2^n; let $\omega(j)$ denote the number of powers of 2 used in the binary representation of j, and let $\upsilon(j)$ denote the smallest power of 2 appearing in the binary representation of j (e.g., if $n = 4$, $j = 28$, then $\mathbf{j} = 12$, $\omega(j) = 3$, $\upsilon(j) = 4$, $\omega(\mathbf{j}) = 2$, $\upsilon(\mathbf{j}) = 4$). Let $A(N)$ denote the number of partitions of N into parts $\equiv 1, 2^n + 1, 2^n + 3, \ldots, 2^n + 2^{n-1} - 1 \pmod{2^{n+1} - 2}$. Let $B(N)$ denote the number of partitions $(\lambda_1\lambda_2,\ldots,\lambda_s)$ of N (s arbitrary) such that for $1 \leqslant i < s$, $\lambda_i - \lambda_{i+1} \geqslant \omega(\boldsymbol{\lambda}_{i+1})\cdot(2^n - 1) + \upsilon(\boldsymbol{\lambda}_{i+1}) - \boldsymbol{\lambda}_{i+1}$. Then $A(N) = B(N)$ for all N.

6. We may deduce Schur's theorem from Example 5 in the case in which $n = 2$.

7. The case in which $n = 3$ of Example 5 asserts that the number of partitions of N into parts $\equiv 1, 9, 11 \pmod{14}$ equals the number of partitions $(\lambda_1\lambda_2\cdots\lambda_s)$ (s arbitrary) of n such that for $1 \leqslant i < s$,

$$\lambda_i - \lambda_{i+1} \geqslant \begin{cases} 7 & \text{if} \quad \lambda_{i+1} \equiv 1, 2, 4 \pmod 7, \\ 12 & \text{if} \quad \lambda_{i+1} \equiv 3 \pmod 7, \\ 10 & \text{if} \quad \lambda_{i+1} \equiv 5, 6 \pmod 7, \\ 15 & \text{if} \quad \lambda_{i+1} \equiv 0 \pmod 7). \end{cases}$$

8. Let $\mathscr{A}_{k,a}(N)$ denote the number of partitions of N into parts $\not\equiv 0, \pm 2a \pmod{4k+2}$. Let $\mathscr{B}_{k,a}(N)$ denote the number of partitions $(1^{f_1}2^{f_2}3^{f_3}\cdots)$ of N such that $f_1 \leqslant 2a - 1$, and

$$f_i + f_{i+1} \leqslant \begin{cases} 2k - 1 & \text{if} \quad f_i \text{ is even,} \\ 2k & \text{if} \quad f_i \text{ is odd.} \end{cases}$$

Then $\mathscr{A}_{k,a}(N) = \mathscr{B}_{k,a}(N)$ for all N.

We may prove this result, which resembles Theorem 7.5, by mimicking the proof of Theorem 7.5 but using

$$J_{k,i}(0; x^2; q^2)(-xq)_\infty$$

instead of $J_{k,i}(0; x; q)$.

9. The number of partitions of n into parts congruent to 2, 3, or 4 modulo 6 equals the number of partitions of n without consecutive integers, with no ones, and with no part repeated three or more times.

The proof here considers

$$J_{2,i+\frac{1}{2}}(0; x^2; q^2)(-xq)_\infty.$$

10. It is a simple matter to prove that for $a = 1$, 2, or 3

$$\prod_{\substack{n=0 \\ n \not\equiv 0, \pm a \pmod 7}}^{\infty} (1 - q^n)^{-1} = (-q)_\infty \sum_{n=0}^{\infty} \frac{q^{2n^2 + 2[(4-a)/2]n}}{(q^2; q^2)_n (-q)_{2n+1-[a/2]}}.$$

This is established by showing that if

$$R_a(x) = R_a(x, q) = (-xq)_\infty \sum_{n \geqslant 0} \frac{x^{2n} q^{2n^2 + 2[(4-a)/2]n}}{(q^2; q^2)_n (-xq)_{2n+1-[a/2]}},$$

then $R_3(x) - R_2(x) = x^2q^2R_1(xq)$; $R_2(x) - R_1(x) = xqR_2(xq)$; $R_1(x) = R_3(xq)$. From there it immediately follows that $R_a(x) = J_{3,a}(0; x; q)$.

11*. The number of partitions of n into parts congruent to 2, 5, or 11 modulo 12 equals the number of partitions $(\lambda_1\lambda_2\cdots\lambda_s)$ (s arbitrary) of n such that $\lambda_s \neq 1$ or 3, and for $1 \leqslant i < s$, $\lambda_i - \lambda_{i+1} \geqslant 6$ with strict inequality for $\lambda_{i+1} \equiv 0, 1$, or $3 \pmod 6$.

This theorem is deduced from the following analytic identities. Define A_n by

$$\frac{(-xq^2; q^6)_\infty(-xq^4; q^6)_\infty(-xq^5; q^6)_\infty}{(x; q^6)_\infty} = \sum_{n=0}^{\infty} \frac{A_n x^n}{(q^6; q^6)_n}.$$

Let $c_m(n)$ denote the number of partitions $(\lambda_1\lambda_2\cdots\lambda_s)$ (s arbitrary) of n such that $m \geqslant \lambda_1$, $\lambda_s \neq 1$ or 3, and for $1 \leqslant i < s$, $\lambda_i - \lambda_{i+1} \geqslant 6$ with strict inequality if $\lambda_{i+1} \equiv 0, 1$ or 3 (mod 6), and let

$$d_m(q) = \sum_{n\geqslant 0} c_m(n)q^n.$$

Then

$$d_{6n-1}(q) = \sum_{0\leqslant 2j\leqslant n} q^{6nj-3j^2+2j} \begin{bmatrix} n-j \\ j \end{bmatrix}_{q^6} A_{n-2j}$$

where $\left[\begin{smallmatrix} a \\ b \end{smallmatrix}\right]_{q^6}$ is $\left[\begin{smallmatrix} a \\ b \end{smallmatrix}\right]$ with q replaced by q^6.

Notes

The material in the introduction comes primarily from Chapters 1 and 6 of *Ramanujan* by G. H. Hardy (1940). The first two continued fractions in the introduction were proved by Watson (1929). The analytic identities in Section 7.2 appear for $a = 0$ in Rogers (1919) and Selberg (1936) and for $a = -q^{-1}$ in Andrews (1967d). Very general identities of this nature are treated in Andrews (1968c, 1974b).

Theorem 7.5 is due to Gordon (1961), our proof is by Andrews (1966; see also 1969d, 1975a, b). Theorem 7.8 is due to Andrews (1974c); a related theorem involving the Alder polynomials is due to V. N. Singh (1957). Theorem 7.11 in the case $k = 2$ is due to H. Göllnitz (1967) and B. Gordon (1965); for arbitrary k the result was obtained by Andrews (1967d). Identity (7.4.4) is also due to Andrews (1975d).

For more detailed accounts of work of this nature there are several survey articles with extensive bibliographies: Alder (1969), Andrews (1970, 1972, 1974a), Gupta (1970). Reviews of recent work in this area occur in Section P68 of LeVeque (1974).

Examples 1–3. Andrews (1968a).
Example 4. Andrews (1968a, 1971a, b), Schur (1926).
Examples 5–7. Andrews (1968b, 1969a, 1972).
Example 8. Andrews (1967b).
Example 9. Andrews (1967c, 1969c).
Example 10. Selberg (1936), Andrews (1968c, 1975a).
Example 11. Göllnitz (1967), Andrews (1969b).

References

Alder, H. L. (1969). "Partition identities—from Euler to the present," *Amer. Math. Monthly* **76**, 733–746.

Andrews, G. E. (1966). "An analytic proof of the Rogers–Ramanujan–Gordon identities," *Amer. J. Math.* **88**, 844–846.

Andrews, G. E. (1967a). "On Schur's second partition theorem," *Glasgow Math. J.* **9**, 127–132.

Andrews, G. E. (1967b). "Partition theorems related to the Rogers–Ramanujan identities," *J. Combinatorial Theory* **2**, 422–430.

Andrews, G. E. (1967c). "Some new partition theorems," *J. Combinatorial Theory* **2**, 431–436.

Andrews, G. E. (1967d). "A generalization of the Göllnitz–Gordon partition theorems," *Proc. Amer. Math. Soc.* **18**, 945–952.

Andrews, G. E. (1968a). "On partition functions related to Schur's second partition theorem," *Proc. Amer. Math. Soc.* **18**, 441–444.

Andrews, G. E. (1968b). "A new generalization of Schur's second partition theorem," *Acta Arith.* **14**, 429–434.

Andrews, G. E. (1968c). "On q-difference equations for certain well-poised basic hypergeometric series," *Quart. J. Math. Oxford Ser.* (2) **19**, 433–447.

Andrews, G. E. (1969a). "A general theorem on partitions with difference conditions," *Amer. J. Math.* **91**, 18–24.

Andrews, G. E. (1969b). "A partition theorem of Göllnitz and related formulae," *J. Reine Angew. Math.* **236**, 37–42.

Andrews, G. E. (1969c). "Some new partition theorems (II)," *J. Combinatorial Theory* **7**, 262–263.

Andrews, G. E. (1969d). "A generalization of the classical partition theorems," *Trans. Amer. Math. Soc.* **145**, 205–221.

Andrews, G. E. (1970). "A polynomial identity which implies the Rogers–Ramanujan identities," *Scripta Math.* **28**, 297–305.

Andrews, G. E. (1971a). "The use of computers in search of identities of the Rogers–Ramanujan type," *Computers in Number Theory* (A. O. L. Atkin and B. J. Birch, eds.), pp. 377–387. Academic Press, London.

Andrews, G. E. (1971b). "On a theorem of Schur and Gleissburg," *Arch. Math.* (Basel) **22**, 165–167.

Andrews, G. E. (1971c). *Number Theory*. Saunders, Philadelphia.

Andrews, G. E. (1972). "Partition identities," *Advances in Math.* **9**, 10–51.

Andrews, G. E. (1974a). "A general theory of identities of the Rogers–Ramanujan type," *Bull. Amer. Math. Soc.* **80**, 1033–1052.

Andrews, G. E. (1974b). "On the general Rogers–Ramanujan theorem," *Mem. Amer. Math. Soc.* **152**.

Andrews, G. E. (1974c). "An analytic generalization of the Rogers–Ramanujan identities for odd moduli," *Proc. Nat. Acad. Sci. USA* **71**, 4082–4085.

Andrews, G. E. (1975a). "On the Alder polynomials and a new generalization of the Rogers–Ramanujan identities," *Trans. Amer. Math. Soc.* **204**, 40–64.

Andrews, G. E. (1975b). "Partially ordered sets and the Rogers–Ramanujan identities," *Aequationes Math.* **12**, 94–107.

Andrews, G. E. (1975c). "On Rogers–Ramanujan type identities related to the modulus 11," *Proc. London Math. Soc.* (3) **30**, 330–346.

Andrews, G. E. (1975d). "Problems and prospects for basic hypergeometric functions," *The Theory and Applications of Special Functions* (R. Askey, ed.), pp. 191–224. Academic Press, New York.

Cheema, M. S. (1969). "Duality in the theory of partitions," *Res. Bull. Panjab Univ.* **20**, 201–206.

Göllnitz, H. (1967). "Partitionen mit Differenzenbedingungen," *J. reine angew. Math.* **225**, 154–190.

Gordon, B. (1961). "A combinatorial generalization of the Rogers–Ramanujan identities," *Amer. J. Math.* **83**, 393–399.

Gordon, B. (1965). "Some continued fractions of the Rogers–Ramanujan type," *Duke Math. J.* **31**, 741–748.

Gupta, H. (1970). "Partitions – a survey," *J. Res. Nat. Bur. Standards* **74B**, 1–29.
Hardy, G. H. (1940). *Ramanujan*. Cambridge Univ. Press, London and New York; reprinted by Chelsea, New York.
Lehner, J. (1941). "A partition function associated with the modulus five," *Duke Math. J.* **8**, 631–655.
LeVeque, W. J. (1974). *Reviews in Number Theory*, Vol. 4. Amer. Math. Soc., Providence, R.I.
Rogers, L. J. (1894). "Second memoir on the expansion of certain infinite products," *Proc. London Math. Soc.* **25**, 318–343.
Rogers, L. J. (1919). "Proof of certain identities in combinatory analysis," *Proc. Cambridge Phil. Soc.* **19**, 211–214.
Schur, I. J. (1917). "Ein Beitrag zur additiven Zahlentheorie und zur Theorie der Kettenbrüche," *S.-B. Preuss. Akad. Wiss. Phys.-Math. Kl.* pp. 302–321. (Reprinted in I. Schur, *Gesammelte Abhandlungen*, Vol. 2, pp. 117–136. Springer, Berlin, 1973.)
Schur, I. J. (1926). "Zur additiven Zahlentheorie," *S.-B. Preuss. Akad. Wiss. Phys.-Math. Kl.* pp. 488–495. (Reprinted in I. Schur, *Gesammelte Abhandlungen*, Vol. 3, pp. 43–50. Springer, Berlin, 1973.)
Selberg, A. (1936). "Über einige arithmetische Identitäten," *Avhl. Norske Vid.* **8**, 23 pp.
Singh, V. N. (1957). "Certain generalized hypergeometric identities of the Rogers–Ramanujan type," *Pacific J. Math.* **7**, 1011–1014.
Watson, G. N. (1929). "Theorems stated by Ramanujan (VII): Theorems on continued fractions," *J. London Math. Soc.* **4**, 39–48.

CHAPTER 8

A General Theory of Partition Identities

8.1 Introduction

The results in Chapter 7 indicate that a general study of partition identities such as the Rogers–Ramanujan identities or the Göllnitz–Gordon identities should help to illuminate these appealing but seemingly unmotivated theorems. In this chapter we shall undertake the foundations of this study. As will become abundantly clear, there are very few truly satisfactory answers to the questions that we shall examine. We shall instead have to settle for partial answers. After presenting the fundamental structure of such problems in the next section, we devote Section 8.3 to "partition ideals of order 1," a topic which we can handle adequately and which suggests the type of answers we would like for our general questions of Section 8.2. The final section of the chapter describes a large class of partition problems wherein the related generating function satisfies a linear homogeneous q-difference equation with polynomial coefficients. In some ways this final section is unsatisfactory, in that the theory of q-difference equations has not been adequately developed to provide answers generally to questions about partition identities and partition asymptotics; however, the theorems of Section 8.4 do suggest that q-difference equations are indeed worthy of future research.

8.2 Foundations

We begin with a simple intuitive observation which forms the basis of our work here. In all the partition identities considered in Chapter 7 (Theorem 7.5, Corollaries 7.6 and 7.7, Theorem 7.11) partition functions were considered that enumerated partitions lying in some subset C of the set of all partitions. For example, C might be the set of all partitions where the difference between parts is at least 2 (see Corollary 7.6), or C might be the set of all partitions with parts congruent to 1, 4, 7 (mod 8) (see Theorem 7.11 when $k = i = 2$). The interesting fact to note is that *all the considered subsets C of partitions have*

ENCYCLOPEDIA OF MATHEMATICS and Its Applications, Gian-Carlo Rota (ed.). 2, George E. Andrews, The Theory of Partitions

the property that if π is a partition in C and one or several parts are removed from π to form a new partition π', then π' is also in C.

This observation suggests that we should define a partial ordering on partitions by $\pi' \leqslant \pi$ whenever any integer i appears at least as often in π as in π'. This approach immediately leads us to a lattice structure on partitions:

DEFINITION 8.1. Let $\mathscr{S}$ denote the set of all sequences $\{f_i\}_{i=1}^{\infty}$ of nonnegative integers wherein only finitely many of the f_i are nonzero.

For reference we note that each sequence $\{f_i\}_{i=1}^{\infty} = \{f_i\}$ in $\mathscr{S}$ uniquely corresponds to the partition $(1^{f_1}2^{f_2}3^{f_3}4^{f_4}\cdots)$.

DEFINITION 8.2. We define a partial order "$\leqslant$" on $\mathscr{S}$ by asserting that $\{f_i\} \leqslant \{g_i\}$ whenever $f_i \leqslant g_i$ for all i.

We note that this definition is consistent with the order structure intuitively suggested at the beginning of this section. Furthermore, $\mathscr{S}$ is actually a lattice whose "meet" and "join" operations are given by $\{f_i\} \cap \{g_i\} = \{\min(f_i, g_i)\}$, $\{f_i\} \cup \{g_i\} = \{\max(f_i, g_i)\}$.

DEFINITION 8.3. A subset C of $\mathscr{S}$ which has the property that whenever $\{f_i\} \in C$ and $\{g_i\} \leqslant \{f_i\}$, then necessarily also $\{g_i\} \in C$, is called a *partition ideal.*

In the notation of lattice theory the sets C would be called "semi-ideals" or "order ideals."

DEFINITION 8.4. We define a function σ which associates with each element of $\mathscr{S}$ the number that is actually being partitioned: $\sigma(\{f_i\}) = \sum_{i=1}^{\infty} f_i \cdot i$.

From the lattice-theoretic point of view, σ is a positive valuation on $\mathscr{S}$ in that $\sigma(\{f_i\} \cap \{g_i\}) + \sigma(\{f_i\} \cup \{g_i\}) = \sigma(\{f_i\}) + \sigma(\{g_i\})$, and if $\{f_i\} > \{g_i\}$, then $\sigma(\{f_i\}) > \sigma(\{g_i\})$.

DEFINITION 8.5. We say two partition ideals C_1 and C_2 are *equivalent*: $C_1 \sim C_2$ if $p(C_1, n) = p(C_2, n)$ for all n.

After this inundation of definitions let us slow down a little and see how these definitions actually fit in with the results obtained in Chapter 7. First, let

$$\mathscr{O} = \{\{f_i\} \in \mathscr{S} | f_i > 0 \text{ implies } i \text{ odd}\}, \tag{8.2.1}$$

$$\mathscr{D} = \{\{f_i\} \in \mathscr{S} | f_i \leqslant 1\}. \tag{8.2.2}$$

Then Euler's theorem (Corollary 1.2) asserts that

$$\mathscr{O} \sim \mathscr{D} \tag{8.2.3}$$

since $p(\mathcal{O}, n)$ is the number of partitions of n with odd parts and $p(\mathcal{D}, n)$ is the number of partitions of n with distinct parts.

Next, let

$$\mathcal{B}_{k,a} = \{\{f_i\} \in \mathcal{S} | f_i + f_{i+1} \leqslant k - 1, f_1 \leqslant a - 1\}, \tag{8.2.4}$$

$$\mathcal{A}_{k,a} = \{\{f_i\} \in \mathcal{S} | f_i > 0 \text{ implies } i \not\equiv 0, \pm a (\text{mod } 2k + 1)\}. \tag{8.2.5}$$

Then Gordon's generalization of the Rogers–Ramanujan identities is immediately seen to be equivalent to the assertion that

$$\mathcal{A}_{k,a} \sim \mathcal{B}_{k,a} \qquad \text{for} \quad 1 \leqslant a \leqslant k. \tag{8.2.6}$$

Finally, if we define

$$\mathcal{D}_{k,a} = \{\{f_i\} \in \mathcal{S} | f_i + f_{i+1} \leqslant k - 1, f_{2i} + f_{2i+1} + f_{2i+2} \leqslant k - 1,$$
$$f_i > 1 \text{ implies } i \text{ even}, f_1 + f_2 \leqslant a - 1\} \tag{8.2.7}$$

and

$$\mathcal{C}_{k,a} = \{\{f_i\} \in \mathcal{S} | f_i > 0 \text{ implies } i \not\equiv 2 (\text{mod } 4) \text{ and}$$
$$i \not\equiv 0, \pm (2a - 1)(\text{mod } 4k)\}, \tag{8.2.8}$$

then the generalization of the Göllnitz–Gordon identities given in Theorem 7.11 asserts that

$$\mathcal{C}_{k,a} \sim \mathcal{D}_{k,a} \qquad \text{for} \quad 1 \leqslant a \leqslant k. \tag{8.2.9}$$

That $\mathcal{A}_{k,a}$, $\mathcal{B}_{k,a}$, $\mathcal{C}_{k,a}$, $\mathcal{D}_{k,a}$ are all partition ideals is immediate from their definitions; however, it requires a moment's calculation to see that truly $p(\mathcal{B}_{k,a}, n) = B_{k,a}(n)$ (and $p(\mathcal{D}_{k,a}, n) = D_{k,a}(n)$). To aid the reader in this matter, we observe that if $(b_1 b_2 \cdots b_3)$ is a partition of n with $b_j \geqslant b_{j+1}, b_j - b_{j+k-1} \geqslant 2$, then such partitions are characterized by the fact that for any i, the total number of appearances of i *and* $i + 1$ (namely, $f_i + f_{i+1}$) in that partition is at most $k - 1$; otherwise, if b_j is chosen as the first i, then $b_j - b_{j+k-1} \leqslant 1$.

Since the equivalence relation in Definition 8.5 can be used to describe the results in Chapter 7 as well as Euler's theorem (Corollary 1.2), we are led to the following question:

Fundamental Problem. Fully characterize the equivalence classes of partition ideals.

This problem is phrased intentionally in a vague manner. A minimally satisfactory answer would be some sort of readily usable algorithm for determining whether two partition ideals are equivalent. We shall, unfortunately, be unable to scratch the surface of this problem; however, we shall consider a refinement in the next section, and for this simpler problem we shall obtain an adequate solution (Theorem 8.4).

8.3 Partition Ideals of Order 1

The last section was based on the observation that a certain order of partitions is inherent in all the partition problems treated in Chapter 7. Next we note a certain "local" property involved in these problems: to determine if $\{f_i\} \in \mathscr{A}_{k,a}$ (see (8.2.5)) we need only examine the sequence *one* term at a time, at each stage checking the condition "$f_i > 0$ implies $i \equiv 0, \pm a \pmod{2k+1}$." However, to determine if $\{f_i\} \in \mathscr{B}_{k,a}$ we must check *consecutive pairs* f_i, f_{i+1} against the condition "$f_i + f_{i+1} \leqslant k - 1$" after having checked the initial condition $f_1 \leqslant a - 1$. The following definition makes precise the notion of how narrowly an examination can be focused on $\{f_i\} \in \mathscr{S}$ to determine whether $\{f_i\} \in C$.

DEFINITION 8.6. We shall say that a partition ideal C has order k if k is the least positive integer such that whenever $\{f_i\} \notin C$, then there exists m such that $\{f_i'\} \notin C$ where

$$f_i' = \begin{cases} f_i & \text{for } i = m, m+1, \ldots, m+k-1, \\ 0 & \text{otherwise.} \end{cases}$$

For the partition ideals considered in Section 8.2 we observe immediately that $\mathscr{O}$, $\mathscr{D}$, $\mathscr{A}_{k,a}$, and $\mathscr{C}_{k,a}$ are all of order 1, while $\mathscr{B}_{k,a}$ is of order 2 and $\mathscr{D}_{k,a}$ is of order 3. Inasmuch as Euler's theorem (Corollary 1.2) asserts the equivalence of two partition ideals of order 1, it is reasonable to ask the following:

Problem. Fully characterize the equivalence classes of partition ideals of order 1.

We shall complete this section by treating this problem in quite a satisfactory manner.

THEOREM 8.1. *Let C be a partition ideal. Then C is of order 1 if and only if there exists $d_1, d_2, d_3, \ldots$ such that*

$$C = \{\{f_i\} \in \mathscr{S} \mid f_i \leqslant d_i \text{ for all } i\} \tag{8.3.1}$$

where each d_i is a nonnegative integer or $+\infty$.

Remark. Once Theorem 8.1 is proved we note that each d_j is determined by $d_j = \sup_{\{f_i\} \in C} f_j$.

Proof. We let $\bar{C}$ denote the set of partitions on the right-hand side of (8.3.1). Clearly any such set is a partition ideal of order 1 since it is clearly a partition ideal and if $\{f_i\} \notin \bar{C}$, then there exists j such that $f_j > d_j$ for some j and hence $\{f_i'\} \notin \bar{C}$ where $f_i' = 0$ if $i \neq j$ and $f_j' = f_j$.

On the other hand, suppose C is of order 1; then define $d_j = \sup_{\{f_i\} \in C} f_j$. With this definition of C it is clear that $C \subset \bar{C}$. Conversely, if $\{f_i\} \notin C$, then there exists j such that $\{f_i'\} \notin C$ where $f_i' = 0$ if $i \neq j$ and $f_j' = f_j$. I claim that $f_j > d_j$, for if not, by the very definition of d_j we see that there must exist $\{h_i\} \in C$ such that $f_j \leqslant h_j \leqslant d_j$ and thus $\{f_i'\} \leqslant \{h_i\} \in C$, which would imply that $\{f_i'\} \in C$, an impossibility. Thus since $f_j > d_j$, we see that $\{f_i\} \notin \bar{C}$, and therefore $C = \bar{C}$, as desired. ∎

Our next result shows that the partition ideals of order 1 have a large amount of algebraic structure:

THEOREM 8.2. *The ideals of $\mathscr{S}$ are just the partition ideals of order* 1.

Proof. Recall the meet and join operations in $\mathscr{S}$: $\{f_i\} \cap \{g_i\} = \{\min(f_i, g_i)\}$ and $\{f_i\} \cup \{g_i\} = \{\max(f_i, g_i)\}$, and recall that an ideal S of $\mathscr{S}$ is merely a partition ideal that is closed under the join operation.

Let us therefore assume that S is an ideal of $\mathscr{S}$. Thus S is a partition ideal. Define $d_j = \sup_{\{f_i\} \in S} f_j$, and let

$$\bar{S} = \{\{f_i\} \in \mathscr{S} \mid f_i \leqslant d_i \text{ for all } i\}.$$

Clearly $S \subset \bar{S}$. If now $\{f_i\} \in \bar{S}$, we may define $\{f_i^{(j)}\}$ by $f_i^{(j)} = 0$ if $i \neq j$, $f_j^{(j)} = f_j$ and we note that $\{f_i\} = \{f_i^{(1)}\} \cup \{f_i^{(2)}\} \cup \cdots \cup \{f_i^{(N)}\}$ where N is chosen so that $f_i = 0$ for all $i > N$. Next we note that for each j there exists $\{h_i\} \in S$ such that $f_j \leqslant h_j \leqslant d_j$ by the definition of the d_j. Consequently, $\{f_i^{(j)}\} \leqslant \{h_i\} \in S$ and so $\{f_i^{(j)}\} \in S$ for each j. Therefore, $\{f_i\} = \{f_i^{(1)}\} \cup \{f_i^{(2)}\} \cup \cdots \cup \{f_i^{(N)}\}$ is in S since S is an ideal of $\mathscr{S}$. Hence $\bar{S} = S$ and so by Theorem 8.1, S is a partition ideal of order 1.

The reverse implication is trivial. For if C is a partition ideal of order 1, then by Theorem 8.1,

$$C = \{\{f_i\} \in \mathscr{S} \mid f_i \leqslant d_i \text{ for all } i\}.$$

Hence, if $\{f_i\} \in C$ and $\{g_i\} \in C$, then so is $\{f_i\} \cup \{g_i\} = \{\max(f_i, g_i)\}$, and so C is an ideal of $\mathscr{S}$. ∎

Our next result provides a very usable representation of the generating function for $P(C, n)$ whenever C is a partition ideal of order 1.

THEOREM 8.3. *Let C be a partition ideal of order* 1 *with $d_j = \sup_{\{f_i\} \in C} f_j$. Then*

$$\sum_{n \geqslant 0} p(C, n) q^n = \frac{\prod_{\substack{j=1 \\ d_j < \infty}}^{\infty} (1 - q^{j(d_j+1)})}{\prod_{j=1}^{\infty} (1 - q^j)}.$$

Proof. By Theorem 8.1, $p(C, n)$ is precisely the number of partitions of n in which each integer i appears as a part at most d_i times. Hence (as in the proof of Theorem 1.1)

$$\sum_{n \leq 0} p(C, n)q^n = \prod_{\substack{j=1 \\ d_j < \infty}}^{\infty} (1 + q^j + q^{2j} + \cdots + q^{d_j j})$$

$$\prod_{\substack{j=1 \\ d_j = \infty}}^{\infty} (1 + q^j + q^{2j} + q^{3j} + \cdots)$$

$$= \prod_{\substack{j=1 \\ d_j < \infty}}^{\infty} \frac{(1 - q^{j(d_j+1)})}{(1 - q^j)} \cdot \prod_{\substack{j=1 \\ d_j = \infty}}^{\infty} \frac{1}{(1 - q^j)}$$

$$= \frac{\prod_{\substack{j=1 \\ d_j < \infty}}^{\infty} (1 - q^{j(d_j+1)})}{\prod_{j=1}^{\infty} (1 - q^j)}. \qquad \blacksquare$$

This last result allows us to provide a relatively simple test for determining the equivalence of two partition ideals of order 1, and in this way we have what might be termed a "minimal" solution to the problem posed at the beginning of this section.

THEOREM 8.4. *Let C and C' be partition ideals of order 1 with $d_j = \sup_{(f_i) \in C} f_j$ and $d_j' = \sup_{(f_i) \in C} f_j$. Then $C \sim C'$ if and only if the two sequences of positive integers $\{j(d_j + 1)\}_{\substack{j=1 \\ d_j < \infty}}^{\infty}$ and $\{j(d_j' + 1)\}_{\substack{j=1 \\ d_j < \infty}}^{\infty}$ are merely reorderings of each other.*

Remark. This theorem may be rephrased in terms of multisets (a topic treated in Section 3.4): A multiset M is a pair (S, f) where S is a set and f is a function from S into the positive integers. For each $a \in S$, $f(a)$ is termed the "multiplicity of a in M." Now each of the sequences $\{j(d_j + 1)\}_{\substack{j=1 \\ d_j < \infty}}^{\infty}$ and $\{j(d_j' + 1)\}_{\substack{j=1 \\ d_j < \infty}}^{\infty}$ uniquely defines a corresponding multiset D (resp. D') where, say, $D = (\Delta, \delta)$ with Δ the set of positive integers of the form $j(d_j + 1)$ for some j and $\delta(r)$ the multiplicity function which counts the number of appearances of r in the sequence $\{j(d_j + 1)\}_{\substack{j=1 \\ d_j < \infty}}^{\infty}$. Theorem 8.4 merely asserts that $C \sim C'$ if and only if $D = D'$.

Proof. If the two sequences in question are merely reorderings of each other, then by Theorem 8.3

$$\sum_{n \geq 0} p(C, n)q^n = \frac{\prod_{\substack{j=1 \\ d_j < \infty}}^{\infty} (1 - q^{j(d_j+1)})}{\prod_{j=1}^{\infty} (1 - q^j)}$$

$$= \frac{\prod_{\substack{j=1 \\ d_j' < \infty}}^{\infty} (1 - q^{j(d_j'+1)})}{\prod_{j=1}^{\infty} (1 - q^j)}$$

$$= \sum_{n \geqslant 0} p(C', n)q^n.$$

Hence $p(C, n) = p(C', n)$ for all n and so $C \sim C'$.

Suppose the two sequences in question are not merely reorderings of each other. Let $\delta(r)$ (resp. $\delta'(r)$) denote the number of times r appears in $\{j(d_j + 1)\}_{\substack{j=1 \\ d_j < \infty}}^{\infty}$ (resp. $\{j(d_j' + 1)\}_{\substack{j=1 \\ d_j' < \infty}}^{\infty}$), and let h be the least integer such that $\delta(h) \neq \delta(h')$. Now let us suppose $C \sim C'$. Hence

$$\frac{\prod_{j=1}^{\infty} (1 - q^j)^{\delta(j)}}{\prod_{j=1}^{\infty} (1 - q^j)} = \sum_{n \geqslant 0} p(C, n)q^n$$

$$= \sum_{n \geqslant 0} p(C', n)q^n$$

$$= \frac{\prod_{j=1}^{\infty} (1 - q^j)^{\delta'(j)}}{\prod_{j=1}^{\infty} (1 - q^j)}.$$

Therefore

$$\prod_{j=h}^{\infty} (1 - q^j)^{\delta(j)} = \prod_{j=h}^{\infty} (1 - q^j)^{\delta'(j)},$$

and this is impossible since the coefficient of q^h on the left-hand side is $-\delta(h)$, which does not equal $-\delta'(h)$, the coefficient of q^h on the right-hand side. This contradiction shows that $C \nsim C'$. ■

A myriad of partition theorems follow easily from Theorem 8.4. We record a few next.

COROLLARY 8.5. *Let $P_1(n)$ denote the number of partitions of n in which each integer i appears at most $i - 1$ times. Let $P_2(n)$ denote the number of partitions of n into nonsquares. Then $P_1(n) = P_2(n)$ for all n.*

Proof. Let $C_1 = \{\{f_i\} \in \mathscr{S} | f_i \leqslant i - 1\}$ and $C_2 = \{\{f_i\} \in \mathscr{S} | f_i > 0 \Rightarrow i$ not a perfect square$\}$. Then the assertion $P_1(n) = P_2(n)$ is just $C_1 \sim C_2$. For C_1 the associated $\{j(d_j + 1)\}_{\substack{j=1 \\ d < \infty}}^{\infty}$ is $\{j(j - 1 + 1)\}_{j=1}^{\infty} = \{j^2\}_{j=1}^{\infty}$, while for C_2 it is also $\{j^2(0 + 1)\}_{j=1}^{\infty} = \{j^2\}_{j=1}^{\infty}$. Hence $C_1 \sim C_2$ by Theorem 8.4. ■

COROLLARY 8.6. *Let M_1 and M_2 be two sets of positive integers. Let $2M_1$ denote the doubles of elements of M_1 (i.e., $2M_1 = \{j | (j/2) \in M_1\}$). Then the number of partitions of n into distinct parts taken from M_1 always equals* *the number of partitions of n into parts taken from M_2 if and only if $2M_1 \subseteq M_1$ and $M_2 = M_1 - 2M_1$.*

Proof. Let $C_1 = \{\{f_i\} \in \mathscr{S} | f_i \leqslant 1$ for all i, $f_i = 1$ implies $i \in M_1\}$; and let $C_2 = \{\{f_i\} \in \mathscr{S} | f_i > 0$ implies $f_i \in M_2\}$. Then we are to show that $C_1 \sim C_2$ if and only if $2M_1 \subseteq M_1$ and $M_2 = M_1 - 2M_1$. The sequence $\{j(d_j + 1)\}_{\substack{j=1 \\ d_j < \infty}}^{\infty}$ associated with C_1 consists of $\{j \cdot 2\}_{\substack{j=1 \\ j \in M_1}}^{\infty}$ together with $\{j\}_{\substack{j=1 \\ j \notin M_1}}^{\infty}$, while that associated with C_2 consists of $\{j\}_{\substack{j=1 \\ j \notin M_2}}^{\infty}$, and these two multisets are identical if and only if $2M_1 \cap M_1{}^c = \emptyset$ (otherwise a second-order multiplicity would occur in C_1's multiset while only first-order multiplicities occur in C_2's) and $2M_1 \cup M_1{}^c = M_2{}^c$. Hence by Theorem 8.4 $C_1 \sim C_2$ if and only if $2M_1 \subseteq M_1$ and $M_2 = M_1 - 2M_1$. ∎

COROLLARY 8.7. *Every integer is uniquely the sum of distinct powers of* 2.

Proof. Let $M_1 = \{1, 2, 4, 8, 16, 32, \ldots\}$ and let $M_2 = \{1\}$. Then $2M_1 = \{2, 4, 8, 16, \ldots\} \subseteq M_1$, and $M_2 = M_1 - 2M_1$. Hence by Corollary 8.6 the number of partitions of n into distinct parts taken from M_1 equals the number of partitions of n into parts taken from M_2, which is exactly 1, since there is obviously a unique partition of n that uses only ones. ∎

8.4 Linked Partition Ideals

The last section provided a nice guide for us concerning how we would like to treat the entire question of equivalence classes of partition ideals. Unfortunately the power of a result like Theorem 8.3 is unavailable to us when the partition ideal C is not of order 1. In this section we shall at least show that a wide class of partition ideals (linked partition ideals) have two-variable generating functions that are the solutions of finite linear homogeneous q-difference equations with polynomial coefficients. The class of linked partition ideals includes both the $\mathscr{B}_{k,k}$ and $\mathscr{D}_{k,k}$ of Chapter 7, as well as most of the partition ideals that appear in known partition identities. It includes partition ideals C of order 1 if and only if the sequence of local suprema $\{d_i\}_{i=1}^{\infty}$ is periodic modulo k for some k and $d_i \neq \infty$ for any i; however, partition ideals of order 1 have been adequately treated in Section 8.3, so that the important applications of linked partition ideals occur when the ideal in question is of order 2 or more.

In this section we shall consider a two-variable generating function for each partition ideal C:

$$f_C(x; q) = \sum_{\{f_i\} \in C} x^{\Sigma f_i} q^{\Sigma f_i i}$$

$$\equiv \sum_{\{f_i\} \in C} x^{\#(\{f_i\})} q^{\sigma(\{f_i\})}, \tag{8.4.1}$$

where the exponent on x is the number of parts of the partition $(1^{f_1}2^{f_2}3^{f_3}\cdots)$ and the exponent of q is the number being partitioned. Note that

$$f_C(1; q) = \sum_{\{f_i\}\in C} q^{\sigma(\{f_i\})} = \sum_{n=0}^{\infty} p(C, n)q^n. \tag{8.4.2}$$

Furthermore, it is not difficult to show (by the means used in Theorem 1.1) that

$$f_{\mathscr{S}}(x; q) = \prod_{n=1}^{\infty} (1 - xq^n)^{-1}, \tag{8.4.3}$$

a product absolutely convergent for $|q| < 1$, $|x| < |q|^{-1}$. Consequently, since the number of partitions of n with m parts in an arbitrary partition ideal C clearly does not exceed the total of such in $\mathscr{S}$, we see by the comparison test (applying (8.4.3) to (8.4.1)) that the series in (8.4.1) is absolutely convergent for $|q| < 1$, $|x| < |q|^{-1}$.

Q-difference equations (like (7.1.1) and (7.2.4)) were very important in Chapter 7. Now we must clear away the specific nature of the problems to determine general qualities of partition ideals that allow q-difference equations to arise. The first important property is the existence of a "modulus."

Definition 8.7. Let C be a partition ideal; we define $C^{(m)} = \{\{f_i\} \in C \mid f_1 = f_2 = \cdots = f_m = 0\}$.

Thus $C^{(m)}$ is the set of partitions in C that have all their parts larger than m.

Definition 8.8. We denote by ϕ the bijection $\phi\colon \mathscr{S} \to \mathscr{S}^{(1)}$ given by $\phi(\{f_i\}) = \{g_i\}$ where

$$g_i = \begin{cases} 0 & \text{if } i = 1, \\ f_{i-1} & \text{if } i > 1. \end{cases}$$

Intuitively the function ϕ merely adds 1 to each part of the partition, since

$$\sigma(\phi\{f_i\}) = \sum_{i=2}^{\infty} if_{i-1} = \sum_{i=1}^{\infty} f_i(i + 1);$$

thus where initially i appeared f_i times now $i + 1$ appears f_i times.

Definition 8.9. We shall say that a partition ideal C has *modulus* m if m is a positive integer such that $\phi^m C = C^{(m)}$. A modulus is not necessarily unique; indeed, any integral multiple of a modulus for C is also one.

Many partition ideals have no modulus (e.g., the C_1 and C_2 considered in Section 8.3 in the proof of Corollary 8.5). The partition ideal $\mathscr{C}$ in (8.2.1) has modulus 2; $\mathscr{Q}$ in (8.2.2) has modulus 1; $\mathscr{B}_{k,a}$ in (8.2.4) has modulus 1;

$\mathscr{A}_{k,a}$ in (8.2.5) has modulus $2k + 1$; $\mathscr{D}_{k,a}$ in (8.2.7) has modulus 2, and $\mathscr{C}_{k,a}$ in (8.2.8) has modulus $4k$.

The relationship between the modulus and q-difference equations is made explicit by the following:

LEMMA 8.8. *If the partition ideal C has modulus m, then $f_C(xq^m; q) = f_{C^{(m)}}(x; q)$.*

Proof.

$$
\begin{aligned}
f_C(xq^m; q) &= \sum_{\{f_i\} \in C} x^{\#(\{f_i\})} q^{m\#(\{f_i\}) + \sigma(\{f_i\})} \\
&= \sum_{\{f_i\} \in C} x^{\#(\phi^m(\{f_i\}))} q^{\sigma(\phi^m(\{f_i\}))} \\
&= \sum_{\{f_i\} \in \phi^m C} x^{\#(\{f_i\})} q^{\sigma(\{f_i\})} \\
&= \sum_{\{f_i\} \in C^{(m)}} x^{\#(\{f_i\})} q^{\sigma(\{f_i\})} = f_{C^{(m)}}(x; q). \qquad \blacksquare
\end{aligned}
$$

DEFINITION 8.10. We define, for each partition ideal C of modulus m,

$$L_C = \{\{f_i\} \in C \mid f_i = 0 \text{ for } i > m\}.$$

Thus L_C is just a subpartition ideal of C consisting of all partitions in C whose parts do not exceed m. Note that L_C depends on which modulus m is chosen.

DEFINITION 8.11. For each $\{f_i\}$ and $\{g_i\}$ in $\mathscr{S}$, define $\{f_i\} \oplus \{g_i\} = \{f_i + g_i\}$.

LEMMA 8.9. *If C is a partition ideal of modulus m, then for each $\pi \in C$ there exists a unique sequence $\pi_1, \pi_2, \pi_3, \pi_4, \ldots$ of elements of L_C such that*

$$\pi = \pi_1 \oplus (\phi^m \pi_2) \oplus (\phi^{2m} \pi_3) \oplus (\phi^{3m} \pi_4) \oplus \cdots. \tag{8.4.4}$$

Proof. We note that since C has modulus m, $\phi^{jm} C = C^{(jm)}$ for each $j = 1, 2, 3, \ldots$. (This is because $\phi: C \to C^{(m)}$ is a bijection and consequently $\phi C^{(m)}$ must contain precisely the elements $\{f_i\}$ of C with $f_1 = f_2 = \cdots = f_{2m} = 0$; that is, $\phi^m C^{(m)} = C^{(2m)}$ and so $\phi^{2m} C = \phi^m \phi^m C = \phi^m C^{(m)} = C^{(2m)}$; the proof for arbitrary j follows similar lines.)

Now for any $\pi = \{f_i\}$, we have uniquely

$$
\begin{aligned}
\pi = \{&f_1, f_2, \ldots, f_m, 0, 0, 0, \ldots\} + \{0, 0, \ldots, 0, f_{m+1}, \\
&f_{m+2}, \ldots, f_{2m}, 0, 0, \ldots\} + \{0, 0, \ldots, 0, f_{2m+1}, \\
&f_{2m+2}, \ldots, f_{3m}, 0, 0, \ldots\} + \cdots \\
= \pi_1 &\oplus (\phi^m \pi_2) \oplus (\phi^{2m} \pi_3) \oplus \cdots
\end{aligned}
$$

where $\pi_j = \{g_i\}_{i=1}^{\infty}$ and

$$g_i = \begin{cases} f_{jm+i} & \text{for} \quad i = 1, 2, \ldots, m, \\ 0 & \text{for} \quad i > m. \end{cases}$$

To conclude our lemma, we must show that when C has modulus m and $\pi \in C$, then each $\pi_j \in L_C$. Clearly, $\phi^{(j-1)m}\pi_j \leqslant \pi \in C$; hence $\phi^{(j-1)m}\pi_j \in C$. But $\phi^{(j-1)m}C = C^{((j-1)m)}$. Therefore since $\phi^{(j-1)m}\pi_j \in C^{((j-1)m)} = \phi^{(j-1)m}C$, we see that $\pi_j \in C$, and so $\pi_j \in L_C$. ∎

DEFINITION 8.12. We say that C is a *linked partition ideal* if

(i) C has a modulus, say m;

(ii) the L_C corresponding to m is a finite set;

(iii) for each $\pi_a \in L_C$ there corresponds a subset $\mathscr{L}_C(\pi_a)$ of L_C (called the linking set of π_a) and a positive integer $l(\pi_a)$ (called the span of π_a) such that for any $\pi \in \mathscr{S}$, $\pi \in C$ if and only if the representation (8.4.4) has the property that for each j, $\pi_{j+1}, \pi_{j+2}, \ldots, \pi_{j+l(\pi_j)-1} = \{0, 0, 0, 0, \ldots\}$ and $\pi_{j+l(\pi_j)} \in \mathscr{L}_C(\pi_j)$.

Since the definition of a linked partition ideal is somewhat complicated, let us consider some examples. From (8.2.4) let us consider

$$\mathscr{B} \equiv \mathscr{B}_{k,k} = \{\{f_i\} \in \mathscr{S} | f_i + f_{i+1} \leqslant k - 1\}.$$

Here $\mathscr{B}$ has modulus 1, and if we let $\pi_i = \{i, 0, 0, 0, \ldots\}$, then $L_{\mathscr{B}} = \{\pi_0, \pi_1, \pi_2, \ldots, \pi_{k-1}\}$, $l(\pi_i) = 1$ for $0 \leqslant i \leqslant k - 1$ and

$$\mathscr{L}_{\mathscr{B}}(\pi_i) = \{\pi_0, \pi_1, \ldots, \pi_{k-i-1}\} \qquad \text{for} \quad 0 \leqslant i \leqslant k - 1.$$

From (8.2.2), we see that $\mathscr{Q}$ has modulus 1, and if $\pi_i = \{i, 0, 0, 0, \ldots\}$, then $L = \{\pi_0, \pi_1\}$, $l(\pi_0) = l(\pi_1) = 1$ and

$$\mathscr{L}_{\mathscr{Q}}(\pi_0) = \mathscr{L}_{\mathscr{Q}}(\pi_1) = \{\pi_0, \pi_1\}.$$

From (8.2.7), we see that $\mathscr{Q}_{k,k} = \mathscr{Q}^*$ has modulus 2, and if $\pi_i = \{0, i, 0, 0, \ldots\}$, $\psi_i = \{1, i, 0, 0, \ldots\}$, then $L_{\mathscr{Q}^*} = \{\pi_0, \pi_1, \ldots, \pi_{k-1}, \psi_0, \psi_1, \ldots, \psi_{k-2}\}$, $l(\pi_i) = l(\psi_i) = l(\pi_{k-1}) = 1$ for $i = 0, 1, \ldots, k - 2$, $\mathscr{L}_{\mathscr{Q}^*}(\psi_i) = \mathscr{L}_{\mathscr{Q}^*}(\pi_i) = \{\pi_0, \pi_1, \ldots, \pi_{k-i-1}, \psi_0, \psi_1, \ldots, \psi_{k-i-2}\}$.

If the linking set $\mathscr{L}_C(\pi) = \{\pi_0\}$ and $l(\pi) = r$, then there actually is a second choice of span and linking set for π, namely, $\mathscr{L}_C(\pi) = L_C$ and $l(\pi) = r + 1$; generally the first choice is best in order to keep the spans as small as possible.

We remark that $\mathscr{B}_{k,a}$ for $1 \leqslant a < k$ and $\mathscr{Q}_{k,a}$ for $1 \leqslant a < k$ are *not* linked partition ideals; however, they are importantly related to $\mathscr{B}_{k,k}$ and $\mathscr{Q}_{k,k}$, respectively, as we now make clear.

DEFINITION 8.13. If C is a linked partition ideal and $\bar{\pi} \in L_C$, define $C_{\bar{\pi}}$ as the subset of C consisting of all π in C for which the representation (8.4.4) has $\pi_1 = \bar{\pi}$.

Thus for $1 \leqslant i < k$

$$\mathscr{B}_{\pi_i} = \pi_i \ominus \phi\mathscr{B}_{k,k-i},$$

$$\mathscr{D}^*_{\pi_i} = \pi_i \oplus \phi^2\mathscr{D}_{k,k-i},$$

$$\mathscr{D}^*_{\psi_i} = \psi_i \oplus \phi^2\mathscr{D}_{k,k-i}.$$

As we shall see, the $C_{\bar{\pi}}$ are very important in the main theorem on linked partition ideals (Theorem 8.11). Crucial to the proof of Theorem 8.11 will be the following result, which is an adaptation to q-difference equations of an algorithm of F. J. Murray and K. S. Miller (1954) for the corresponding problem in differential equations.

LEMMA 8.10. *Suppose we have a system of q-difference equations*

$$y_j(xq^m) = \sum_{k=1}^{n} p_{jk}(x)y_k(x), \qquad 1 \leqslant j \leqslant n, \tag{8.4.5$_j$}$$

valid for $|q| < 1$, $|x| < |q|^{-1}$, where the $y_j(x)$ are analytic in x and q inside this domain and the $p_{jk}(x)$ are rational functions in x and q. Then there exists $r \leqslant n$ such that

$$\sum_{k=0}^{r} \bar{p}_k(x)y_1(xq^{km}) = 0, \tag{8.4.6}$$

where the $\bar{p}_k(x)$ are rational functions of x and q and $\bar{p}_r(x) = 1$.

Proof. Throughout this proof we shall be replacing the rational functions $p_{jk}(x)$ by new rational functions, say $p^*_{ik}(x)$; rather than constantly change notation (and wind up with $\bar{p}_{jk}^{*\dagger\#1}(x)$), we shall write the new functions as $p_{jk}(x)$ also. The reader must be warned, then, that the $p_{jk}(x)$ in one line may not be the same as the $p_{jk}(x)$ in the next line.

We begin by considering $(8.4.5)_1$.

$$y_1(xq^m) = p_{11}(x)y_1(x) + (p_{12}(x)y_2(x) + \cdots + p_{1n}(x)y_n(x)).$$

Either $p_{12}(x), \ldots, p_{in}(x)$ are all identically 0 or not. If they are, we have (8.4.6) with $r = 1$. If they are not identically 0, we may (through a suitable renumbering of the $y_2(x), \ldots, y_n(x)$) assume then that $p_{12}(x)$ is not identically 0. Define

$$w_2(x) = p_{12}(x)y_2(x) + \cdots + p_{1n}(x)y_n(x) \tag{8.4.7}$$

and now we may eliminate $y_2(x)$ from our system of equations via

$$y_2(x) = (w_2(x) - p_{13}(x)y_3(x) - \cdots - p_{1n}(x)y_n(x))/p_{12}(x). \tag{8.4.8}$$

Our new system is

$$y_1(xq^m) = p_{11}(x)y_1(x) + w_2(x)$$

$$\begin{aligned} w_2(xq^m) &= \sum_{k=2}^{n} p_{1k}(xq^m)y_k(xq^m) \\ &= \sum_{k=2}^{n} p_{1k}(xq^m) \sum_{l=1}^{n} p_{kl}(x)y_l(x) \\ &= \sum_{\substack{l=1 \\ l\neq 2}}^{n} y_l(x)\left(\sum_{k=2}^{n} p_{1k}(xq^m)p_{kl}(x)\right) \\ &\quad + \frac{1}{p_{12}(x)}\left(\sum_{k=2}^{n} p_{1k}(xq^m)p_{k2}(x)\right) \\ &\quad \times (w_2(x) - p_{13}(x)y_3(x) - \cdots - p_{1n}(x)y_n(x)), \end{aligned} \qquad (8.4.9)_1$$

and following our initially stated convention we write

$$\begin{aligned} w_2(xq^m) = p_{21}(x)y_1(x) &+ p_{22}(x)w_2(x) + p_{23}(x)y_3(x) \\ &+ \cdots + p_{2n}(x)y_n(x). \end{aligned} \qquad (8.4.9)_2$$

After substituting (8.4.8) into $(8.4.5)_j$ for $3 \leqslant j \leqslant n$, we see that the resulting equations are again of the same form, namely

$$\begin{aligned} y_j(xq^m) = p_{j1}(x)y_1(x) &+ p_{j2}(x)w_2(x) \\ &+ \sum_{k=3}^{n} p_{jk}(x)y_k(x), \qquad 3 \leqslant j \leqslant n. \end{aligned} \qquad (8.4.9)_j$$

The foregoing procedure can be repeated; this time we examine in $(8.4.9)_2$

$$w_3(x) = p_{23}(x)y_3(x) + \cdots + p_{2n}(x)y_n(x). \qquad (8.4.10)$$

If the $p_{23}(x), \ldots, p_{2n}(x)$ are all identically zero, then $(8.4.9)_1$ and $(8.4.9)_2$ imply

$$\begin{aligned} y_1(xq^{2m}) - p_{11}(xq^m)y_1(xq^m) = p_{21}(x)y_1(x) &+ p_{22}(x)(y_1(xq^m) \\ &- p_{11}(x)y_1(x)), \end{aligned}$$

which is (8.4.6) with $r = 2$, and we would be done. If not all the $p_{23}(x), \ldots, p_{2n}(x)$ are identically zero, then the elimination $y_3(x)$ (in the same manner in which $y_2(x)$ was eliminated) yields

$$y_1(xq^m) = p_{11}(x)y_1(x) + w_2(x),$$

$$w_2(xq^m) = p_{21}(x)y_1(x) + p_{22}(x)w_2(x) + w_3(x),$$

$$w_3(xq^m) = p_{31}(x)y_1(x) + p_{32}(x)w_2(x) + p_{33}(x)w_3(x)$$

$$+ \sum_{k=4}^{n} p_{3k}(x)y_k(x),$$

$$y_j(xq^m) = p_{j1}(x)y_1(x) + p_{j2}(x)w_2(x) + p_{j3}(x)w_3(x)$$

$$+ \sum_{k=4}^{n} p_{jk}(x)y_k(x).$$

We may repeat this process until either all of the $y_2(x), y_3(x), \ldots, y_n(x)$ have been eliminated or until one of the resulting

$$w_{j+1}(x) = p_{j,u+1}(x)y_{j+1}(x) + \cdots + p_{jn}(x)y_n(x)$$

has $p_{j,j+1}(x), \ldots, p_{jn}(x)$ all identically zero. In any event, there exists $r \leqslant n$ such that

$$y_1(xq^m) = p_{11}(x)y_1(x) + w_2(x),$$

$$w_2(xq^m) = p_{21}(x)y_1(x) + p_{22}(x)w_2(x) + w_3(x),$$

$$\vdots$$

$$w_r(xq^m) = p_{r1}(x)y_1(x) + p_{r2}(x)w_2(x) + \cdots + p_{r,r}(x)w_r(x).$$

The first of these equations may be used to eliminate $w_2(x)$ from the system. Hence

$$y_1(xq^{2m}) - p_{11}(xq^m)y_1(xq^m) = p_{21}(x)y_1(x) + p_{22}(x)(y_1(xq^m)$$

$$- p_{11}(x)y_1(x)) + w_3(x);$$

$$w_3(xq^m) = p_{31}(x)y_1(x) + p_{32}(x)(y_1(xq^m) - p_{11}(x)y_1(x))$$

$$+ p_{33}(x)w_3(x) + w_4(x),$$

$$\vdots$$

$$w_r(xq^m) = p_{r1}(x)y_1(x) + p_{r2}(x)(y_1(xq^m) - p_{11}(x)y_1(x))$$

$$+ p_{r3}(x)w_3(x) + \cdots + p_{rr}(x)w_r(x).$$

The next step produces

$$y_1(xq^{3m}) + \alpha(x)y_1(xq^{2m}) + \beta(x)y_1(xq^m) + \gamma(x)y_1(x) = w_4(x),$$

and so on. Finally the elimination of $w_r(x)$ yields (8.4.6) as desired. ■

THEOREM 8.11. *Let C be a linked partition ideal, then $f_C(x; q) = \sum_{\{f_i\}\in C} x^{\Sigma f_i} q^{\Sigma f_i \cdot i}$ satisfies a finite linear homogeneous q-difference equation with coefficients that are polynomials in x and q.*

Proof. We begin by numbering the elements of L_C, say $L_C = \{\pi_0, \pi_1, \pi_2, \ldots, \pi_k\}$, and since for all C, $\{0, 0, 0, \ldots\} \in L_C$, we make the convention that $\pi_0 = \{0, 0, 0, \ldots\}$. We should quickly remark that the π_i here are *not necessarily* the same as the π_i used in treating the examples considered prior to this theorem.

Let us define for $0 \leqslant i \leqslant k$

$$H_i(x) = H_i(x, q) = \sum_{\pi\in C_{\pi_i}} x^{\#(\pi)} q^{\sigma(\pi)}. \tag{8.4.11}$$

Now since $\pi_0 = \{0, 0, 0, \ldots\}$, it follows immediately that $C_{\pi_0} = C^{(m)} = \phi^m C$. Hence by Lemma 8.8

$$H_0(x) = \sum_{\pi\in C_{\pi_0}} x^{\#(\pi)} q^{\sigma(\pi)}$$

$$= \sum_{\pi\in C^{(m)}} x^{\#(\pi)} q^{\sigma(\pi)} = f_C(xq^m; q). \tag{8.4.12}$$

Therefore we need only prove that $H_0(x)$ satisfies a linear homogeneous q-difference equation with polynomial coefficients, since then (8.4.12) will automatically imply the same for $f_C(x; q)$.

Our *first* goal in this proof is to show that the $H_i(x)$ satisfy a system of linear homogeneous q-difference equations. To this end we note that if $\pi \in C_{\pi_i}$, then by part (iii) of Definition 8.11

$$\pi = \pi_i \oplus \underbrace{\pi_0 \oplus \pi_0 \oplus \cdots \oplus \pi_0}_{l(\pi_i)-1 \text{ times}} \oplus \phi^{l(\pi_i)m}\pi'$$

where

$$\pi' \in C_{\pi_{j_1}} \cup C_{\pi_{j_2}} \cup C_{\pi_{j_3}} \cup \cdots \cup C_{\pi_{j_r}}$$

with $\mathscr{L}_C(\pi_i) = \{\pi_{j_1}, \pi_{j_2}, \pi_{j_3}, \ldots, \pi_{j_r}\}$. Obviously the $C_{\pi_{j_1}}, C_{\pi_{j_2}}, \ldots, C_{\pi_{j_r}}$ are disjoint sets of partitions. Hence for $0 \leqslant i \leqslant k$

$$H_i(x) = \sum_{\pi\in C_{\pi_i}} x^{\#(\pi)} q^{\sigma(\pi)}$$

$$= \sum_{\pi'\in C_{\pi_{j_1}} \cup\cdots\cup C_{\pi_{j_r}}} x^{\#(\pi_i)} q^{\sigma(\pi_i)} x^{\#(\phi^{l(\pi_i)m}\pi')} q^{\sigma(\phi^{l(\pi_i)m}\pi')}$$

$$= x^{\#(\pi_i)} q^{\sigma(\pi_i)} \sum_{h=1}^{r} \sum_{\pi' \in C_{\pi_{j_h}}} x^{\#(\pi')} q^{l(\pi_i)m\#(\pi') + \sigma(\pi')}$$

$$= x^{\#(\pi_i)} q^{\sigma(\pi_i)} \sum_{h=1}^{r} H_{j_h}(xq^{l(\pi_i)m}). \tag{8.4.13}$$

To conclude our theorem we need only show that our system of q-difference equations can be reduced to one higher-order equation in $H_0(x)$, and this requires the utilization of Lemma 8.10.

Define

$$h_{ij}(x) = H_i(x^{-1}q^{-jm}), \qquad 0 \leqslant i \leqslant k, \quad 0 \leqslant j < l(\pi_i). \tag{8.4.14}$$

The system (8.4.13) is easily seen to be equivalent to the following system of $\Omega = l(\pi_0) + l(\pi_i) + \cdots + l(\pi_k)$ equations:

$$h_{ij}(xq^m) = h_{i,j+1}(x), \qquad 0 \leqslant i \leqslant k, \quad 0 \leqslant j < l(\pi_i) - 1,$$

$$h_{i,l(\pi_i)-1}(xq^m) = x^{-\#(\pi_i)} q^{\sigma(\pi_i) - ml(\pi_i)\#(\pi_i)} \sum_{h=1}^{r} h_{j_h,0}(x). \tag{8.4.15}$$

Thus by Lemma 8.10, $h_{00}(x) = H_0(x^{-1})$ satisfies

$$\sum_{k=0}^{\Omega} \bar{p}_k(x) H_0(x^{-1}q^{-km}) = 0;$$

hence

$$0 = \sum_{k=0}^{\Omega} \bar{p}_k(x^{-1}q^{-\Omega m}) H_0(xq^{(\Omega-k)m}); \tag{8.4.16}$$

furthermore, we may assume that the $\bar{p}_k(x^{-1}q^{-\Omega m})$ are polynomials in x and q since we may clear denominators by multiplication of (8.4.16) by a suitably chosen polynomial. By the remark following Eq. (8.4.12) we see that the establishment (8.4.16) is sufficient to prove Theorem 8.11. ■

Examples

1. For each $k \geqslant 2$, every integer has a unique representation to the base k; that is, every integer is uniquely representable as a sum of powers of k where no part is repeated more than $k - 1$ times.

2. A *perfect partition* of n is a partition $\pi = (\lambda_1, \ldots, \lambda_s)$ such that for each $m \leqslant n$ there is exactly one $\pi' \leqslant \pi$ such that π' is a partition of m. The number of perfect partitions of n is the same as the number of ordered factorizations of $n + 1$. To see this, we remark that all perfect partitions of n may be written

$$(1^{g_1-1}g_1^{g_2-1}(g_1g_2)^{g_3-1}\cdots(g_1g_2\cdots g_{k-1})^{g_k-1});$$

hence

$$\begin{aligned}n &= (g_1-1)\cdot 1+(g_2-1)\cdot g_1+(g_3-1)\cdot g_1g_2+\cdots+(g_k-1)\cdot(g_1g_2\cdots g_{k-1})\\ &= g_1g_2\cdots g_k-1.\end{aligned}$$

3. MacMahon, who introduced perfect partitions, generalized the idea to a *partition of infinity*, by which he meant the formal expression

$$\pi_\infty = (1^{g_1-1}g_1^{g_2-1}(g_1g_2)^{g_3-1}\cdots(g_1g_2\cdots g_{k-1})^{g_k-1}\cdots)$$

defined for each sequence $\{g_i\}_{i=1}^{\infty}$ of integers each larger than 1. From Example 2 it is obvious that for each integer n there is a unique π with $\pi < \pi_\infty$ such that π is a partition of n. A moment's reflection shows that if $\mathscr{M}$ is the equivalence class of ideals of order 1 that contains $\{\{f_i\} | f_i = 0 \text{ if } i > 1\}$, then

$$p(\mathscr{M}; n) = 1 \qquad \text{for all } n,$$

and each "partition of infinity" is essentially a partition ideal in $\mathscr{M}$.

4. The constant sequences $\{g_i\}_{i=1}^{\infty}$ with $g_i = k \geqslant 2$ produce the partitions of infinity (or elements of $\mathscr{M}$) that occur in Example 1.

5. It is a simple matter to show that every integer is uniquely representable as a sum of nonconsecutive Fibonacci numbers (the Fibonacci numbers are $1, 2, 3, 5, 8, 13, 21, \ldots, u_n (= u_{n-1} + u_{n-2}), \ldots$). This shows that not every element of $\mathscr{M}$ corresponds to a partition of infinity.

6. Let $\mathscr{D}(r; b_1, b_2, \ldots, b_m; m)$ denote the set of all partitions $(\lambda_1\lambda_2\cdots\lambda_s)$ (s arbitrary) such that $\lambda_{i-r} - \lambda_i \geqslant b_j$ whenever $r < i \leqslant s$ and $\lambda_i \equiv j \pmod{m}$. Then $\mathscr{D}(r; b_1, \ldots, b_m; m)$ is a linked partition ideal.

7. We let $\mathscr{E}$ denote the set of all partitions $(\lambda_1\lambda_2\cdots\lambda_s)$ (s arbitrary) such that $\lambda_{i-1} - \lambda_i \geqslant 9$, $(1 < i \leqslant s)$ if λ_i odd, $\lambda_{i-2} - \lambda_i \geqslant 5$, $(2 < i \leqslant s)$ if $\lambda_i \equiv 2 \pmod 4$, and $\lambda_{i-2} - \lambda_i > 0$ $(2 < i \leqslant s)$ always. Then $\mathscr{E}$ is a linked partition ideal. A modulus of $\mathscr{E}$ is 4. The set $L_{\mathscr{E}}$ has nine elements:

$$\pi_0 = \{0,0,0,0,0,\ldots\}, \quad \pi_1 = \{1,0,0,0,0,\ldots\}, \quad \pi_2 = \{0,0,1,0,0,\ldots\},$$

$$\pi_3 = \{0,1,1,0,0,\ldots\}, \quad \pi_4 = \{0,1,0,1,0,\ldots\}, \quad \pi_5 = \{0,2,0,0,0,\ldots\},$$

$$\pi_6 = \{0,1,0,0,0,\ldots\}, \quad \pi_7 = \{0,0,0,1,0,\ldots\}, \quad \pi_8 = \{0,0,0,2,0,\ldots\}.$$

The spans of π_1, π_2, π_3 are 2, while the remaining π_i have span 1. The linking sets are $\mathscr{L}_{\mathscr{E}}(\pi_0) = L_{\mathscr{E}}$; $\mathscr{L}_{\mathscr{E}}(\pi_1) = L_{\mathscr{E}} - \{\pi_1\}$; $\mathscr{L}_{\mathscr{E}}(\pi_2) = \{\pi_7, \pi_8\} = \mathscr{L}_{\mathscr{E}}(\pi_3)$; $\mathscr{L}_{\mathscr{E}}(\pi_4) = \{\pi_2, \pi_7, \pi_8\} = \mathscr{L}_{\mathscr{E}}(\pi_5)$; $\mathscr{L}_{\mathscr{E}}(\pi_6) = L_{\mathscr{E}} = \mathscr{L}_{\mathscr{E}}(\pi_7) = \mathscr{L}_{\mathscr{E}}(\pi_8)$.

Notes

The material presented here originally appeared in Andrews (1972, 1974, 1975). The presentation of linked partition ideals is expanded and improved over that in Andrews (1974) and was first exposed in this form in lectures at the University of Erlangen in 1975. The pair of sets (M_1, M_2) in Corollary

8.6 is called an *Euler pair*; this result was originally given in Andrews (1969b), and was extended by Subbarao (1971). Reviews of work related to the material in this chapter occur in LeVeque (1974), Section P68.

Example 1. This is just the ancient radix representation theorem (see Andrews, 1969a, b for this treatment).

Examples 2–4. MacMahon (1886, 1891, 1923).

Example 5. This is Zeckendorf's theorem. See Daykin (1960) for a more extensive treatment.

References

Andrews, G. E. (1969a). "On radix representation and the Euclidean algorithm," *Amer. Math. Monthly* **76**, 66–68.

Andrews, G. E. (1969b). "Two theorems of Euler and a general partition theorem," *Proc. Amer. Math. Soc.* **20**, 499–502.

Andrews, G. E. (1971). *Number Theory*. Saunders, Philadelphia.

Andrews, G. E. (1972). "Partition identities," *Advances in Math.* **9**, 10–51.

Andrews, G. E. (1974). "A general theory of identities of the Rogers–Ramanujan type," *Bull. Amer. Math. Soc.* **80**, 1033–1052.

Andrews, G. E. (1975). "Problems and prospects for basic hypergeometric functions," *Theory and Application of Special Functions*. (R. Askey, ed.), pp. 191–224. Academic Press, New York.

Daykin, D. E. (1960). "Representations of natural numbers as sums of generalized Fibonacci numbers," *J. London Math. Soc.* **35**, 143–160.

LeVeque, W. J. (1974). *Reviews in Number Theory*, Vol. 4. Amer. Math. Soc., Providence, R.I.

MacMahon, P. A. (1886). "Certain special partitions of numbers," *Quart. J. Math. Oxford Ser.* **21**, 367–373.

MacMahon, P. A. (1891). "The theory of perfect partitions and the compositions of multipartite numbers," *Messenger of Math.* **20**, 103–119.

MacMahon, P. A. (1915). *Combinatory Analysis*, Vol. 1. Cambridge Univ. Press, London and New York (reprinted by Chelsea, New York, 1960).

MacMahon, P. A. (1923). "The partitions of infinity with some arithmetic and algebraic consequences," *Proc Cambridge Phil. Soc.* **21**, 642–650.

Miller, K. S., and Murray, F. J. (1954). *Existence Theorems for Ordinary Differential Equations*. New York Univ. Press, New York.

Subbarao, M. V. (1971). "Partition theorems for Euler pairs," *Proc. Amer. Math. Soc.* **28**, 330–336.

CHAPTER 9

Sieve Methods Related to Partitions

9.1 Introduction

Up to this point we have established identities between partition functions either by exhibiting a bijection between the sets of partitions enumerated (e.g., Theorems 1.4–1.6) or by proving that the related generating functions were identical (e.g., Theorems 7.5 and 7.11). In this chapter we shall illustrate how various types of *inclusion–exclusion* arguments, or *sieves*, may be used to prove many results of interest in the theory of partitions. In Section 9.2, we shall prove Euler's theorem (Corollary 1.2) and an identity of Sylvester using the inclusion–exclusion principle. In Section 9.3, we study an unusual sieve related to Atkin's successive ranks of a partition, and we shall prove a new partition identity (Theorem 9.12) related to the $A_{k,i}(n)$ in Theorem 7.5. This last sieve introduces a number of tantalizing unanswered questions about partitions.

9.2 Inclusion-Exclusion

We begin with a reexamination of Corollary 1.2 that will yield a purely combinatorial treatment of Euler's theorem.

THEOREM 9.1 (see Corollary 1.2). $p(\mathscr{D}, n) = p(\mathscr{O}, n)$ *for all* n.

Proof. Recall the main argument in the proof of Corollary 1.2

$$\sum_{n=0}^{\infty} p(\mathscr{D}, n)q^n = \prod_{n=1}^{\infty} (1 + q^n) = \prod_{n=1}^{\infty} \frac{(1 - q^{2n})}{(1 - q^n)} = \prod_{n=1}^{\infty} \frac{1}{1 - q^{2n-1}} = \sum_{n=0}^{\infty} p(\mathscr{O}, n)q^n.$$

The key to our proof lies in a combinatorial interpretation of $\prod_{n=1}^{\infty}\cdot$ $(1 - q^{2n})/(1 - q^n)$. Namely, we write

$$E(n) = \sum_{\substack{a_1 > \cdots > a_r \\ b_1 \geqslant \cdots \geqslant b_s}} (-1)^r$$

where the summation is over all representations of n of the form

ENCYCLOPEDIA OF MATHEMATICS and Its Applications, Gian-Carlo Rota (ed.). 2, George E. Andrews, The Theory of Partitions

$$n = a_1 + \cdots + a_r + b_1 + \cdots + b_s, \qquad a_1 > \cdots > a_r, \quad b_1 \geqslant \cdots \geqslant b_s, \quad (9.2.1)$$

and each a_i is even. Now each ordinary partition $\pi = (C_1 C_2 \cdots C_t)$ may be rearranged to yield several representations (9.2.1). In fact, if m is the number of *different* even C_i, then the contribution of π to $E(n)$ is precisely

$$1 - \binom{m}{1} + \binom{m}{2} - \binom{m}{3} + \cdots + (-1)^m \binom{m}{m} = \begin{cases} 1 & \text{if } m = 0, \\ (1-1)^m = 0 & \text{if } m > 0. \end{cases}$$

Consequently π contributes 1 if all parts are odd and 0 otherwise. Therefore

$$E(n) = p(\mathcal{O}, n).$$

Next we note that since each a_i is even in (9.2.1), we may write $a_i = 2\alpha_i$, and thus our representation is equivalent to

$$n = \alpha_1 + \cdots + \alpha_r + \alpha_1 + \cdots + \alpha_r + b_1 + \cdots + b_s,$$

$$\alpha_1 > \cdots > \alpha_r, \quad b_1 \geqslant \cdots \geqslant b_s. \quad (9.2.2)$$

Now each ordinary partition $\pi = (C_1 C_2 \cdots C_t)$ may be rearranged to yield several representations (9.2.2). In fact, if m is the number of parts of π that occur with repetition, then the contribution of π to $E(n)$ is precisely

$$1 - \binom{m}{1} + \binom{m}{2} - \binom{m}{3} + \cdots + (-1)^m \binom{m}{m} = \begin{cases} 1 & \text{if } m = 0, \\ (1-1)^m = 0 & \text{if } m > 0. \end{cases}$$

Consequently, π contributes 1 if all parts are distinct and 0 otherwise. Hence

$$E(n) = p(\mathcal{D}, n),$$

and thus $p(\mathcal{O}, n) = p(\mathcal{D}, n)$ $(= E(n))$, as desired. ∎

The foregoing argument may be easily extended to numerous other results, such as Corollary 8.6. Our next result utilizes a truncated inclusion–exclusion argument.

THEOREM 9.2 (Sylvester)

$$\sum_{n=0}^{\infty} \frac{(-1)^n x^n q^{\frac{1}{2}n(3n+1)}(1 - xq^{2n+1})}{(q)_n (xq^{n+1})_\infty} = 1. \quad (9.2.3)$$

Remark. Theorem 9.2 is easy to establish from the results in Chapter 7. In fact, the left-hand side of (9.2.3) is just $J_{1,1}(0; x; q)$ and by Lemma 7.2, $J_{1,1}(0; x; q) = J_{1,1}(0; xq; q)$. Hence

$$J_{1,1}(0; x; q) = J_{1,1}(0; xq^N; q) \to J_{1,1}(0; 0; q) = 1 \quad \text{as} \quad N \to \infty.$$

Proof. We shall examine the left-hand side of (9.2.3) as the generating function for

$$C(M, N) = \sum_{\substack{\pi \\ \#(\pi)=M \\ \sigma(\pi)=N}} \alpha_\pi$$

and our object will be to determine the weighting factor α_π.

The left-hand side of (9.2.3) is

$$\sum_{n=0}^{\infty} (-1)^n x^n q^{n^2} \cdot \frac{q^{\frac{1}{2}n(n+1)}}{(q)_n} \cdot \frac{1}{(xq^{n+1})_\infty}$$

$$+ \sum_{n=0}^{\infty} (-1)^{n+1} x^{n+1} q^{(n+1)^2} \cdot \frac{q^{\frac{1}{2}n(n+1)}}{(q)^n} \cdot \frac{1}{(xq^{n+1})_\infty}, \tag{9.2.4}$$

and the approach used in the combinatorial proof of Eq. (2.2.9) at the end of Chapter 2 shows that $x^n q^{n^2} \cdot q^{\frac{1}{2}n(n+1)}/(q)_n$ is the generating function for all partitions with n parts, all distinct and each $> n$. Hence

$$x^n q^{n^2} \cdot \frac{q^{\frac{1}{2}n(n+1)}}{(q)_n} \cdot \frac{1}{(xq^{n+1})_\infty}$$

generates representations of integers N as follows:

$$N = a_1 + a_2 + \cdots + a_n + b_1 + \cdots + b_s \tag{9.2.5}$$

where $a_1 > a_2 > \cdots > a_n > n$, $b_1 \geqslant \cdots \geqslant b_s > n$ with s arbitrary. Now each ordinary partition π whose smallest part is m and that contains t different integers can be broken into representation (9.2.5) in numerous ways merely by selecting a subset of size n of the t possible choices for the a_i. Hence, the $(-1)^n$ factor in the first sum in (9.2.4) means that the total contribution of π to the first sum is

$$1 - \binom{t}{1} + \binom{t}{2} - \cdots + (-1)^{m-1}\binom{t}{m-1} = (-1)^{m-1}\binom{t-1}{m-1}.$$

The second sum in (9.2.4) is very much the same as the first except now the representations of N generated are

$$N = a_1 + a_2 + \cdots + a_n + (n+1) + b_1 + \cdots + b_s \tag{9.2.6}$$

where $a_1 > a_2 > \cdots > a_n > n + 1$, $b_1 \geqslant \cdots \geqslant b_s > n$ with s arbitrary.

In this instance, each ordinary partition π whose smallest part is m and that contains t different integers (some or all of these t integers may occur in π with repetition) can be broken into several representations (9.2.6); namely, provided that $m = n + 1$, exactly

$$\binom{t-1}{n} = \binom{t-1}{m-1}$$

choices may be made for the a_i. Hence π contributes only to the term of the second sum in (9.2.4) with $n = m - 1$ and its contribution to this term is

$$(-1)^m \binom{t-1}{m-1}.$$

Hence the total contribution of a partition π with smallest part m is

$$1 - \binom{t}{1} + \binom{t}{2} - \cdots + (-1)^{m-1}\binom{t-1}{m-1} + (-1)^m \binom{t-1}{m-1} = 0.$$

Hence for

$$C(M, N) = \sum_{\substack{\pi \\ \#(\pi) = M \\ \sigma(\pi) = N}} \alpha_\pi$$

we see that the weighting factor α_π is zero if π has a smallest part m. Thus the only partition that can have a positive weighting factor is the empty partition of zero whose weighting factor is clearly 1. Therefore

$$\sum_{n=0}^{\infty} \frac{(-1)^n x^n q^{\frac{1}{2}n(3n+1)}(1 - xq^{2n+1})}{(q)_n (xq^{n+1})_\infty} = \sum_{M \geq 0, N \geq 0} C(M, N) x^M q^N$$

$$= C(0, 0) = 1. \qquad \blacksquare$$

9.3 A Sieve for Successive Ranks

As we know, any partition may be represented by a Ferrers graph. For example, (7 7 5 3 3 1 1 1) has the representation

. . . .

. . . .

. . . . .

. . .

. . .

.

.

.

In 1944, F. J. Dyson defined the *rank* of a partition as the largest part minus the number of parts. Thus the rank of the preceding partition is $7 - 8 = -1$.

Later, Atkin defined the *successive ranks* of a partition. One can view the graphical representation of a partition as a set of nested right angles of nodes. The partition (7 7 5 3 3 1 1 1) may be viewed as three such right angles

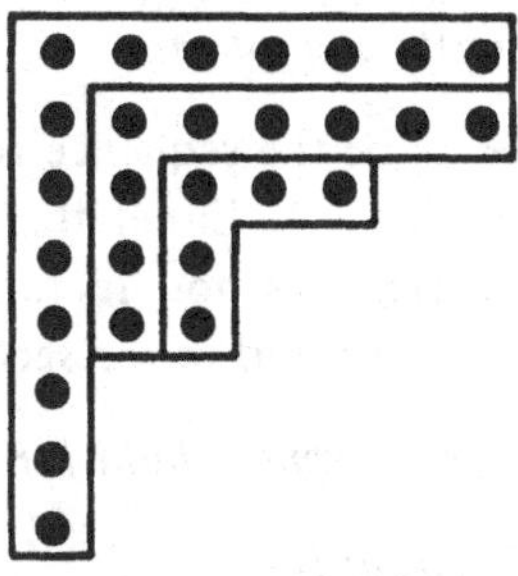

If π is a partition, we define $r_i(\pi)$ as the number of nodes in the horizontal part of the ith right angle in the graph of π minus the number of nodes in the vertical part of the ith right angle. Thus if π is (7 7 5 3 3 1 1 1), then $r_i(\pi) = 7 - 8 = -1$, $r_2(\pi) = 6 - 4 = 2$, $r_3(\pi) = 3 - 3 = 0$.

Our interest lies in the "oscillation" of the ranks between certain positive and negative bounds.

DEFINITION 9.1. Let h be the largest integer for which there exists a sequence $j_1 < j_2 < \cdots < j_h$ such that $r_{j_1}(\pi) > 2k - i - 1$, $r_{j_2}(\pi) \leqslant -(i-1)$, $r_{j_3}(\pi) > 2k - i - 1$, $r_{j_4}(\pi) \leqslant -(i-1)$, and so on. We define h to be the (k, i)-*positive oscillation* of π.

DEFINITION 9.2. Let g be the largest integer for which there exists a sequence $j_1 < j_2 < \cdots < j_g$ such that $r_{j_1}(\pi) \leqslant -(i-1)$, $r_{j_2}(\pi) > 2k - i - 1$, $r_{j_3}(\pi) \leqslant -(i-1)$, $r_{j_4}(\pi) > 2k - i - 1$, and so on. We define g to be the (k, i)-*negative oscillation* of π.

If π denotes the partition (7 7 5 3 3 1 1 1), then the (1, 1)-positive oscillation of π is 2, the (2, 1)-negative oscillation of π is 1.

Our interest centers upon partition functions related to these oscillations.

DEFINITION 9.3. Let $p_{k,i}(a, b; \mu; N)$ (resp. $m_{k,i}(a, b; \mu; N)$) denote the number of partitions of N with at most b parts, with largest part at most a, and with (k, i)-positive (resp. (k, i)-negative) oscillation at least μ.

To these partition functions we associate the related generating functions

$$P_{k,i}(a, b; \mu; q) = \sum_{N \geqslant 0} p_{k,i}(a, b; \mu; N) q^N, \tag{9.3.1}$$

$$M_{k,i}(a, b; \mu; q) = \sum_{N \geqslant 0} m_{k,i}(a, b; \mu; N) q^N. \tag{9.3.2}$$

We next shall derive recurrences for these functions that will allow us to identify them with expressions involving Gaussian polynomials. We start by noting the obvious close relationship between negative and positive oscillation.

LEMMA 9.3. *The (k, i)-positive oscillation of π is either 1 larger or 1 smaller than the (k, i)-negative oscillation of π.*

Proof. Let $j_1 < j_2 < \cdots < j_h$ be the sequence related to the (k, i)-positive oscillation as given in Definition 9.1. Either there does exist a $j_0 < j_1$ such that $r_{j_0}(\pi) \leqslant -(i-1)$ or there does not. In the first case, we see that the (k, i)-negative oscillation is 1 larger and in the second case it is 1 smaller. ■

LEMMA 9.4. *The following recurrences hold for $\mu \geqslant 1$:*

$$\begin{aligned} &m_{k,i}(a, b; \mu; N) - m_{k,i}(a-1, b; \mu; N) \\ &\qquad - m_{k,i}(a, b-1; \mu; N) + m_{k,i}(a-1, b-1; \mu; N) \\ &\quad = \begin{cases} m_{k,i}(a-1, b-1; \mu; N-a-b+1) & \text{if } a-b > -(i-1), \\ p_{k,i}(a-1, b-1; \mu-1; N-a-b+1) & \text{if } a-b \leqslant -(i-1); \end{cases} \end{aligned} \tag{9.3.3}$$

$$\begin{aligned} &p_{k,i}(a, b; \mu; N) - p_{k,i}(a-1, b; \mu; N) \\ &\qquad - p_{k,i}(a, b-1; \mu; N) + p_{k,i}(a-1, b-1; \mu; N) \\ &\quad = \begin{cases} m_{k,i}(a-1, b-1; \mu-1; N-a-b+1) & \text{if } a-b > 2k-i-1, \\ p_{k,i}(a-1, b-1; \mu; N-a-b+1) & \text{if } a-b \leqslant 2k-i-1. \end{cases} \end{aligned} \tag{9.3.4}$$

Proof. Let us start by examining the left-hand side of (9.3.3). The expression $m_{k,i}(a, b; \mu; N) - m_{k,i}(a-1, b; \mu; N)$ denotes the number of partitions of N with at most b parts, with (k, i)-negative oscillation at least μ, and with largest part exactly equal to a. Therefore, the expression

$$\begin{aligned} &(m_{k,i}(a, b; \mu; N) - m_{k,i}(a-1, b; \mu; N)) - (m_{k,i}(a, b-1; \mu; N) \\ &\qquad - m_{k,i}(a-1, b-1; \mu; N)) \end{aligned}$$

(which is the left-hand side of (9.3.3)) denotes the number of partitions of N with exactly b parts, with largest part exactly a, and with (k, i)-negative oscillation at least μ.

We now transform these partitions that are enumerated by the left-hand side of (9.3.3) by deleting the largest part (namely a) and subtracting 1 from each of the $b-1$ remaining parts. In terms of the graphical representation, we have removed the outer right angle of nodes from each partition. The transformed partitions are partitions of $N-a-b+1$ into at most $b-1$ parts with largest part at most $a-1$.

Suppose first that $a - b > -(i - 1)$; then the removal of the outer right angle of nodes from the graphical representation has no effect on the (k, i)-negative oscillation, which is still equal to μ. Consequently our transformed partition is of the type enumerated by $m_{k,i}(a - 1, b - 1; \mu; N - a - b + 1)$. Since the foregoing procedure is clearly reversible, we see that if $a - b > -(i - 1)$, then

$$m_{k,i}(a, b; \mu; N) - m_{k,i}(a - 1, b; \mu; n) - m_{k,i}(a, b - 1; \mu; N)$$

$$+ m_{k,i}(a - 1, b - 1; \mu; N)$$

$$= m_{k,i}(a - 1, b - 1; \mu; N - a - b + 1).$$

Therefore the top half of (9.3.3) is established.

Now suppose that $a - b \leqslant -(i - 1)$; in this case, the removal of the outer right angle of nodes from the graphical representation may affect the (k, i)-negative oscillation. In fact, the resulting partition is now characterized by the fact that it has (k, i)-positive oscillation at least $\mu - 1$. Thus the transformed partitions in this case are of the type enumerated by $p_{k,i}(a - 1, b - 1; \mu - 1; N - a - b + 1)$. Again the transformation used is reversible. Therefore if $a - b \leqslant -(i - 1)$, then

$$m_{k,i}(a, b; \mu; N) - m_{k,i}(a - 1, b; \mu; N) - m_{k,i}(a, b - 1; \mu; N)$$

$$+ m_{k,i}(a - 1, b - 1; \mu; N)$$

$$= p_{k,i}(a - 1, b - 1; \mu - 1; N - a - b + 1).$$

Therefore the bottom half of (9.3.3) is established.

We omit the proof of (9.3.4) since it perfectly parallels the proof of (9.3.3) with the roles of (k, i)-positive and (k, i)-negative oscillation interchanged. ■

It is now a simple matter to translate Lemma 9.4 into relations among the generating functions.

COROLLARY 9.5. *For each* $\mu \geqslant 1$,

$$M_{k,i}(a, b; \mu; q) - M_{k,i}(a - 1, b; \mu; q)$$

$$- M_{k,i}(a, b - 1; \mu; q) + M_{k,i}(a - 1, b - 1; \mu; q)$$

$$= q^{a+b-1} \begin{cases} M_{k,i}(a - 1, b - 1; \mu; q) & \text{if } a - b > -(i - 1), \\ P_{k,i}(a - 1, b - 1; \mu - 1; q) & \text{if } a - b \leqslant -(i - 1); \end{cases}$$

(9.3.5)

$$P_{k,i}(a, b; \mu; q) - P_{k,i}(a-1, b; \mu; q)$$
$$- P_{k,i}(a, b-1; \mu; q) + P_{k,i}(a-1, b-1; \mu; q)$$
$$= q^{a+b-1} \begin{cases} M_{k,i}(a-1, b-1; \mu-1; q) & \text{if } a-b > 2k-i-1, \\ P_{k,i}(a-1, b-1; \mu; q) & \text{if } a-b \leqslant 2k-i-1. \end{cases} \tag{9.3.6}$$

Proof. Comparing coefficients of q^N in these identities, we can establish these results directly from Lemma 9.4. ∎

LEMMA 9.6. *The following relations hold:*

$$P_{k,i}(a, b; 0; q) = M_{k,i}(a, b; 0; q) = \begin{bmatrix} a+b \\ a \end{bmatrix}, \tag{9.3.7}$$

and for $\mu \geqslant 1$

$$P_{k,i}(0, b; \mu; q) = P_{k,i}(a, 0; \mu; q) = M_{k,i}(0, b; \mu; q)$$
$$= M_{k,i}(a, 0; \mu; q) = 0, \tag{9.3.8}$$

where $\begin{bmatrix} a+b \\ a \end{bmatrix}$ *is the Gaussian polynomial discussed in Section* 3.2.

Proof. For (9.3.7), we observe that all partitions have (k, i)-positive and (k, i)-negative oscillation at least 0. Hence $P_{k,i}(a, b; 0; q)$ and $M_{k,i}(a, b; 0; q)$ are each the generating function for partitions with largest part at most a and with at most b parts. By Theorem 3.1, this generating function is $\begin{bmatrix} a+b \\ a \end{bmatrix}$; thus (9.3.7) is established.

As for (9.3.8), we observe that only the empty partition of 0 has either no parts or no largest part. Since the empty partition of 0 has 0 for all of its (k, i)-oscillations, we see that the generating functions appearing in (9.3.8) must all be identically 0. ∎

LEMMA 9.7.

$$q^L \begin{bmatrix} A+B \\ B-X \end{bmatrix} - q^L \begin{bmatrix} A+B-1 \\ B-X \end{bmatrix} - q^L \begin{bmatrix} A+B-1 \\ B-X-1 \end{bmatrix} + q^L \begin{bmatrix} A+B-2 \\ B-X-1 \end{bmatrix}$$
$$= q^{A+B-1} q^L \begin{bmatrix} A+B-2 \\ B-X-1 \end{bmatrix}.$$

Proof. By (3.3.3), we see that

$$q^L \begin{bmatrix} A+B \\ B-X \end{bmatrix} - q^L \begin{bmatrix} A+B-1 \\ B-X \end{bmatrix} - q^L \begin{bmatrix} A+B-1 \\ B-X-1 \end{bmatrix} + q^L \begin{bmatrix} A+B-2 \\ B-X-1 \end{bmatrix}$$

$$= q^{L+A+X}\begin{bmatrix} A+B-1 \\ B-X-1 \end{bmatrix} - q^{L+A+X}\begin{bmatrix} A+B-2 \\ B-X-2 \end{bmatrix}$$

$$= q^{L+A+B-1}\begin{bmatrix} A+B-2 \\ B-X-1 \end{bmatrix} \qquad \text{(by (3.3.4))}. \qquad \blacksquare$$

LEMMA 9.8. *For each integer $r \geqslant 0$, let*

$$f(A, r; X, Y, L) = q^L \frac{(1-q^{(r+1)(A+Y+1)})}{(1-q^{A+Y+1})}\begin{bmatrix} 2A+X \\ A+Y \end{bmatrix} + \sum_{j=0}^{r-2} q^{L+A+Y+j+2}\frac{(1-q^{(r-j-1)(A+Y+1)})}{(1-q^{A+Y+1})}\begin{bmatrix} 2A+X+j \\ A+Y-1 \end{bmatrix}.$$

Then for each integer $r \geqslant 1$,

$$f(A, r; X, Y, L) - f(A, r-1; X, Y, L) - f(A-1, r+1; X, Y, L) + f(A-1, r; X, Y, L) = q^{L+2A+2Y+r}\begin{bmatrix} 2A+X+r-2 \\ A+Y \end{bmatrix}.$$

Proof.

$$f(A, r; X, Y, L) - f(A, r-1; X, Y, L) = q^{L+r(A+Y+1)}\begin{bmatrix} 2A+X \\ A+Y \end{bmatrix} + \sum_{j=0}^{r-2} q^{L+A+Y+j+2+(r-j-2)(A+Y+1)}\begin{bmatrix} 2A+X+j \\ A+Y-1 \end{bmatrix}.$$

Also, by repeated application of (3.3.4), we see that

$$q^{L+2A+2Y+r}\begin{bmatrix} 2A+X+r-2 \\ A+Y \end{bmatrix} = \sum_{j=0}^{r-3} q^{j(A+Y)+L+2A+2Y+r}\begin{bmatrix} 2A+X+r-3-j \\ A+Y-1 \end{bmatrix} + q^{L+2A+2Y+r+(r-2)(A+Y)}\begin{bmatrix} 2A+X \\ A+Y \end{bmatrix}$$

$$= \sum_{j=0}^{r-3} q^{(r-1-j)(A+Y)+L+r}\begin{bmatrix} 2A+X+j \\ A+Y-1 \end{bmatrix} + q^{L+r(A+Y+1)}\begin{bmatrix} 2A+X \\ A+Y \end{bmatrix}$$

$$= \sum_{j=0}^{r-3} q^{L+A+Y+j+2+(r-j-2)(A+Y+1)} \begin{bmatrix} 2A+X+j \\ A+Y-1 \end{bmatrix}$$

$$+ q^{L+r(A+Y+1)} \begin{bmatrix} 2A+X \\ A+Y \end{bmatrix}.$$

Therefore, combining these two results, we derive that

$$f(A, r; X, Y, L) - f(A, r-1; A, Y, L) - q^{L+2A+2Y+r} \begin{bmatrix} 2A+X+r-2 \\ A+Y \end{bmatrix}$$

$$+ f(A-1, r; X, Y, L)$$

$$= q^{L+A+Y+r} \begin{bmatrix} 2A+X+r-2 \\ A+Y-1 \end{bmatrix} + f(A-1, r; X, Y, L)$$

$$= q^{L+A+Y+r} \left\{ \sum_{j=0}^{r-1} q^{j(A+Y-1)} \begin{bmatrix} 2A+X+r-3-j \\ A+Y-2 \end{bmatrix} \right.$$

$$\left. + q^{r(A+Y-1)} \begin{bmatrix} 2A+X-2 \\ A+Y-1 \end{bmatrix} \right\}$$

$$+ f(A-1, r; X, Y, L) \qquad \text{(by repeated application of (3.3.4))}$$

$$= \sum_{j=0}^{r-1} q^{L+A+Y+r+(r-1-j)(A+Y-1)} \begin{bmatrix} 2A+X-2+j \\ A+Y-2 \end{bmatrix}$$

$$+ q^{L+(r+1)(A+Y)} \begin{bmatrix} 2A+X-2 \\ A+Y-1 \end{bmatrix} + f(A-1, r; X, Y, L)$$

$$= q^L \frac{(1-q^{(r+2)(A+Y)})}{(1-q^{A+Y})} \begin{bmatrix} 2(A-1)+X \\ A-1+Y \end{bmatrix}$$

$$+ \sum_{j=0}^{r-2} q^{L+A+Y+j+1} \frac{(1-q^{(r-j)(A+Y)})}{(1-q^{A+Y})} \begin{bmatrix} 2(A-1)+X+j \\ A-1+Y-1 \end{bmatrix}$$

$$= f(A-1, r+1; X, Y, L). \qquad \blacksquare$$

We are now prepared to prove the main theorem on generating functions that will be essential to our sieve.

Theorem 9.9. *If $b = a$ or $a-1$ and $k \geqslant i > 0$, then*

$$M_{k,i}(a, b; 2\mu; q) = q^{\mu((4k+2)\mu+(2k-2i+1))} \begin{bmatrix} a+b \\ b-(2k+1)\mu \end{bmatrix}, \tag{9.3.9}$$

$$M_{k,i}(a,b;2\mu-1;q) = q^{(2\mu-1)((2k+1)\mu-(2k-i+1))}\begin{bmatrix} a+b \\ b-(2k+1)\mu+2k-i+1 \end{bmatrix}, \tag{9.3.10}$$

$$P_{k,i}(a,b;2\mu;q) = q^{\mu((4k+2)\mu-(2k-2i+1))}\begin{bmatrix} a+b \\ b+(2k+1)\mu \end{bmatrix}, \tag{9.3.11}$$

$$P_{k,i}(a,b;2\mu-1;q) = q^{(2\mu-1)((2k+1)\mu-i)}\begin{bmatrix} a+b \\ a-(2k+1)\mu+i \end{bmatrix}. \tag{9.3.12}$$

Proof. We first note that the recurrence relations (9.3.5) and (9.3.6) together with the initial conditions (9.3.7) and (9.3.8) uniquely determine $M_{k,i}(a,b;\mu;q)$ and $P_{k,i}(a,b;\mu;q)$. We now define new functions $M^*_{k,i}(a,b;\mu;q)$ and $P^*_{k,i}(a,b;\mu;q)$ in terms of Gaussian polynomials. Our object is to identify these new functions with the generating functions being considered.

$$M^*_{k,i}(a,b;2\mu;q) = q^{\mu((4k+2)\mu+(2k-2i+1))}\begin{bmatrix} a+b \\ b-(2k+1)\mu \end{bmatrix} \tag{9.3.13}$$

for $a-b \geqslant -(i-1)$;

$$M^*_{k,i}(a,b;2\mu-1;q) = q^{(2\mu-1)((2k+1)\mu-(2k-i+1))}\begin{bmatrix} a+b \\ b-(2k+1)\mu+2k-i+1 \end{bmatrix} \tag{9.3.14}$$

for $a-b \geqslant -(i-1)$;

$$P^*_{k,i}(a,b;2\mu;q) = q^{\mu((4k+2)\mu-(2k-2i+1))}\begin{bmatrix} a+b \\ a-(2k+1)\mu \end{bmatrix} \tag{9.3.15}$$

for $a-b \leqslant 2k-i$;

$$P^*_{k,i}(a,b;2\mu-1;q) = q^{(2\mu-1)((2k+1)\mu-i)}\begin{bmatrix} a+b \\ a-(2k+1)\mu+i \end{bmatrix} \tag{9.3.16}$$

for $a-b \leqslant 2k-i$;

$$M^*_{k,i}(a-1,a+i-1+r;2\mu;q)$$

$$= q^{\mu((4k+2)\mu+(2k-2i+1))}\frac{\left(1-q^{(r+2)(a+i-1-(2k+1)\mu)}\right)}{1-q^{a+i-1-(2k+1)\mu}}$$

$$\times \begin{bmatrix} 2a+i-3 \\ a+i-2-(2k+1)\mu \end{bmatrix} + \sum_{j=0}^{r-1} q^{\mu((4k+2)\mu+(2k-2i+1))+a+i-(2k+1)\mu+j}$$

$$\times \frac{(1-q^{(r-j)(a+i-1-(2k+1)\mu)})}{1-q^{a+i-1-(2k+1)\mu}} \begin{bmatrix} 2a+i-3+j \\ a+i-3-(2k+1)\mu \end{bmatrix} \tag{9.3.17}$$

for $r \geqslant -1$, $a \geqslant 1$, $\mu \geqslant 1$;

$$M^*_{k,i}(a-1, a+i-1+r; 2\mu-1; q)$$

$$= q^{(2\mu-1)((2k+1)\mu-(2k-i+1))} \frac{(1-q^{(r+2)(a-(2k+1)\mu+2k)})}{(1-q^{a-(2k+1)\mu+2k})}$$

$$\times \begin{bmatrix} 2a+i-3 \\ a-(2k+1)\mu+2k-1 \end{bmatrix}$$

$$+ \sum_{j=0}^{r-1} q^{(2\mu-1)((2k+1)\mu-(2k-i+1))+a-(2k+1)\mu+2k+1+j}$$

$$\times \frac{(1-q^{(r-j)(a-(2k+1)\mu+2k)})}{(1-q^{a-(2k+1)\mu+2k})} \begin{bmatrix} 2a+i-3+j \\ a-(2k+1)\mu+2k-2 \end{bmatrix} \tag{9.3.18}$$

for $r \geqslant -1$, $a \geqslant 1$, $\mu \geqslant 1$;

$$P^*_{k,i}(b+2k-i-1+r, b-1; 2\mu; q)$$

$$= q^{\mu((4k+2)\mu-(2k-2i+1))} \frac{(1-q^{(r+1)(b+2k-i-(2k+1)\mu)})}{1-q^{b+2k-i-(2k+1)\mu}}$$

$$\times \begin{bmatrix} 2b+2k-i-2 \\ b+2k-i-1-(2k+1)\mu \end{bmatrix}$$

$$+ \sum_{j=0}^{r-2} q^{\mu((4k+2)\mu-(2k-2i+1))-(2k+1)\mu+b+j+2k-i+1}$$

$$\times \frac{(1-q^{(r-j-1)(b+2k-i-(2k+1)\mu)})}{1-q^{b+2k-i-(2k+1)\mu)}} \begin{bmatrix} 2b+2k-i-2+j \\ b+2k-i-2-(2k+1)\mu \end{bmatrix} \tag{9.3.19}$$

for $r \geqslant 0$, $b \geqslant 1$, $\mu \geqslant 1$;

$$P^*_{k,i}(b+2k-i-1+r, b-1; 2\mu-1; q)$$

$$= q^{(2\mu-1)((2k+1)\mu-i)} \frac{(1-q^{(r+1)(b+2k-(2k+1)\mu)})}{1-q^{b+2k-(2k+1)\mu}}$$

$$\times \begin{bmatrix} 2b+2k-i-2 \\ b+2k-1-(2k+1)\mu \end{bmatrix}$$

$$+\sum_{j=0}^{r-2} q^{(2\mu-1)((2k+1)\mu-i)-(2k+1)\mu+b+j+2k+1}$$

$$\times \frac{(1-q^{(r-j-1)(b+2k-(2k+1)\mu)})}{1-q^{b+2k-(2k+1)\mu}} \begin{bmatrix} 2b+2k-i-2+j \\ b+2k-2-(2k+1)\mu \end{bmatrix} \quad (9.3.20)$$

for $r \geqslant 0$, $b \geqslant 1$, $\mu \geqslant 1$.

Finally

$$M^*_{k,i}(a, b; 0; q) = P^*_{k,i}(a, b; 0; q) = \begin{bmatrix} a+b \\ a \end{bmatrix}. \quad (9.3.21)$$

We first observe that the initial condition (9.3.7) is prescribed in (9.3.21). Furthermore, if either a or b is set equal to zero in $M^*_{k,i}(a, b; \mu; q)$ or $P^*_{k,i}(a, b; \mu; q)$, with $\mu \geqslant 1$, then inspection of the appropriate defining equation among (9.3.13)–(9.3.20) shows that the resulting function is identically zero. Hence the initial conditions prescribed in our Lemma 9.6 are satisfied by $M^*_{k,i}(a, b; \mu; q)$ and $P^*_{k,i}(a, b; \mu; q)$.

We now shall examine the recurrence relations (9.3.5) and (9.3.6). The top line of (9.3.5) and the bottom line of (9.3.6) follow in each case by applying Lemma 9.7 to the appropriate expressions among (9.3.13)–(9.3.16). Finally, in the notation of Lemma 9.8, we see that

$$M^*_{k,i}(a-1, a+i-1+r; 2\mu; q)$$
$$= f(a, r+1; i-1, i-2-(2k+1)\mu, (4k+2)\mu^2+(2k-2i+1)\mu) \quad (9.3.22)$$

for $a \geqslant 1$, $r \geqslant -1$, $\mu > 0$;

$$M^*_{k,i}(a-1, a+i-1+r; 2\mu-1; q)$$
$$= f(a, r+1; i-3, 2k-1-(2k+1)\mu, (2\mu-1)((2k+1)\mu-2k+i-1)) \quad (9.3.23)$$

for $a \geqslant 1$, $r \geqslant -1$, $\mu > 0$;

$$P^*_{k,i}(b+2k-i-1+r, b-1; 2\mu; q)$$
$$= f(b, r; 2k-i-2, 2k-i-1-(2k+1)\mu, (4k+2)\mu^2-(2k-2i+1)\mu) \quad (9.3.24)$$

for $b \geqslant 1$, $r \geqslant 0$, $\mu > 0$;

$$P^*_{k,i}(b + 2k - i - 1 + r, b - 1, 2\mu - 1; q)$$
$$= f(b, r; 2k - i - 2, 2k - 1 - (2k + 1)\mu, (2\mu - 1)((2k + 1)\mu - i)), \tag{9.3.25}$$

for $b \geqslant 1$, $r \geqslant 0$, $\mu > 0$.

It is now simply a matter of inspection to verify that Lemma 9.8 implies that the bottom line of (9.3.5) and the top line of (9.3.6) hold for $M^*_{k,i}(a, b; \mu; q)$ and $P^*_{k,i}(a, b; \mu; q)$ if $\mu > 0$.

Thus we have observed that $M^*_{k,i}(a, b; \mu; q)$ and $P^*_{k,i}(a, b; \mu; q)$ satisfy (9.3.5), (9.3.6), (9.3.7), and (9.3.8); and since these four equations uniquely define $M_{k,i}(a, b; \mu; q)$ and $P_{k,i}(a, b; \mu; q)$, we see that

$$P_{k,i}(a, b; \mu; q) = P^*_{k,i}(a, b; \mu; q) \tag{9.3.26}$$

and

$$M_{k,i}(a, b; \mu; q) = M^*_{k,i}(a, b; \mu; q). \tag{9.3.27}$$

Equation (9.3.9) now follows from (9.3.13) and (9.3.27); (9.3.10) follows from (9.3.14) and (9.3.27); (9.3.11) follows from (9.3.15) and (9.3.25); finally, (9.3.12) follows from (9.3.16) and (9.3.26). ■

DEFINITION 9.4. Let $p_{k,i}(\mu; N)$ (resp. $m_{k,i}(\mu; N)$) denote the number of partitions of N with (k, i)-positive (resp. (k, i)-negative) oscillation at least μ.

DEFINITION 9.5. Let

$$P_{k,i}(\mu; q) = \sum_{N \geqslant 0} p_{k,i}(\mu; N) q^N, \qquad M_{k,i}(\mu; q) = \sum_{N \geqslant 0} m_{k,i}(\mu; N) q^N.$$

THEOREM 9.10. *The following relations hold for* $|q| < 1$.

$$M_{k,i}(2\mu; q) = \frac{q^{\mu((4k+2)\mu + 2k - 2i + 1)}}{(q)_\infty}, \qquad \mu \geqslant 0; \tag{9.3.28}$$

$$M_{k,i}(2\mu - 1; q) = \frac{q^{(2\mu - 1)((2k+1)\mu - 2k + i - 1)}}{(q)_\infty}, \qquad \mu > 0; \tag{9.3.29}$$

$$P_{k,i}(2\mu; q) = \frac{q^{\mu((4k+2)\mu - 2k + 2i - 1)}}{(q)_\infty}, \qquad \mu \geqslant 0; \tag{9.3.30}$$

$$P_{k,i}(2\mu - 1; q) = \frac{q^{(2\mu - 1)((2k+1)\mu - i)}}{(q)_\infty}, \qquad \mu > 0. \tag{9.3.31}$$

Proof. First we observe that for $|q| < 1$,

$$|P_{k,i}(\mu; q) - P_{k,i}(a, a; \mu; q)| \leqslant \sum_{n=a+1}^{\infty} p(n)|q|^n \to 0 \quad \text{as} \quad a \to \infty$$

and

$$|M_{k,i}(\mu; q) - M_{k,i}(a, a; \mu; q)| \leq \sum_{n=a+1}^{\infty} p(n)|q|^n \to 0 \quad \text{as} \quad a \to \infty$$

where $p(n)$ is the ordinary partition function. Finally

$$\lim_{a\to\infty} q^x \begin{bmatrix} 2a + y \\ a - z \end{bmatrix} = \frac{q^x}{(q)_\infty}.$$

Consequently (9.3.28) follows from (9.3.9), (9.3.29) from (9.3.10), (9.3.30) from (9.3.11), and (9.3.31) from (9.3.12). ∎

Now we introduce the partition function that arises from our sieve technique.

DEFINITION 9.6. Let $Q_{k,i}(n)$ denote the number of partitions π of n such that $-(i-2) \leqslant r_j(\pi) \leqslant 2k - i - 1$ for each of the successive ranks of π.

The following lemma is the inclusion–exclusion aspect of our sieve.

LEMMA 9.11. *For each integer* $n \geqslant 0$,

$$Q_{k,i}(n) = p_{k,i}(0; n) + \sum_{\mu=1}^{\infty} (-1)^\mu m_{k,i}(\mu; n) + \sum_{\mu=1}^{\infty} (-1)^\mu p_{k,i}(\mu; n). \quad (9.3.32)$$

Proof. First we remark that $Q_{k,i}(n)$ counts the set of all partitions of n that have 0 as (k, i)-positive oscillation *and* 0 as (k, i)-negative oscillation.

On the other hand, the right-hand side of (9.3.32) is a weighted count of the partitions of n. First suppose π is a partition of n with 0 as (k, i)-positive and (k, i)-negative oscillation. Then π is counted once by $p_{k,i}(0; n)$ and not at all by each of $m_{k,i}(\mu; n)$ and $p_{k,i}(\mu; n)$ for each $\mu > 0$. Next suppose that π is a partition of n with (k, i)-positive oscillation $r > 0$. By Lemma 9.3, the (k, i)-negative oscillation of π is either $r - 1$ or $r + 1$; if $r - 1$, then the weighted count for π is

$$1 + \sum_{\mu=1}^{r-1} (-1)^\mu + \sum_{\mu=1}^{r} (-1)^\mu = 1 + \sum_{\mu=1}^{r-1} (-1)^\mu - \sum_{\mu=0}^{r-1} (-1)^\mu = 0;$$

if $r + 1$, then the weighted count for π is

$$1 + \sum_{\mu=1}^{r+1} (-1)^\mu + \sum_{\mu=1}^{r} (-1)^\mu = 1 - \sum_{\mu=0}^{r} (-1)^\mu + \sum_{\mu=1}^{r} (-1)^\mu = 0.$$

Finally, if the $(k - i)$-positive oscillation of π is 0 and the (k, i)-negative oscillation is 1, then the weighted count of π is $1 - 1 = 0$.

Thus we see that the right-hand expression in (9.3.32) counts once each partition of n with 0 as (k, i)-positive and (k, i)-negative oscillation, while it counts 0 for each of the other partitions of n. Consequently the right-hand expression in (9.3.32) is just equal to $Q_{k,i}(n)$. ■

THEOREM 9.12. *Recall that $A_{k,i}(n)$ denotes the number of partitions of n into parts that are not congruent to $0, \pm i \pmod{2k+1}$. Then for $0 < i \leqslant k$ and for each $n \geqslant 0$*

$$A_{k,i}(n) = Q_{k,i}(n).$$

Proof. By Lemma 9.11, we see that

$$\begin{aligned}
&\sum_{n=0}^{\infty} Q_{k,i}(n)q^n \\
&\quad = \sum_{n=0}^{\infty} p_{k,i}(0; n)q^n + \sum_{\mu=1}^{\infty}(-1)^{\mu}\sum_{n=0}^{\infty} m_{k,i}(\mu; n)q^n + \sum_{\mu=1}^{\infty}(-1)^{\mu}\sum_{n=0}^{\infty} p_{k,i}(\mu; n)q^n \\
&\quad = P_{k,i}(0; q) + \sum_{\mu=1}^{\infty}(-1)^{\mu}M_{k,i}(\mu; q) + \sum_{\mu=1}^{\infty}(-1)^{\mu}P_{k,i}(\mu; q) \\
&\quad = \sum_{\mu=1}^{\infty} M_{k,i}(2\mu; q) - \sum_{\mu=1}^{\infty} M_{k,i}(2\mu-1; q) + \sum_{\mu=0}^{\infty} P_{k,i}(2\mu; q) - \sum_{\mu=1}^{\infty} P_{k,i}(2\mu-1; q) \\
&\quad = (q)_{\infty}^{-1}\left\{\sum_{\mu=1}^{\infty} q^{\mu((4k+2)\mu+2k-2i+1)} - \sum_{\mu=1}^{\infty} q^{(2\mu-1)((2k+1)\mu-2k+i-1)}\right. \\
&\qquad \left. + \sum_{\mu=0}^{\infty} q^{\mu((4k+2)\mu-2k+2i-1)} - \sum_{\mu=1}^{\infty} q^{(2\mu-1)((2k+1)\mu-i)}\right\} \\
&\quad = (q)_{\infty}^{-1}\left\{\sum_{\mu=-\infty}^{\infty} q^{\frac{1}{2}2\mu((2k+1)2\mu-2k+2i-1)}\right. \\
&\qquad \left. - \sum_{\mu=-\infty}^{\infty} q^{\frac{1}{2}(2\mu-1)((2k+1)(2\mu-1)-2k+2i-1)}\right\} \\
&\quad = (q)_{\infty}^{-1}\sum_{n=-\infty}^{\infty}(-1)^n q^{\frac{1}{2}n((2k+1)n-2k+2i-1)} \\
&\quad = (q)_{\infty}^{-1}\prod_{m=0}^{\infty}(1-q^{(m+1)(2k+1)})(1-q^{(2k+1)m+i})(1-q^{(2k+1)(m+1)-i})
\end{aligned}$$

(by Jacobi's identity, Theorem 2.8)

$$\begin{aligned}
&\quad = \prod_{\substack{n=1 \\ n \not\equiv 0, \pm i \pmod{2k+1}}}^{\infty} (1-q^n)^{-1} \\
&\quad = \sum_{n=0}^{\infty} A_{k,i}(n)q^n.
\end{aligned}$$

Hence comparing coefficients of q^n in the identities above, we see that

$$A_{k,i}(n) = Q_{k,i}(n)$$

for each $n \geqslant 0$. ■

It is very easy to show that $Q_{2,2}(n) = B_{2,2}(n)$ and $Q_{2,1}(n) = B_{2,1}(n)$, and then Corollaries 7.6 and 7.7 (the Rogers–Ramanujan identities) follow directly from Theorem 9.12.

To prove that $Q_{2,2}(n) = B_{2,2}(n)$, let us consider a partition π of the type enumerated by $B_{2,2}(n)$, say $\pi = (C_1 C_2 \cdots C_t)$, where $C_i - C_{i+1} \geqslant 2$. We form a graphical representation of π as follows. The ith part of π is represented as the ith right angle in a graphical representation where if $C_i = 2s + 1$, the angle is

s + 1

s + 1

and if $C_i = 2s$, the angle is

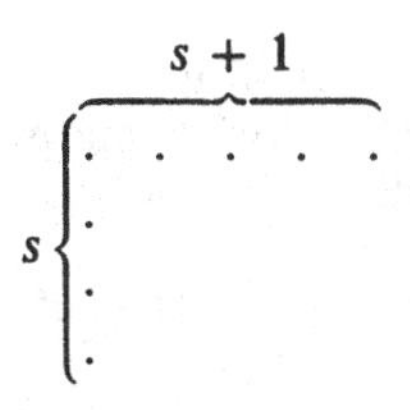

Since $C_i - C_{i+1} \geqslant 2$, these angles do indeed form the successive angles of a graphical representation of a partition, and clearly the successive ranks are either zero or one. Thus a partition of the type enumerated by $Q_{2,2}(n)$ is produced. The foregoing procedure clearly establishes a one-to-one correspondence between the two types of partitions, and therefore $B_{2,2}(n) = Q_{2,2}(n)$.

To prove that $B_{2,1}(n) = Q_{2,1}(n)$ we follow the same procedure except that now for parts $C_i = 2s + 1$, the angle is

s + 2

s

The argument proceeds as before with the only difference being that now an angle with only one node in it cannot arise, and this is precisely the difference between $B_{2,2}(n)$ and $B_{2,1}(n)$. Thus $B_{2,1}(n) = Q_{2,1}(n)$.

There are several questions of interest that arise naturally from this work.

Question 1. Can a direct graphical proof be found to show that $B_{k,a}(n) = Q_{k,a}(n)$ in general?

Question 2. Are there sieves that can be used in studying some or all of the partition functions found in Chapters 7 and 8?

Question 3. We can directly deduce from Theorem 9.10 that

$$m_{k,i}(2\mu; n) = p(n - \mu((4k + 2)\mu + 2k - 2i + 1)),$$

$$m_{k,i}(2\mu - 1; n) = p(n - (2\mu - 1)((2k + 1)\mu - 2k + i - 1)),$$

$$p_{k,i}(2\mu; n) = p(n - \mu((4k + 2)\mu - 2k + 2i - 1)),$$

$$p_{k,i}(2\mu - 1; n) = p(n - (2\mu - 1)((2k + 1)\mu - i)).$$

Are there direct combinatorial proofs of these relations?

Examples

1. The inclusion–exclusion method used in the proof of Theorem 9.1 may be employed to provide combinatorial interpretations and proofs of many elementary infinite product identities, such as

$$\frac{\prod_{n=1}^{\infty}(1 - q^n)}{\prod_{n=1}^{\infty}(1 - q^n)} = 1; \qquad \frac{\prod_{n=1}^{\infty}(1 - q^{n+m})}{\prod_{n=1}^{\infty}(1 - q^n)} = \prod_{n=1}^{m}(1 - q^n)^{-1}.$$

2. Let $\varepsilon_e(n)$ (resp. $\varepsilon_o(n)$) denote the number of partitions of n into distinct nonnegative parts with smallest part even and an even (resp. odd) number of even parts. Then

$$\varepsilon_o(n) - \varepsilon_e(n) = \begin{cases} 0 & \text{if } n \text{ is not a square,} \\ 1 & \text{if } n \text{ is a square.} \end{cases}$$

For example, when $n = 9$ the five partitions enumerated by $\varepsilon_e(9)$ are $8 + 1 + 0$, $7 + 2 + 0$, $6 + 3 + 0$, $5 + 4 + 0$, $4 + 3 + 2$, while the six partitions enumerated by $\varepsilon_o(9)$ are $9 + 0$, $7 + 2$, $6 + 2 + 1 + 0$, $5 + 4$, $5 + 3 + 1 + 0$, $4 + 3 + 2 + 0$.

The related generating function identity is

$$\sum_{n=0}^{\infty} q^{2n}(q^{2n+2}; q^2)_\infty(-q^{2n+1}; q^2)_\infty = \sum_{m=0}^{\infty} q^{m^2}.$$

This analytic assertion is quite easily proved (e.g., in Corollary 2.3, replace q by q^2, then set $c = -q$, $a = 0$, $t = q^2$, and let $b \to 0$). Actually we can combinatorially prove each step in the following analytic argument:

$$\sum_{n=0}^{\infty} (\varepsilon_o(n) - \varepsilon_e(n))q^n$$

$$= \sum_{n=0}^{\infty} q^{2n}(q^{2n+2}; q^2)_\infty(-q^{2n+1}; q^2)_\infty$$

$$= (q^2; q^2)_\infty \sum_{n=0}^{\infty} q^{2n} \frac{1}{(q^2; q^2)_n}(-q^{2n+1}; q^2)_\infty \qquad \text{(use Example 1)}$$

$$= (q^2; q^2)_\infty \sum_{n=0}^{\infty} \frac{q^{2n}}{(q^2; q^2)_n} \sum_{m=0}^{\infty} \frac{q^{m^2+2nm}}{(q^2; q^2)_m} \qquad \text{(see Example 17, Chapter 2)}$$

$$= (q^2; q^2)_\infty \sum_{m=0}^{\infty} \frac{q^{m^2}}{(q^2; q^2)_m} \frac{1}{(q^{2m+2}; q^2)_\infty} \qquad \text{(see Example 17, Chapter 2)}$$

$$= \sum_{m=0}^{\infty} q^{m^2} \qquad \text{(use Example 1).}$$

3. The work in Example 2 can be generalized by means of the identity

$$\sum_{n=0}^{\infty} q^{2kn}(q^{2kn+2k}; q^{2k})_\infty(-q^{2kn+1}; q^2)_\infty$$
$$= 1 + \sum_{n=1}^{\infty} q^{n^2} \prod_{j=1}^{n} \left(1 + q^{2j} + q^{4j} + \cdots + q^{2(k-1)j}\right).$$

4. The identities (here $[x]$ is the greatest integer function)

$$\sum_{j \geq 0} q^{j^2+ja} \begin{bmatrix} n+1-a-j \\ j \end{bmatrix}$$
$$= \sum_{\lambda=-\infty}^{\infty} (-1)^\lambda q^{\frac{1}{2}\lambda(5\lambda+1)-2a\lambda} \begin{bmatrix} n+1 \\ [\frac{1}{2}(n+1-5\lambda)]+a \end{bmatrix} \qquad (a = 0 \text{ or } 1),$$

and

$$\sum_{\lambda=-\infty}^{\infty} (-1)^\lambda q^{\frac{1}{2}\lambda(3\lambda+1)} \begin{bmatrix} n+1 \\ [\frac{1}{2}(n+1-3\lambda)] \end{bmatrix} = 1$$

are both deducible from Theorem 9.9. The sieve used in Theorem 9.12 is now applied directly to the generating functions in Theorem 9.9. The case in which $a + b = n + 1$, $k = 2$, $i = 2 - a$ essentially yields the top result; the case in which $a + b = n + 1$, $k = i = 1$ yields the second result immediately.

Notes

In the late nineteenth and early twentieth centuries, K. T. Vahlen (1893) and L. von Schrutka (1916, 1917) undertook an extensive study of partitions from the point of view taken in Section 9.2. The theorems we have chosen come from Andrews (1969, 1975).

The work in Section 9.3 also comes from Andrews (1971, 1972a). Such results were discovered in an effort to obtain a combinatorial interpretation of Schur's (1917) analytic proof of the Rogers–Ramanujan–Schur identities. Reviews of recent work related to this area may be found in Section P68 of LeVeque (1974).

Example 1. Vahlen (1893), Andrews (1969).

Example 2. Andrews (1972b), Stenger (1973).

Example 4. Schur (1917), Andrews (1970, 1971, 1972a).

Question 1. W.H. Burge has found such a proof. See *Discrete Math.*, 34 (1981), 9–15.

Question 3. D.M. Bressoud has given a combinatorial account of these identities as well as having observed that Theorem 9.12 is also true when k is half an odd integer. See *J. Number Th.*, 12 (1980), 87–100. Indeed the proof given above works in this case as well.

References

Andrews, G. E. (1969). "On a calculus of partition functions," *Pacific J. Math.* **31**, 555–562.

Andrews, G. E. (1970). "A polynomial identity which implies the Rogers–Ramanujan identities," *Scripta Math.* **28**, 297–305.

Andrews, G. E. (1971). "Sieves for theorems of Euler, Rogers and Ramanujan," *The Theory of Arithmetic Functions* (A. A. Gioia and D. L. Goldsmith, eds.) (Lecture Notes in Math. No. 251), pp. 1–20. Springer, Berlin.

Andrews, G. E. (1972a). "Sieves in the theory of partitions," *Amer. J. Math.* **94**, 1214–1230.

Andrews, G. E. (1972b). "Problem 5865," *Amer. Math. Monthly* **79**, 668.

Andrews, G. E. (1975). "Partially ordered sets and the Rogers–Ramanujan identities," *Aequationes Math.* **12**, 94–107.

Atkin, A. O. L. (1966). "A note on ranks and conjugacy of partitions," *Quart. J. Math. Oxford Ser.* (2) **17**, 335–338.

Dyson, F. J. (1944). "Some guesses in the theory of partitions," *Eureka* (Cambridge) **8**, 10–15.

LeVeque, W. J. (1974). *Reviews in Number Theory*, Vol. 4. Amer. Math. Soc., Providence, R.I.

Schrutka, L. von (1916). "Zur Systematik der additiven Zahlentheorie," *J. Reine Angew. Math.* **146**, 245–254.

Schrutka, L. von (1917). "Zur additiven Zahlentheorie," *S.-B. Kaiserl. Akad. Wiss. Wien, Abt. IIa* **125**, 1081–1163.

Schur, I. J. (1917). "Ein Beitrag zur additiven Zahlentheorie und zur Theorie der Kettenbrüche," *S.-B. Preuss. Akad. Wiss. Phys.-Math. Kl.* pp. 302–321. (Reprinted in I. Schur, *Gesammelte Abhandlungen*, Vol. 2, pp. 117–136. Springer, Berlin, 1973.)

Stenger, A. (1973). "Solution to Problem 5865," *Amer. Math. Monthly* **80**, 1148.

Vahlen, K. T. (1893). "Beiträge zu einer additiven Zahlentheorie," *J. Reine Angew. Math.* **112**, 1–36.

CHAPTER 10

Congruence Properties of Partition Functions

10.1 Introduction

In Chapter 5 we observed what remarkable fruits arose from the Hardy–Ramanujan collaboration. Out of this effort also arose Ramanujan's amazing discoveries about divisibility properties of $p(n)$. The following words of Ramanujan describe his famous conjecture on this topic:

> A recent paper by Mr. Hardy and myself contains a table, calculated by Major MacMahon, of the values of $p(n)$, the number of unrestricted partitions of n, for all values of n from 1 to 200. On studying the numbers in this table I observed a number of curious congruence properties, apparently satisfied by $p(n)$. Thus
>
> $$\begin{array}{rlllll} (1) & p(4), & p(9), & p(14), & p(19), \ldots & \equiv 0 \pmod{5}, \\ (2) & p(5), & p(12), & p(19), & p(26), \ldots & \equiv 0 \pmod{7}, \\ (3) & p(6), & p(17), & p(28), & p(39), \ldots & \equiv 0 \pmod{11}, \\ (4) & p(24), & p(49), & p(74), & p(99), \ldots & \equiv 0 \pmod{25}, \\ (5) & p(19), & p(54), & p(89), & p(124), \ldots & \equiv 0 \pmod{35}, \\ (6) & p(47), & p(96), & p(145), & p(194), \ldots & \equiv 0 \pmod{49}, \\ (7) & p(39), & p(94), & p(149), \ldots & & \equiv 0 \pmod{55}, \\ (8) & p(61), & p(138), \ldots & & & \equiv 0 \pmod{77}, \\ (9) & p(116), \ldots & & & & \equiv 0 \pmod{121}, \\ (10) & p(99), \ldots & & & & \equiv 0 \pmod{125}. \end{array}$$
>
> From these data I conjectured the truth of the following theorem: If $\delta = 5^a 7^b 11^c$ and $24\lambda \equiv 1 \pmod{\delta}$, then
>
> $$p(\lambda),\ p(\lambda + \delta),\ p(\lambda + 2\delta), \ldots \equiv 0 \pmod{\delta}.$$
>
> This theorem is supported by all the available evidence; but I have not yet been able to find a general proof.

Ramanujan then presents in this same paper proofs of the congruences

$$p(5n + 4) \equiv 0 \pmod{5}, \qquad (10.1.1)$$

ENCYCLOPEDIA OF MATHEMATICS and Its Applications, Gian-Carlo Rota (ed.). 2, George E. Andrews, The Theory of Partitions

$$p(7n + 5) \equiv 0 \pmod 7, \tag{10.1.2}$$

and he sketches proofs of the results:

$$p(25n + 24) \equiv 0 \pmod{25} \quad \text{and} \quad p(49n + 47) \equiv 0 \pmod{49}.$$

In connection with these congruences, Ramanujan proves a number of identities related to the partition function, such as

$$\sum_{n=0}^{\infty} p(5n + 4)q^n = 5 \prod_{n=1}^{\infty} \frac{(1 - q^{5n})^5}{(1 - q^n)^6}, \tag{10.1.3}$$

a result considered by G. H. Hardy to be an example of Ramanujan's best work.

The consideration of Ramanujan's full conjecture on the congruence properties of $p(n)$ produced a number of partial results, such as those of Ramanujan described earlier. In the 1930s, S. Chowla noticed (from the examination of an extended table of values of $p(n)$ due to H. Gupta) that

$$p(243) = 133978259344888 \not\equiv 0 \pmod{7^3};$$

however,

$$24 \cdot 243 \equiv 1 \pmod{7^3}.$$

Hence Ramanujan's conjecture as it stands is false. G. N. Watson showed in 1938 that a modified version of the conjecture was true for all powers of 5 and 7, and in 1967 (48 years after Ramanujan originally published his conjecture) A. O. L. Atkin proved the full modified conjecture:

If $\delta = 5^a 7^b 11^c$ *and* $24\lambda \equiv 1 \pmod{\delta}$, *then* $p(\lambda) \equiv 0 \pmod{5^a 7^{[(b+2)/2]} 11^c}$.

In this chapter, our object is to obtain an understanding of the analytic methods that have been most effective in attacking such problems. Some form of the *Hecke operator* almost always appears in the analytic considerations, and in Section 10.2, we shall see that such operators suffice to yield Rödseth's proof of the Churchhouse conjecture on binary partitions.

As for Ramanujan's conjecture, we shall content ourselves with the powers of 5, and we shall present a simplified version of Watson's original proof due to Atkin. This proof has received an excellent presentation in M. I. Knopp's book, *Modular Functions in Analytic Number Theory* (1970); our version differs primarily in the fact that it is less complete. In particular, properties of $p(n)$ that rely primarily on the fact that $\sum_{n \geq 0} p(n)e^{2\pi i n\tau} = e^{\pi i\tau/12}(\eta(\tau))^{-1}$ (where $\eta(\tau)$ is a modular form) would require extensive preliminary work, which would be much more appropriate in a book on analytic number theory or elliptic function theory (Knopp's book cited above, e.g.). Thus in Chapter 5 we assumed the truth of (5.2.2), which is a disguised form of the relation describing the behavior of $\eta(\tau)$ under the action of the modular group. In this chapter we shall assume (10.3.1), the modular equation of fifth degree. After obtaining Ramanujan's conjecture for powers of 5, we shall conclude this chapter with a brief discussion of the Dyson conjectures (proved by Atkin and Swinnerton–Dyer in the 1950s) which provide combinatorial interpretations of (10.1.1) and (10.1.2).

10.2 Rödseth's Theorem for Binary Partitions

We begin with some elementary remarks about certain operators closely related to the Hecke operators: These operators U_m are defined on all functions $f(q) = \sum_{n=-\infty}^{\infty} a_n q^n$ meromorphic around $q = 0$ and are given by

$$U_m\{f(q)\} = \sum_{n=-\infty}^{\infty} a_{mn} q^n.$$

Thus if $\rho = e^{2\pi i/m}$, then

$$U_m f(q) = \frac{1}{m}\left(f(q^{1/m}) + f(\rho q^{1/m}) + \cdots + f(\rho^{m-1} q^{1/m})\right), \tag{10.2.1}$$

since

$$\sum_{j=0}^{m-1} \rho^{rj} = \begin{cases} m & \text{if } r \equiv 0 \pmod{m}, \\ 0 & \text{if } r \not\equiv 0 \pmod{m}. \end{cases}$$

Clearly U_m is a linear operator.

The type of partitions we shall be considering in this section are *binary partitions*, that is, partitions in which the parts are all powers of 2. Thus the six binary partitions of 7 are (124), $(1^3 4)$, (12^3), $(1^3 2^2)$, (1^7), (21^5).

We denote by $b(n)$ the number of binary partitions of n, and we let

$$B(q) = \sum_{n=0}^{\infty} b(n) q^n = \prod_{n=0}^{\infty} (1 - q^{2^n})^{-1}.$$

Next we define a sequence of functions for $m \geqslant 1$,

$$\begin{aligned} \mathscr{F}_m(q) &= (U_2{}^{m+1} - U_2{}^{m-1}) B(q) \\ &= \sum_{n=0}^{\infty} (b(2^{m+1} n) - b(2^{m-1} n)) q^n. \end{aligned} \tag{10.2.2}$$

First we have

$$\begin{aligned}
\mathscr{F}_1(q^4) &= (U_2{}^4 - U_2{}^0)B(q^4) = \sum_{n=0}^{\infty} (b(4n) - b(n))q^{4n} \\
&= \tfrac{1}{4}\sum_{j=0}^{3} B(i^j q) - B(q^4) \\
&= \prod_{n=2}^{\infty} \frac{1}{(1-q^{2n})} \left\{ \frac{1}{4}\left[\sum_{j=0}^{3} \frac{1}{(1-i^j q)(1-i^{2j}q^2)} \right] - 1 \right\} \\
&= \prod_{n=2}^{\infty} \frac{1}{(1-q^{2n})} \left\{ \frac{1}{4}\left[\sum_{j=0}^{3} \frac{(1+i^j q)(1+i^{2j}q^2)^2}{(1-q^4)^2} \right] - 1 \right\} \\
&= \frac{1}{(1-q^4)^2} \prod_{n=2}^{\infty} \frac{1}{(1-q^{2n})} (1 + q^4 - 1 + 2q^4 - q^8) \\
&= \frac{q^4(3-q^4)}{(1-q^4)^3} \prod_{n=3}^{\infty} \frac{1}{(1-q^{2n})}.
\end{aligned}$$

Hence

$$\mathscr{F}_1(q) = \frac{q(3-q)}{(1-q)^3} G(q) \tag{10.2.3}$$

where

$$G(q) = \prod_{n=1}^{\infty} \frac{1}{(1-q^{2n})} = (1-q)B(q).$$

In general, we have

$$\mathscr{F}_{m+1}(q^2) = U_2 \mathscr{F}_m(q^2) = \tfrac{1}{2}(\mathscr{F}_m(q) + \mathscr{F}_m(-q)) \tag{10.2.4}$$

(by (10.2.1)). Thus

$$\begin{aligned}
\mathscr{F}_2(q^2) &= \tfrac{1}{2}(\mathscr{F}_1(q) + \mathscr{F}_1(-q)) \\
&= \tfrac{1}{2} q G(q) \left[\frac{(3-q)}{(1-q)^3} - \frac{(3+q)}{(1+q)^3} \right] \\
&= \frac{8q^2 G(q)}{(1-q^2)^3}.
\end{aligned}$$

Therefore

$$\mathscr{F}_2(q) = \frac{8qG(q)}{(1-q)^4}. \tag{10.2.5}$$

We now derive similar formulas for $\mathscr{F}_3(q)$ and $\mathscr{F}_4(q)$.

$$\begin{aligned}\mathscr{F}_3(q^2) &= U_2\mathscr{F}_2(q^2) = \tfrac{1}{2}(\mathscr{F}_2(q) + \mathscr{F}_2(-q)) \\ &= \frac{4qG(q)}{(1-q^2)^4}((1+q)^4 - (1-q)^4) = \frac{32q^2G(q)(1+q^2)}{(1-q^2)^4}.\end{aligned}$$

Thus

$$\mathscr{F}_3(q) = \frac{32qG(q)(1+q)}{(1-q)^5}. \tag{10.2.6}$$

This implies

$$\begin{aligned}\mathscr{F}_3(q) + 4\mathscr{F}_2(q) &= \frac{32qg(q)}{(1-q)^5}((1+q) + (1-q)) \\ &= \frac{64qG(q)}{(1-q)^5}.\end{aligned} \tag{10.2.7}$$

Next, applying U_2 to (10.2.7), we find that

$$\begin{aligned}\mathscr{F}_4(q^2) + 4\mathscr{F}_3(q^2) &= U_2(\mathscr{F}_3(q^2) + 4\mathscr{F}_2(q^2)) = \frac{32qG(q)}{(1-q^2)^5}((1+q)^5 - (1-q)^5) \\ &= \frac{64q^2G(q)}{(1-q^2)^5}((1-q^2)^2 - 12(1-q^2) + 16) \\ &= \frac{64q^2G(q)}{(1-q^2)^3} - \frac{768q^2G(q)}{(1-q^2)^4} + \frac{1024q^2G(q)}{(1-q^2)^5}.\end{aligned}$$

Thus

$$\mathscr{F}_4(q) + 4\mathscr{F}_3(q) = 8\mathscr{F}_2(q) - 12(\mathscr{F}_3(q) + 4\mathscr{F}_2(q)) + \frac{1024qG(q)}{(1-q)^6}$$

or

$$\mathscr{F}_4(q) + 16\mathscr{F}_3(q) + 40\mathscr{F}_2(q) = \frac{1024qG(q)}{(1-q)^6}. \tag{10.2.8}$$

It is now possible to prove Theorem 10.1, which generalizes (10.2.5), (10.2.7), and (10.2.8).

THEOREM 10.1. *There exist integers $c_j(m)$ such that for $m \geqslant 2$*

$$\sum_{j=0}^{m-2} c_j(m)\mathscr{F}_{m-j}(q) = \frac{2^{\binom{m+1}{2}}qG(q)}{(1-q)^{m+2}};$$

furthermore, $c_0(m) = 1$, $16|c_1(m)$ *if* $m \geqslant 4$, $8|c_2(m)$, $16\nmid c_2(m)$, *and* $2^{2j}|c_j(m)$, *for* $3 \leqslant j \leqslant m-2$.

Proof. From (10.2.5), (10.2.7), and (10.2.8) we see that Theorem 10.1 is true for $m = 2, 3$, and 4. We now proceed by mathematical induction on m, and we assume $m \geqslant 5$.

By applying U_2 to the equation stated in the theorem (with q replaced by q^2) we find

$$\sum_{j=0}^{m-2} c_j(m)\mathscr{F}_{m+1-j}(q^2) = \frac{2^{\binom{m+1}{2}}qG(q)}{(1-q^2)^{m+2}} \cdot \frac{1}{2}((1+q)^{m+2} - (1-q)^{m+2}).$$

Now we define integers $\delta_k(m+2)$ for $0 \leqslant k \leqslant [(m+1)/2]$ by

$$\tfrac{1}{2}((1+q)^{m+2} - (1-q)^{m+2}) = q \sum_{k=0}^{[(m+1)/2]} \delta_k(m+2)(1-q^2)^k;$$

this equation truly defines integers $\delta_k(m+2)$, since the left-hand side of the equation is an odd polynomial. Furthermore, we see that

$$\delta_0(m+2) = 2^{m+1}$$

and

$$\begin{aligned}\delta_1(m+2) &= \lim_{q\to 1}(1-q^2)^{-1}(\tfrac{1}{2}((1+q)^{m+2} - (1-q)^{m+2}) - q2^{m+1}) \\ &= -m2^{m-1}.\end{aligned}$$

Thus

$$\begin{aligned}\sum_{j=0}^{m-2} c_j(m)\mathscr{F}_{m+1-j}(q) = {} & \frac{2^{\binom{m+2}{2}}qG(q)}{(1-q)^{m+3}} - m2^{m-1}\sum_{j=0}^{m-2} c_j(m)\mathscr{F}_{m-j}(q) \\ & + \sum_{k=2}^{[(m+1)/2]} \delta_k(m+2)2^{\binom{m+1}{2} - \binom{m-k+2}{2}} \\ & \times \sum_{h=0}^{m-k-1} c_h(m-k+1)\mathscr{F}_{m-k-h+1}(q).\end{aligned}$$

Thus we may derive expressions for the $c_j(m+1)$:

$$c_0(m+1) = c_0(m) = 1;$$

$$\begin{aligned}c_1(m+1) &= c_1(m) + m2^{m-1}c_0(m) \\ &= c_1(m) + m2^{m-1}.\end{aligned}$$

Now since $16|c_1(m)$ and $m > 4$, we see that $16|c_1(m+1)$.

$$c_2(m+1) = c_2(m) + m2^{m-1}c_1(m) + \delta_2(m+2)2^m c_0(m-1).$$

Since $8|c_2(m)$, and $m \geqslant 5$, we see that $8|c_2(m+1)$. Also since $16 \nmid c_2(m)$, we see that $16 \nmid c_2(m+1)$, because all of the remaining terms on the right-hand side of the equation above are divisible by 16.

Now for $2 < j \leqslant m-1$

$$c_j(m+1) = c_j(m) + (m+2)2^{m-1}c_{j-1}(m)$$
$$+ \sum_{\substack{k+h=j\\ k\geqslant 2\\ h\geqslant 0}} \delta_k(m+2)2^{\binom{m+1}{2}-\binom{m-k+2}{2}}c_h(m-k+1).$$

By hypothesis, $2^{2j}|c_j(m)$ and $2^{2j}|2^{m-1}c_{j-1}(m)$. Thus to prove that $2^{2j}|c_j(m+1)$, we need only establish that

$$\binom{m+1}{2} - \binom{m-k+2}{2} + 2h \geqslant 2j$$

when $k+h=j$, $[(m+1)/2] \geqslant k \geqslant 2$, $h \geqslant 0$.

$$\binom{m+1}{2} - \binom{m-k+2}{2} + 2h - 1 = -\binom{k+1}{2} + km - 2 + 2j$$
$$\geqslant km - \tfrac{1}{2}(k^2+k) - 2 + 2j$$
$$= k(m - \tfrac{1}{2}(k+1)) - 2 + 2j$$
$$\geqslant 2j,$$

since $2 \leqslant k \leqslant [(m+1)/2]$; thus $2^{2j}|c_j(m+1)$, and Theorem 10.1 is established. ∎

The theorem above allows us to prove certain congruences for $b(4n) - b(n)$ originally conjectured by R. F. Churchhouse and proved independently by O. Rödseth and H. Gupta.

THEOREM 10.2. *If $k \geqslant 1$ and $t \equiv 1 \pmod 2$, then*

$$b(2^{2k+2}t) - (b^{2k}t) \equiv 0 \pmod{2^{3k+2}}, \tag{10.2.9}$$

$$b(2^{2k+1}t) - b(2^{2k-1}t) \equiv 0 \pmod{2^{3k}}; \tag{10.2.10}$$

furthermore, (10.2.9) *and* (10.2.10) *are best possible in that no higher power of* 2 *divides* $b(4n) - b(n)$.

Proof. We shall show that

$$2^{-3k-2}\mathscr{F}_{2k+1}(q) \quad \text{and} \quad 2^{-3k}\mathscr{F}_{2k}(q)$$

have integral coefficients and that the coefficients of the odd powers of q are themselves odd. This is equivalent to Theorem 10.2.

If $k = 1$, then (10.2.5) implies

$$\tfrac{1}{8}\mathscr{F}_2(q) = \frac{qG(q)}{(1-q)^4} \equiv \frac{qB(q)}{(1-q)^3} \equiv \frac{q}{(1-q)^2} \equiv \sum_{j=0}^{\infty} q^{2j+1} (\text{mod } 2).$$

Also by (10.2.6),

$$\frac{1}{32}\mathscr{F}_3(q) = \frac{qG(q)(1+q)}{(1-q)^5} \equiv \frac{qB(q)}{(1-q)^3} \equiv \sum_{j=0}^{\infty} q^{2j+1} (\text{mod } 2).$$

Thus Theorem 2 is established for $k = 1$.

Assume that Theorem 10.2 is true for all integers less than k. Then by Theorem 10.1

$$2^{-3k}\mathscr{F}_{2k}(q) + 2^{-3k}c_1(2k)\mathscr{F}_{2k-1}(q) + 2^{-3k}c_2(2k)\mathscr{F}_{2k-2}(q)$$
$$+ \sum_{j=3}^{2k-1} 2^{-3k}c_j(2k)\mathscr{F}_{2k-j}(q) = \frac{2^{\binom{2k+1}{2}-3k}qG(q)}{(1-q)^{2k+2}}.$$

By Theorem 10.1, $16|c_1(2k)$, $8|c_2(2k)$, and $2^{2j}|c_j(2k)$, also for $k > 1$,

$$\binom{2k+1}{2} - 3k > 0.$$

Hence from the induction hypothesis we see that $2^{-3k}\mathscr{F}_{2k}(q)$ has integral coefficients. Furthermore since $16 \nmid c_2(2k)$,

$$2^{-3k}\mathscr{F}_{2k}(q) \equiv 2^{-3k+3}\mathscr{F}_{2k-2}(q) \equiv \sum_{j=0}^{\infty} q^{2j+1} \ (\text{mod } 2).$$

Now by Theorem 10.1,

$$2^{-3k-2}\mathscr{F}_{2k+1}(q) + 2^{-3k-2}c_1(2k+1)\mathscr{F}_{2k}(q)$$
$$+ 2^{-3k-2}c_2(2k+1)\mathscr{F}_{2k-1}(q) + \sum_{j=3}^{2k-1} 2^{-3k-3}c_j(2k+1)\mathscr{F}_{2k+1-j}(q)$$
$$= \frac{2^{\binom{2k+2}{2}-3k-2}qG(q)}{(1-q)^{2k+3}}.$$

Again by Theorem 10.1, $16|c_1(2k+1)$, $8|c_2(2k+1)$, $2^{2j}|c_j(2k)$, also for $k>1$,

$$\binom{2k+2}{2} - 3k - 2 > 0.$$

Hence from the induction hypothesis we see that $2^{-3k-2}\mathscr{F}_{2k+1}(q)$ has integral coefficients. Furthermore, since $16\nmid c_2(2k+1)$,

$$2^{-3k-2}\mathscr{F}_{2k+1}(q) \equiv 2^{-3k+1}\mathscr{F}_{2k-1}(q) \equiv \sum_{j=0}^{\infty} q^{2j+1} \pmod 2.$$

Thus we have Theorem 10.2. ■

10.3 Ramanujan's Conjecture for 5^n

Our first task is to translate Ramanujan's conjecture into terms amenable to the use of generating functions. In particular, we shall obtain a general family of identities of which (10.1.3) is the simplest case. Besides the Hecke operator U_5, we also require the modular equation of fifth degree, which we state in the form

$$\frac{q^5(q^{25};q^{25})_\infty^5}{(q)_\infty^5} = \frac{q^5(q^{25};q^{25})_\infty^6}{(q^5;q^5)_\infty^6}\left(5^2\frac{q^4(q^{25};q^{25})_\infty^4}{(q)_\infty^4} + 5^2\frac{q^3(q^{25};q^{25})_\infty^3}{(q)_\infty^3}\right.$$
$$\left. + 3\cdot 5\frac{q^2(q^{25};q^{25})_\infty^2}{(q)_\infty^2} + 5\frac{q(q^{25};q^{25})_\infty}{(q)_\infty} + 1\right). \quad (10.3.1)$$

For a proof of this result we refer the reader to Knopp (1970, p. 119, Eq. (2.5)). To simplify notation, we define

$$\phi(q) = \frac{q(q^{25};q^{25})_\infty}{(q)_\infty}, \quad (10.3.2)$$

and

$$g(q) = \frac{q(q^5;q^5)_\infty^6}{(q)_\infty^6}. \quad (10.3.3)$$

Thus we may rewrite (10.3.1) as

$$\phi(q)^5 = g(q^5)\{5^2\phi(q)^4 + 5^2\phi(q)^3 + 3\cdot 5\phi(q)^2 + 5\phi(q) + 1\}. \quad (10.3.4)$$

In the following theorems it is useful to have

$$S_r = 5U_5\phi^r(q) = \sum_{i=0}^{4}\phi^r(q^{1/5}\rho^i) \quad (10.3.5)$$

where $\rho = e^{2\pi i/5}$. Also we require the 5-adic valuation:

DEFINITION 10.1. If r is a rational number, then by the fundamental theorem of arithmetic r may be uniquely written as $\pm p_1^{a_1} p_2^{a_2} \cdots p_r^{a_r}$ where the p_i are prime numbers and the a_i are integers (positive, negative, or zero). The *5-adic valuation* $\upsilon(r)$ is defined to be the exponent of 5 in this prime decomposition of r.

LEMMA 10.3. *For each positive integer r, S_r is a polynomial in $g(q)$ of the form*

$$S_r = \sum_{j=1}^{\infty} a_{rj} g^j(q)$$

where 5 divides each a_{rj}, also $\upsilon(a_{rj}) \geqslant [(5j - r + 1)/2]$, and $a_{rj} = 0$ unless $[(r + 4)/5] \leqslant j \leqslant r$, where $[x]$ is the greatest integer function.

Proof. We begin by fixing q and considering the fifth-degree polynomial equation in u:

$$u^5 - g(q)(5^2u^4 + 5^2u^3 + 3 \cdot 5u^2 + 5u + 1) = 0. \qquad (10.3.6)$$

From (10.3.4) we see that $u = \phi(q^{1/5})$ is one root; in fact, $u = \phi(q^{1/5}\rho^j)$, $j = 0, 1, 2, 3, 4$ (where $\rho = e^{2\pi i/5}$), are roots of (10.3.6) since the replacement of q by $\rho^j q$ in (10.3.4) only alters $\phi(q)$. Furthermore, since 1 is the coefficient of q in $\phi(q)$ (by (10.3.2)), we see that the $\phi(q^{1/5}\rho^j)$ are all different functions of q and are indeed distinct as complex numbers for q sufficiently close to zero, since $\phi(q) \sim q$ as $q \to 0$. Hence we have determined five distinct roots of (10.3.6) when q is a sufficiently small complex number, and so

$$\prod_{i=0}^{4} (u - \phi(q^{1/5}\rho^i)) = u^5 - g(q)(5^2u^4 + 5^2u^3 + 3 \cdot 5u^2 + 5u + 1). \quad (10.3.7)$$

By analytic continuation (10.3.7) is valid for all q with $|q| < 1$.

Now let us recall Newton's formula that connects the sums of like powers of the roots of a polynomial with the coefficients of the polynomial. Namely, if

$$f(x) = x^n - a_1x^{n-1} + a_2x^{n-2} - \cdots + (-1)^n a_n,$$
$$= (x - r_1)(x - r_2) \cdots (x - r_n)$$

and if

$$s_m = r_1^m + r_2^m + \cdots + r_n^m,$$

then for $1 \leqslant j \leqslant n$

$$s_j - a_1 s_{j-1} + a_2 s_{j-2} - \cdots + (-1)^{j-1} a_{j-1} s_1 + (-1)^j a_j j = 0,$$

and for $j > n$

$$s_j - a_1 s_{j-1} + a_2 s_{j-2} - \cdots + (-1)^j a_j s_{j-n} = 0.$$

Applying Newton's formula to $f(x)$, we find that since S_r is the sum of the rth powers of the roots of the polynomial in (10.3.6),

$$S_1 = 5^2 g(q), \tag{10.3.8)$_1$}$$

$$S_2 = 5^4 g^2(q) + 2\cdot 5^2 g(q), \tag{10.3.8)$_2$}$$

$$S_3 = 5^6 g^3(q) + 3\cdot 5^4 g^2(q) + 9\cdot 5g(q), \tag{10.3.8)$_3$}$$

$$S_4 = 5^8 g^4(q) + 4\cdot 5^6 g^3(q) + 22\cdot 5^3 g^2(q) + 4\cdot 5g(q), \tag{10.3.8)$_4$}$$

$$S_5 = 5^{10} g^5(q) + 5\cdot 5^8 g^4(q) + 40\cdot 5^5 g^3(q) + 20\cdot 5^3 g^2(q) + 5g(q), \tag{10.3.8)$_5$}$$

and for $r > 5$

$$S_r = 5^2 g(q) S_{r-1} + 5^2 g(q) S_{r-2} + 3\cdot 5g(q) S_{r-3} + 5g(q) S_{r-4} + g(q) S_{r-5}. \tag{10.3.8)$_r$}$$

We may now easily check all the assertions of Theorem 10.3 directly; the only one that requires any real care is the inequality for $\upsilon(a_{rj})$, which is directly verified by induction on r. For $1 \leqslant r \leqslant 5$, we need only examine the explicit form of the identity for S_r. For $r > 5$, assume $\upsilon(a_{sj}) \geqslant [(5j - s + 1)/5]$ for $s < r$ and then examine (10.3.8)$_r$:

$$a_{rj} = 5^2 a_{r-1,j-1} + 5^2 a_{r-2,j-1} + 15 a_{r-3,j-1} + 5 a_{r-4,j-1} + a_{r-5,j-1}.$$

Therefore

$$\begin{aligned}\upsilon(a_{rj}) &\geqslant \min(\upsilon(a_{r-1,j-1}) + 2, \upsilon(a_{r-2,j-1}) + 2, \\ &\qquad \upsilon(a_{r-3,j-1}) + 1, \upsilon(a_{r-4,j-1}) + 1, \upsilon(a_{r-5,j-1})) \\ &= \min\left(\left[\frac{5j - r + 1}{5}\right], \left[\frac{5j - r + 2}{5}\right], \left[\frac{5j - r + 1}{2}\right],\right. \\ &\qquad \left.\left[\frac{5j - r + 2}{5}\right], \left[\frac{5j - r + 1}{5}\right]\right) \\ &= \left[\frac{5j - r + 1}{5}\right],\end{aligned}$$

and Lemma 10.3 is established. ■

We are now prepared for the proof of Ramanujan's conjecture for powers of 5. Let us restate the conjecture for powers of 5 only:

If $24\lambda \equiv 1 \pmod{5^a}$, then $p(\lambda) \equiv 0 \pmod{5^a}$.

We begin by remarking that the smallest positive integral solution of $24\lambda \equiv 1 \pmod{5^a}$ is

$$c_a = \begin{cases} \frac{1}{24}(19\cdot 5^{2r-1} + 1) & \text{if } a = 2r - 1, \\ \frac{1}{24}(23\cdot 5^{2r} + 1) & \text{if } a = 2r. \end{cases}$$

Consequently we wish to show that

$$p(5^a m + c_a) \equiv 0 \pmod{5^a}$$

for each $m \geqslant 0$.

In light of our work in Section 10.2, it would be natural to guess that our interest should center on

$$\sum_{m=0}^{\infty} p(5^a m + c_a)q^m = U_5 \sum_{m=0}^{\infty} p(5^{a-1}m + c_a)q^m;$$

however, as we shall see in the utilization of the modular equation, we must instead consider

$$L_{2n-1}(q) = (q^5; q^5)_\infty \sum_{m=0}^{\infty} p(5^{2n-1}m + c_{2n-1})q^{m+1} \tag{10.3.9}$$

and

$$L_{2n}(q) = (q)_\infty \sum_{m=0}^{\infty} p(5^{2n}m + c_{2n})q^{m+1}. \tag{10.3.10}$$

The following theorem provides us with strong information concerning the coefficients of $L_n(q)$, and Ramanujan's conjecture for powers of 5 becomes a simple corollary.

THEOREM 10.4. *Each $L_n(q)$ is a polynomial in $g(q)$ (see (10.3.3) for the definition of $g(q)$) with integral coefficients. Furthermore, if we write this polynomial as $L_n(q) = \sum_{s=0}^{\infty} b_{ns}g^s(q)$, then*

$$b_{n0} = 0, \tag{10.3.11}$$

$$\upsilon(b_{n1}) = n, \tag{10.3.12}$$

$$\upsilon(b_{ns}) \geqslant n \quad \textit{for all} \quad s \geqslant 1. \tag{10.3.13}$$

Proof. The proof proceeds by mathematical induction on n, much in the manner of the proof of Theorem 10.1. The primary complication arises from the peculiarities of the two-line definition of $L_n(q)$ (Eqs. (10.3.9) and (10.3.10) are forced on us by the varying form of the solutions to the congruence $24\lambda \equiv 1 \pmod{5^n}$). Also, we are forced to prove more than (10.3.13), so we replace (10.3.13) by

$$\upsilon(b_{ns}) \geqslant n + [\tfrac{1}{4}(10s - 9 + (-1)^n)]. \tag{10.3.13$'$}$$

When $n = 1$,

$$\begin{aligned} L_1(q) &= (q^5; q^5) \sum_{m=0}^{\infty}{}' p(5m+4)q^{m+1} \\ &= (q^5; q^5)_\infty U_5 \sum_{m=1}^{\infty} p(m-1)q^m \\ &= (q^5; q^5)_\infty U_5 \frac{q}{(q)_\infty} \\ &= U_5 \frac{q(q^{25}; q^{25})_\infty}{(q)_\infty} \end{aligned}$$

(since $U_5 f_1(q^5) f_2(q) = f_1(q) U_5 f_2(q)$)

$$\begin{aligned} &= U_5 \phi(q) \\ &= \tfrac{1}{5} S_1 \\ &= 5g(q) \end{aligned} \tag{10.3.14}$$

(by $(10.3.8)_1$). Hence, $b_{10} = 0$, $b_{11} = 5$, $b_{1s} = 0$ for $s > 1$, and Eqs. (10.3.11), (10.3.12), and (10.3.13)$'$ are all clearly valid when $n = 1$.

Let us now assume that our theorem is valid for every integer $n \leqslant 2r - 1$. We must show that this implies the result is valid for $n = 2r$ and $2r + 1$. By (10.3.9)

$$\begin{aligned} L_{2r}(q) &= (q)_\infty \sum_{m=0}^{\infty} p(5^{2r}m + c_{2r})q^{m+1} \\ &= (q)_\infty \sum_{m=0}^{\infty} p(5^{2r}m + 4\cdot 5^{2r-1} + c_{2r-1})q^{m+1} \\ &= (q)_\infty \sum_{m=0}^{\infty} p(5^{2r-1}(5m+4) + c_{2r-1})q^{m+1} \\ &= U_5 L_{2r-1}(q) \end{aligned}$$

$$= U_5 \sum_{s=0}^{\infty} b_{2r-1,s} g^s(q)$$

$$= \sum_{s=0}^{\infty} b_{2r-1,s} U_5(g^s(q))$$

$$= \sum_{s=0}^{\infty} b_{2r-1,s} U_5\left(\frac{q^s(q^5;q^5)_\infty^{6s}}{(q)_\infty^{6s}}\right)$$

$$= \sum_{s=0}^{\infty} b_{2r-1,s} \frac{(q)_\infty^{6s}}{q^s(q^5;q^5)_\infty^{6s}} U_5\phi^{6s}(q)$$

$$= \sum_{s=0}^{\infty} b_{2r-1,s} \frac{1}{g^s(q)} \frac{1}{5} \sum_{j=1}^{\infty} a_{6s,j} g^j(q) \qquad \text{(by Theorem 10.3)}$$

$$= \sum_{t=-\infty}^{\infty} g^t(q) \sum_{s=0}^{\infty} \tfrac{1}{5} b_{2r-1,s} a_{6s,t+s}$$

$$= \sum_{t=-\infty}^{\infty} g^t(q) b_{2r,t}, \tag{10.3.15}$$

where

$$b_{2r,t} = \tfrac{1}{5} \sum_{s=0}^{\infty} b_{2r-1,s} a_{6s,t+s}. \tag{10.3.16}$$

We may now deduce the truth of (10.3.11), (10.3.12), and (10.3.13)′ for $b_{2r,t}$ from (10.3.16) using Theorem 10.3 (for properties of the $a_{6s,t+s}$) and the induction hypothesis (for properties of the $b_{2r-1,s}$). In particular, if $t \leqslant 0$, then $t + s \leqslant s$, and so by Theorem 10.3 $a_{6s,t+s} = 0$ for $s > 0$ since

$$\left[\frac{6s+4}{5}\right] > s \geqslant t + s.$$

By the induction hypothesis $b_{2r-1,0} = 0$. Hence we see that $b_{2r,t} = 0$ if $t \leqslant 0$, confirming (10.3.11) when $n = 2r$.

Furthermore, since $L_{2r-1}(q)$ is (by hypothesis) a polynomial in $g(q)$, we see that $b_{2r-1,s} = 0$ for $s \geqslant s_0(r)$. Also by Theorem 10.3, $a_{6s,t+s} = 0$ unless $t \leqslant 5s$. Hence, if $t \geqslant 5s_0(r)$, then one of the factors of each term in (10.3.16) is zero; so $b_{2r,t} = 0$ for t sufficiently large. Therefore $L_{2r}(q)$ is a polynomial in $g(q)$. Next

$$b_{2r,1} = \tfrac{1}{5} \sum_{s=0}^{\infty} b_{2r-1,s} \cdot a_{6s,s+1}$$

$$= \tfrac{1}{5} b_{2r-1,1} \cdot a_{6,2}$$

$$= b_{2r-1,1} \cdot 5 \cdot 63$$

(by evaluation of $a_{6,2}$ from Eq. $(10.3.8)_6$). Hence

$$\upsilon(b_{2r,1}) = \upsilon(b_{2r-1,1}) + 1 = 2r,$$

by the induction hypothesis, and (10.3.12) is valid for $n = 2r$.

Finally for $t \geqslant 1$, by (10.3.16), Theorem 10.3, and (10.3.13)′, with $n = 2r - 1$,

$$\upsilon(b_{2r,t}) \geqslant -1 + \min_{\substack{[(s+4)/5] \leqslant t \leqslant 5s \\ s \geqslant 1}} \upsilon(b_{2r-1,s} \cdot a_{6s,t+s})$$

$$= -1 + \min_{\substack{[(s+4)/5] \leqslant t \leqslant 5s \\ s \geqslant 1}} \left\{ 2r - 1 + \left[\frac{5s-5}{2}\right] + \left[\frac{5t-s+1}{2}\right] \right\}$$

$$\geqslant -1 + \left\{ 2r - 1 + \left[\frac{5s-5}{2}\right] + \left[\frac{5t-s+1}{2}\right] \right\}_{\text{at } s=1}$$

$$= 2r - 2 + \left[\frac{5t}{2}\right]$$

$$\geqslant 2r + \left[\frac{5t-4}{2}\right],$$

which establishes (10.3.13), with $n = 2r$.

To conclude our proof by induction we must now prove our assertions for $n = 2r + 1$ using the induction hypotheses and the case in which $n = 2r$. Since the steps resemble those above, we omit most of the details; we derive the counterpart of (10.3.16), and we leave to the reader the necessary deductions from this result.

$$L_{2r+1}(q) = (q^5; q^5)_\infty \sum_{m=0}^{\infty} p(5^{2r+1}m + c_{2r+1}) q^{m+1}$$

$$= (q^5; q^5)_\infty \sum_{m=0}^{\infty} p(5^{2r}m + 3 \cdot 5^{2r} + c_{2r}) q^{m+1}$$

$$= (q^5; q^5)_\infty U_5 \sum_{m=0}^{\infty} p(5^{2r}m + c_{2r}) q^{m+2}$$

$$= U_5 \left\{ (q^{25}; q^{25}) \sum_{m=0}^{\infty} p(5^{2r}m + c_{2r}) q^{m+2} \right\}$$

$$= U_5\{\phi(q)L_{2r}(q)\}$$

$$= U_5 \sum_{s=0}^{\infty} b_{2r,s}\phi(q)g^s(q)$$

$$= \sum_{s=0}^{\infty} b_{2r,s}U_5(\phi(q)g^s(q))$$

$$= \sum_{s=0}^{\infty} b_{2r,s}\frac{(q)_\infty^{6s}}{q^s(q^5;q^5)_\infty^{6s}}U_5(\phi^{6s+1}(q))$$

$$= \sum_{s=0}^{\infty} b_{2r,s}\frac{1}{g^s(q)}\frac{1}{5}S_{6s+1}$$

$$= \sum_{s=0}^{\infty} b_{2r,s}\frac{1}{g^s(q)}\frac{1}{5}\sum_{j=1}^{\infty} a_{6s+1,j}g^j(q) \qquad \text{(by Theorem 10.3)}$$

$$= \sum_{t=-\infty}^{\infty} g^t(q)\sum_{s=0}^{\infty} \tfrac{1}{5}b_{2r,s}a_{6s+1,j} \tag{10.3.17}$$

whence

$$b_{2r+1,t} = \tfrac{1}{5}\sum_{s=0}^{\infty} b_{2r,s}a_{6s+1,t+s}. \tag{10.3.18}$$

The remainder of the proof follows just as in the treatment of (10.3.16). ∎

COROLLARY 10.5 (Ramanujan's conjecture for powers of 5). *If* $24\lambda \equiv 1 \pmod{5^n}$, *then*

$$p(\lambda) \equiv 0 \pmod{5^n}.$$

Proof. By the remarks preceding Theorem 10.4, we see that we must show that

$$p(5^a m + c_a) \equiv 0 \pmod{5^a}$$

for each $m \geqslant 0$. If $a = 2n - 1$, then

$$\sum_{m=0}^{\infty} p(5^{2n-1}m + c_{2n-1})q^{m+1} = \frac{L_{2n-1}(q)}{(q^5;q^5)_\infty}$$

$$= \sum_{s=1}^{\infty} b_{2n-1,s}\frac{g^s(q)}{(q^5;q^5)_\infty}. \tag{10.3.19}$$

Now $g^s(q)/(q^5;q^5)_\infty$ is always a power series in q with integral coefficients, and by Theorem 10.4 $5^{2n-1}|b_{2n-1,s}$ for each s. Hence the coefficient of

q^{m+1} on the right-hand side of (10.3.19) is divisible by 5^{2n-1}; therefore

$$5^{2n-1} | p(5^{2n-1}m + c_{2n-1}).$$

The same argument handles the case in which $a = 2n$; the only change is the replacement of $(q^5; q^5)_\infty$ by $(q)_\infty$. ■

COROLLARY 10.6.

$$\sum_{n=0}^{\infty} p(5n + 4)q^n = \frac{5(q^5; q^5)_\infty^5}{(q)_\infty^6}.$$

Proof. This is just a restatement of Eq. (10.3.14). ■

A natural question to ask at this stage is: Are there any combinatorial interpretations for these congruences? Actually the two congruences (10.1.1) and (10.1.2) have combinatorial interpretations; however, none of the other congruences have known interpretations. The following theorem (conjectured by F. J. Dyson and proved by A. O. L. Atkin and H. P. F. Swinnerton-Dyer) provides the known interpretations:

THEOREM 10.7. *Let $R(k, a; n)$ denote the number of partitions of n whose first rank (see Section 9.3) is congruent to a modulo k. Then*

$$R(5, 0; 5m + 4) = R(5, 1; 5m + 4) = R(5, 2; 5m + 4) = R(5, 3; 5m + 4)$$
$$= R(5, 4; 5m + 4);$$
$$R(7, 0; 7m + 5) = R(7, 1; 7m + 5) = R(7, 2; 7m + 5) = R(7, 3; 7m + 5)$$
$$= R(7, 4; 7m + 5) = R(7, 5; 7m + 5) = R(7, 6; 7m + 5).$$

■

The only known proof of this would require too much space for its presentation here. We should point out that it is primarily an analytic proof which relies heavily on the properties of modular functions. No combinatorial proof of Theorem 10.7 is known.

Examples

In Examples 1–6, we consider problems for the number $b(n; m)$ of partitions of n into powers of m. The functions we shall consider are

$$\Phi(m; q) = \sum_{n=0}^{\infty} b(m; n)q^n = \prod_{n=0}^{\infty} (1 - q^{m^n})^{-1},$$

$$\Gamma(m; q) = (1 - q)\Phi(m; q) = \Phi(m; q^m),$$

$$\Psi_r(m; q) = \sum_{n=0}^{\infty} (b(m; m^{r+1}n) - b(m; m^r n))q^n;$$

note

$$\Psi_{r+1}(m; q) = U_m \Psi_r(m; q) \qquad \text{for} \quad r > 0.$$

1. $\Psi_0(m; q) = \dfrac{q\Gamma(m; q)}{(1 - q)^2}$.

2. $\Psi_1(m; q) = \dfrac{m\Gamma(m; q)q}{(1 - q)^3}$.

3. $\Psi_2(m; q) + \dbinom{m}{2}\Psi_1(m; q) = \dfrac{m^3\Gamma(m, q)q}{(1 - q)^4}$.

4. $$\Psi_3(m; q) + \binom{m}{2}(2m + 1)\Psi_2(m; q) + \left(2m\binom{m}{2}^2 - m^3\binom{m}{2}\right.$$

$$\left. + 2m^2\binom{m+1}{3}\right)\Psi_1(m; q) = \frac{m^6\Gamma(m; q)q}{(1 - q)^5}.$$

5. There exist integers $a_i(r)$ $(= a_i(m; r))$ such that

$$\sum_{i=0}^{r-1} a_i(r)\Psi_{r-i}(m; q) = \frac{m^{\binom{r+1}{2}}G(m; q)q}{(1 - q)^{r+2}},$$

where $a_0(r) = 1$, $m | a_1(r)$ if m is odd, $\frac{1}{2}m | a_1(r)$ if m is even, and $m^{2j-2} | a_j(r)$ for $2 \leqslant j \leqslant r - 1$.

6. For each m, we let $\mu = m$ if m is odd and $\mu = \frac{1}{2}m$ if m is even. Then

$$b(m; m^{r+1}n) - b(m; m^r n) \equiv 0 \pmod{\mu^r}.$$

In Examples 7–13, the simplest known proof that $5 | p(5n + 4)$ is sketched.

7. It is a simple matter to deduce from Jacobi's triple product identity (Theorem 2.8) that

$$(q)_\infty^3 = \sum_{m=0}^{\infty} (-1)^m(2m + 1)q^{\frac{1}{2}m(m+1)}.$$

8. For $-\infty < r < \infty$, $s \geqslant 0$, the expression $1 + \frac{1}{2}r(3r + 1) + \frac{1}{2}s(s + 1)$ is divisible by 5 if and only if $r \equiv 4 \pmod 5$ and $s \equiv 2 \pmod 5$.

9. From Examples 7 and 8 and the trivial identity

$$q(q)_\infty^4 = q(q)_\infty^3(q)_\infty,$$

it follows that the coefficient of q^{5m+5} in $q(q)_\infty^4$ is divisible by 5.

10. We may define $f(q) \equiv g(q) \pmod 5$ if $a_n \equiv b_n \pmod 5$ for all n where $f(q) = \sum_{n\geq 0} a_n q^n$ and $g(q) = \sum_{n \geq 0} b_n q^n$. It is easy to show, using the binomial series, that

$$\frac{1-q^5}{(1-q)^5} \equiv 1 \pmod 5.$$

11. From Example 10, we see that

$$\frac{(q^5;q^5)_\infty}{(q)_\infty^5} \equiv 1 \pmod 5.$$

12. Examples 9 and 11 imply that the coefficient of q^{5m+5} in $(q(q^5;q^5)_\infty)/(q)_\infty$ is divisible by 5.

13. From Example 12 it follows that $5|p(5m+4)$ for all $m \geqslant 0$.

Notes

As was mentioned in Section 10.1, the most readable and complete recent work on the divisibility properties of $p(n)$ is the book of M. I. Knopp (1970). Ramanujan's contributions are primarily contained in Paper 25 of his *Collected Papers* (1927). The paper by Atkin (1967) contains a full account of the 11^n case; G. N. Watson (1938) was the first person to treat fully the cases 5^n and 7^n.

Atkin (1969), and others (see Lehner, 1969) have greatly extended the study of arithmetic properties of modular forms; for example, Atkin (1969) has also shown that

$$p(206839n + 2623) \equiv 0 \pmod{17}.$$

The material on binary partitions is due to Rödseth (1970) and Gupta (1971); our treatment is patterned along the lines of Andrews (1971). Theorem 10.2 was originally conjectured by Churchhouse (1969).

Kolberg (1957, 1960) has devised elegant elementary techniques for proving some of the Ramanujan congruences, such as (10.1.1) and (10.1.2); Rademacher (1973) gives a nice exposition of some of Kolberg's work.

Of Corollary 10.6 (due to Ramanujan), G. H. Hardy (Ramanujan, 1927, p. xxxv) wrote: "It would be difficult to find more beautiful formulae than the 'Rogers–Ramanujan' identities...; but here Ramanujan must take second place to Prof. Rogers; and, if I had to select one formula from all Ramanujan's work, I would agree with Major MacMahon in selecting..., viz.

$$p(4) + p(9)x + p(14)x^2 + \cdots = \frac{5\{(1-x^5)(1-x^{10})(1-x^{15})\cdots\}^5}{\{(1-x)(1-x^2)(1-x^3)\cdots\}^6},$$

where $p(n)$ is the number of partitions of n."

Reviews of recent work on congruences for partition functions occur in Section P76 of LeVeque (1974).

Examples 1–6. Churchhouse (1969), Rödseth (1970), Andrews (1971), Gupta (1972).

Examples 7–13. Paper 25 of Ramanujan (1927).

References

Andrews, G. E. (1971). "Congruence properties of the m-ary partition function," *J. Number Theory* **3**, 104–110.

Atkin, A. O. L. (1967). "Proof of a conjecture of Ramanujan," *Glasgow Math. J.* **8**, 14–32.

Atkin, A. O. L. (1969). "Congruence Hecke operators," *Proc. Symp. Pure Math.* **12**, 33–40.

Atkin, A. O. L. and Swinnerton-Dyer, H. P. F. (1953). "Some properties of partitions," *Proc. London Math. Soc.* (3) **4**, 84–106.

Churchhouse, R. F. (1969). "Congruence properties of the binary partition function," *Proc. Cambridge Phil. Soc.* **66**, 371–376.

Dyson, F. J. (1944). "Some guesses in the theory of partitions," *Eureka* (Cambridge) **8**, 10–15.

Gupta, H. (1971). "Proof of the Churchhouse conjecture concerning binary partitions," *Proc. Cambridge Phil. Soc.* **70**, 53–56.

Gupta, H. (1972). "On m-ary partitions," *Proc. Cambridge Phil. Soc.* **71**, 343–345.

Knopp, M. I. (1970). *Modular Functions in Analytic Number Theory*. Markham, Chicago.

Kolberg, O. (1957). "Some identities involving the partition function," *Math. Scand.* **5**, 77–92.

Kolberg, O. (1960). "Congruences involving the partition function for the moduli 17, 19, and 23," *Univ. Bergen Årbok Naturvit. Rekke* 1959, No. 15, 10 pp.

Lehner, J. (1969). *Lectures on Modular Forms* (Appl. Math. Ser. No. 61). Nat. Bur. Standards, Washington, D.C.

LeVeque, W. J. (1974). *Reviews in Number Theory*, Vol. 4. Amer. Math. Soc., Providence, R.I.

Rademacher, H. (1973). *Topics in Analytic Number Theory*. Springer, Berlin.

Ramanujan, S. (1927). *Collected Papers of S. Ramanujan*. Cambridge Univ. Press, London and New York (reprinted by Chelsea, New York).

Rödseth, O. (1970). "Some arithmetical properties of m-ary partitions," *Proc. Cambridge Phil. Soc.* **68**, 447–453.

Watson, G. N. (1938). "Ramanujans Vermutung über Zerfällungsanzahlen," *J. reine angew. Math.* **179**, 97–128.

CHAPTER 11

Higher-Dimensional Partitions

11.1 Introduction

The previous chapters have (with the exception of the multipartite compositions treated in Chapter 4) treated partitions as a linear array whose sum is prescribed:

$$n = n_1 + n_2 + \cdots + n_s = \sum_{i=1}^{s} n_i, \qquad n_i \geqslant n_{i+1}.$$

In this chapter we shall look at *higher-dimensional partitions*, that is, arrays whose sum is n:

$$n = \sum_{i_1,\ldots,i_r \geqslant 0} n_{i_1 i_2 \cdots i_r} \qquad \text{where} \quad n_{i_1 i_2 \cdots i_r} \geqslant n_{j_1 j_2 \cdots j_r} \tag{11.1.1}$$

whenever $i_1 \leqslant j_1, i_2 \leqslant j_2, \ldots, i_r \leqslant j_r$ (all $n_{i_1 i_2 \cdots i_r}$ nonnegative integers).

Surprisingly, there is much of interest when the dimension is 1 or 2, and very little when the dimension exceeds 2. In Section 11.2, we shall follow L. Carlitz's proof of MacMahon's fundamental theorem utilizing determinant representations of the appropriate generating functions; the recurrence method of attack used here also arises in Chapter 12 when we consider vector partitions.

In Section 11.3, we study a fundamental algorithm of Knuth that provides the method of proof for numerous theorems concerning plane partitions.

In Section 11.4, we consider questions concerning higher-dimensional partitions. Unfortunately, the most important results in this section are negative. In particular, the conjectured form for the generating function for higher-dimensional partitions is shown to be invalid.

11.2 Plane Partitions

From the comments in the preceding section we see that the plane partitions of n are two-dimensional arrays of nonnegative integers in the first quadrant

---

ENCYCLOPEDIA OF MATHEMATICS and Its Applications, Gian-Carlo Rota (ed.). 2, George E. Andrews, The Theory of Partitions

subject to a "nonincreasing" condition along rows and columns. For example, there are six plane partitions of 3:

$$\begin{matrix} 0 & 0 & 0 & \dots \\ 0 & 0 & 0 & \dots \\ 3 & 0 & 0 & \dots \end{matrix} \qquad \begin{matrix} 0 & 0 & 0 \\ 0 & 0 & 0 \\ 2 & 1 & 0 \end{matrix} \qquad \begin{matrix} 0 & 0 & 0 \\ 1 & 0 & 0 \\ 2 & 0 & 0 \end{matrix}$$

$$\begin{matrix} 0 & 0 & 0 \\ 0 & 0 & 0 \\ 1 & 1 & 1 \end{matrix} \qquad \begin{matrix} 0 & 0 & 0 \\ 1 & 0 & 0 \\ 1 & 1 & 0 \end{matrix} \qquad \begin{matrix} 1 & 0 & 0 \\ 1 & 0 & 0 \\ 1 & 0 & 0 \end{matrix}$$

Often we omit the zeros and write

$$3, \quad 21, \quad \begin{matrix} 1 \\ 2, \end{matrix} \quad 111, \quad \begin{matrix} 1 \\ 11, \end{matrix} \quad \begin{matrix} 1 \\ 1 \\ 1 \end{matrix}$$

We remark that many times plane partitions are represented in the fourth quadrant rather than the first; thus the plane partitions of 3 are written

$$3, \quad 21, \quad \begin{matrix} 2, \\ 1 \end{matrix} \quad 111, \quad \begin{matrix} 11, \\ 1 \end{matrix} \quad \begin{matrix} 1 \\ 1 \\ 1 \end{matrix}$$

We shall follow the latter custom to keep our results consistent with the current literature; however, we point out that the use of the first quadrant simplifies notation when investigating higher-dimensional partitions.

DEFINITION 11.1. Let $\pi_r(n_1, n_2, \dots, n_k; q)$ denote the generating function for plane partitions with at most r columns, at most k rows, and with n_i the first entry in the ith row.

Our work on $\pi_r(n_1, n_2, \dots, n_k; q)$ will require an extensive use of Gaussian polynomials and we refer the reader to Section 3.3 for the appropriate background material.

We start by noting that the $\pi_r(n_1, \dots, n_k; q)$ are completely determined by the following recurrence and initial condition:

$$\pi_{r+1}(n_1, n_2, \dots, n_k; q) = q^{n_1+n_2+\cdots+n_k} \sum_{m_k=0}^{n_k} \sum_{m_{k-1}=m_k}^{n_{k-1}} \cdots \sum_{m_1=m_2}^{n_1} \pi_r(m_1, \dots, m_k; q), \tag{11.2.1}$$

$$\pi_1(n_1, n_2, \dots, n_k; q) = q^{n_1+n_2+\cdots+n_k}. \tag{11.2.2}$$

Equation (11.2.2) is obvious, and Eq. (11.2.1) is easily seen by examining the array left after the initial column $(n_1, n_2, \dots, n_k)$ has been removed from an $(r+1)$-column array.

We now observe that $\pi_r(n_1,\ldots,n_k;q)$ is easily computed using (11.2.1), (11.2.2). For example

$$\begin{aligned}\pi_2(n,m;q) &= q^{n+m}\sum_{m_1=0}^{m}\sum_{n_1=m_1}^{n} q^{n_1+m_1}\\ &= q^{n+m}\sum_{m_1=0}^{m} q^{m_1}\frac{q^{m_1}-q^{n+1}}{1-q}\\ &= q^{n+m}\left(\frac{1-q^{2m+2}}{1-q^2} - q^{n+1}\frac{1-q^{m+1}}{1-q}\right)(1-q)^{-1}\\ &= q^{n+m}\left[\frac{(1-q^{n+1})(1-q^{m+1})}{(1-q)^2} - q\frac{(1-q^{m+1})(1-q^m)}{(1-q)(1-q^2)}\right]\\ &= q^{n+m}\left(\begin{bmatrix} n+1\\ 1\end{bmatrix}\begin{bmatrix} m+1\\ 1\end{bmatrix} - q\begin{bmatrix} m+1\\ 2\end{bmatrix}\right)\\ &= q^{n+m}\begin{Vmatrix} \begin{bmatrix} n+1\\ 1\end{bmatrix} & q\begin{bmatrix} m+1\\ 2\end{bmatrix}\\ \begin{bmatrix} n+1\\ 0\end{bmatrix} & \begin{bmatrix} m+1\\ 1\end{bmatrix}\end{Vmatrix}. \end{aligned} \tag{11.2.3}$$

We may now apply (11.2.3) to the right-hand side of (11.2.1) with $r = 2$; using (3.3.9) we may easily deduce that

$$\pi_3(n,m;q) = q^{n+m}\begin{Vmatrix} \begin{bmatrix} n+2\\ 2\end{bmatrix} & q\begin{bmatrix} m+2\\ 3\end{bmatrix}\\ \begin{bmatrix} n+2\\ 1\end{bmatrix} & \begin{bmatrix} m+2\\ 2\end{bmatrix}\end{Vmatrix}.$$

From here we immediately conjecture that

$$\pi_r(n,m;q) = q^{n+m}\begin{Vmatrix} \begin{bmatrix} n+r-1\\ r-1\end{bmatrix} & q\begin{bmatrix} m+r-1\\ r\end{bmatrix}\\ \begin{bmatrix} n+r-1\\ r-2\end{bmatrix} & \begin{bmatrix} m+r-1\\ r-1\end{bmatrix}\end{Vmatrix},$$

and we easily prove this by induction on r using (3.3.9).

After a few calculations with $k = 3$, we can make a reasonable guess concerning the representation of $\pi_r(n_1, n_2,\ldots, n_k; q)$ as a determinant:

THEOREM 11.1

$$\pi_r(n_1,n_2,\ldots,n_k;q) = q^{n_1+\cdots+n_k}\det\left(q^{\frac{1}{2}(i-j)(i-j-1)}\begin{bmatrix} n_j+r-1\\ r-i+j-1\end{bmatrix}\right)_{1\leqslant i,j\leqslant k}. \tag{11.2.4}$$

Proof. When $r = 1$, the determinant is upper triangular with ones down the main diagonal. Hence (11.2.4) reduces to the trivial (11.2.2) when $r = 1$.

Assuming the result for a particular r, we proceed using (11.2.1):

$$q^{-n_1-\cdots-n_k}\pi_{r+1}(n_1,\ldots,n_k;q)$$

$$= \sum_{m_k=0}^{n_k}\sum_{m_{k-1}=m_k}^{n_{k-1}}\cdots\sum_{m_1=m_2}^{n_1} q^{m_1+m_2+\cdots+m_k}$$

$$\times \det\left(q^{\frac{1}{2}(i-j)(i-j-1)}\begin{bmatrix} m_j+r-1 \\ r-i+j-1 \end{bmatrix}\right). \qquad (11.2.5)$$

We may now sum the inner sum, using (3.3.9). The result transforms the right-hand side by leaving a $(k-1)$-fold summation and replacing the ith entry in the first column of the determinant in the summand by

$$q^{\frac{1}{2}(i-1)(i-4)}\left(\begin{bmatrix} n_1+r \\ r-i+1 \end{bmatrix} - \begin{bmatrix} m_2+r-1 \\ r-i+1 \end{bmatrix}\right).$$

Now the unaltered ith entry of the second column is

$$q^{\frac{1}{2}(i-2)(i-3)}\begin{bmatrix} m_2+r-1 \\ r-i+1 \end{bmatrix},$$

so if we multiply the second column by q^{-1} and add the result to our first column, we obtain as ith entry for the transformed first column

$$q^{\frac{1}{2}(i-1)(i-4)}\begin{bmatrix} n_1+r \\ r-i+1 \end{bmatrix}.$$

We now sum with respect to m_2, then we multiply the third column by q^{-2} and add the result to our new second column, whose ith entry becomes

$$q^{\frac{1}{2}(i-2)(i-5)}\begin{bmatrix} n_2+r \\ r-i+2 \end{bmatrix}.$$

At the jth step, we sum with respect to m_j; we then multiply the $(j+1)$st column by q^{-j} and add the result to our jth column. The resulting jth column has as ith entry

$$q^{\frac{1}{2}(i-j)(i-j-3)}\begin{bmatrix} n_j+r \\ r-i+j \end{bmatrix}.$$

Hence, after all k summations are completed, we have

$$q^{-n_1-\cdots-n_k}\pi_{r+1}(n_1,\ldots,n_k;q) = \det\left(q^{\frac{1}{2}(i-j)(i-j-3)}\begin{bmatrix} n_j+r \\ r-i+j \end{bmatrix}\right). \qquad (11.2.6)$$

If we multiply the ith row of the determinant in (11.2.6) by q^{i-1} and divide the jth column by q^{j-1} (and as a result do not alter the value of the determinant), we obtain

$$q^{-n_1-\cdots-n_k}\pi_{r+1}(n_1,\ldots,n_k;q) = \det\left(q^{\frac{1}{2}(i-j)(i-j-1)}\begin{bmatrix} n_j + r \\ r - i + j\end{bmatrix}\right), \quad (11.2.7)$$

which is (11.2.4) with r replaced by $r+1$. Hence our theorem is established by mathematical induction. ∎

DEFINITION 11.2. Let $p_{k,r}(m, n)$ denote the number of plane partitions of m with at most r columns, at most k rows, and with each entry $\leqslant n$, and let

$$\pi_{k,r}(n;q) = \sum_{m=0}^{\infty} p_{k,r}(m,n)q^m.$$

We observe immediately that

$$\begin{aligned}\pi_{k,r}(n;q) &= \sum_{n_k\leqslant\cdots\leqslant n_1\leqslant n} \pi_r(n_1,\ldots,n_k;q) \\ &= q^{-kn}\pi_{r+1}(n,\ldots,n;q) \qquad (11.2.8)\end{aligned}$$

(by (11.2.1)). Hence

$$\pi_{k,r}(n;q) = \det\left(q^{\frac{1}{2}(i-j)(i-j-1)}\begin{bmatrix} n + r \\ r - i + j\end{bmatrix}\right)_{1\leqslant i,j\leqslant k}. \quad (11.2.9)$$

It is possible to represent the polynomial $\pi_{k,r}(n;q)$ as a simple quotient of products of expressions $(q)_j$. This is easy to do by direct computation for small k, and the results lead us to conjecture:

THEOREM 11.2

$$\pi_{k,r}(n;q) = \frac{(q)_1(q)_2\cdots(q)_{k-1}}{(q)_r(q)_{r+1}\cdots(q)_{r+k-1}}\cdot\frac{(q)_{n+r}(q)_{n+r+1}\cdots(q)_{n+r+k-1}}{(q)_n(q)_{n+1}\cdots(q)_{n+k-1}}. \quad (11.2.10)$$

Proof. We proceed by using another ingenious device of L. Carlitz. Let

$$W(k,r) = \det\left(q^{ri+\frac{1}{2}i(i-1)}\begin{bmatrix} j \\ i\end{bmatrix}\right)_{0\leqslant i,j\leqslant k-1}.$$

Hence by the rule for multiplication of determinants

$$\pi_{k,r}(n;q)W(k,r) = \det(c_{ij})_{0\leqslant i,j\leqslant k-1},$$

where

$$c_{ij} = \sum_{s=0}^{j} q^{\frac{1}{2}s(s-1)} \begin{bmatrix} j \\ s \end{bmatrix} q^{rs+\frac{1}{2}(i-s)(i-s-1)} \begin{bmatrix} n+r \\ n+i-s \end{bmatrix}$$

$$= q^{\frac{1}{2}i(i-1)} \begin{bmatrix} n+r+j \\ n+i \end{bmatrix}$$

(by (3.3.10)). Therefore

$$\pi_{k,r}(n;q)W(k,r) = \frac{(q)_{n+r}(q)_{n+r+1}\cdots(q)_{n+r+k-1}}{(q)_n(q)_{n+1}\cdots(q)_{n+k-1}} \det\left(\frac{q^{\frac{1}{2}i(i-1)}}{(q)_{r-i+j}}\right)$$

$$= \frac{(q)_{n+r}(q)_{n+r+1}\cdots(q)_{n+r+k-1}}{(q)_n(q)_{n+1}\cdots(q)_{n+k-1}} C(k,r) \qquad (11.2.11)$$

where neither $W(k, r)$ (which is upper triangular with no zeros on the main diagonal) nor $C(k, r)$ depends on n. We may therefore determine $C(k, r)/W(k, r)$ by setting $n = 0$ in (11.2.11). Thus

$$\frac{C(k,r)}{W(k,r)} = \frac{(q)_1(q)_2\cdots(q)_{k-1}}{(q)_r(q)_{r+1}\cdots(q)_{r+k-1}}, \qquad (11.2.12)$$

and therefore

$$\pi_{k,r}(n;q) = \frac{(q)_1(q)_2\cdots(q)_{k-1}}{(q)_r(q)_{r+1}\cdots(q)_{r+k-1}} \cdot \frac{(q)_{n+r}(q)_{n+r+1}\cdots(q)_{n+r+k-1}}{(q)_n(q)_{n+1}\cdots(q)_{n+k-1}}, \qquad (11.2.13)$$

as desired. ∎

Theorem 11.2 directly implies MacMahon's famous formulas for the generating function of k-rowed plane partitions $\pi_{k,\infty}(\infty; q)$:

COROLLARY 11.3. *For* $|q| < 1$,

$$\sum_{m=0} p_{k,\infty}(m,\infty)q^m = \prod_{j=1}^{\infty}(1-q^j)^{-\min(k,j)}; \qquad (11.2.14)$$

$$\sum_{m=0} p_{\infty,\infty}(m,\infty)q^m = \prod_{j=1}^{\infty}(1-q^j)^{-j}. \qquad (11.2.15)$$

Proof. Equation (11.2.14) follows from Theorem 11.2 by letting r and $n \to \infty$. Equation (11.2.15) follows from (11.2.14) by letting $k \to \infty$ in (11.2.14). ∎

11.3 The Knuth-Schensted Correspondence

In the 1960s a correspondence between certain matrices and plane partitions was developed by Schensted and later extended by Knuth. This combinatorial mapping allowed a great simplification in the deduction of many known restricted plane partition generating functions. Furthermore, many new generating functions can be treated using this elegant procedure.

THEOREM 11.4. *There is a one-to-one correspondence between*

(i) *the set of $k \times k$ matrices A with nonnegative integer entries;*

(ii) *the set of all lexicographically ordered sequences of ordered pairs of integers, each $\leqslant k$;*

(iii) *the set of ordered pairs (π, π') of plane partitions in which there is strict decrease along columns, each entry does not exceed k, and the corresponding rows of π and π' are of the same length.*

Furthermore the desired correspondence is such that the ith column sum (resp. *ith row sum*) *of a matrix A from* (i) *equals the number of times i appears in the corresponding plane partition π* (resp. π') *in* (iii).

Proof. The correspondence between (i) and (ii) is simple; to each sequence described in (ii) associate the incidence matrix $A = (a_{ij})$, that is, a_{ij} is the number of times the pair (i, j) (written for our purposes as $\binom{i}{j}$) occurs in the corresponding sequence of (ii). For example

$$\begin{pmatrix} 3 & 2 & 2 \\ 1 & 1 & 1 \\ 2 & 3 & 1 \end{pmatrix} \leftrightarrow \begin{matrix} 3 & 3 & 3 & 3 & 3 & 3 & 2 & 2 & 2 & 1 & 1 & 1 & 1 & 1 & 1 & 1 \\ 3 & 2 & 2 & 2 & 1 & 1 & 3 & 2 & 1 & 3 & 3 & 2 & 2 & 1 & 1 & 1 \end{matrix} \qquad (11.3.1)$$

We now produce the ordered pair of plane partitions (π, π'); the entries in π arise from the second elements (or bottom row) in the sequence of ordered pairs; those in π' arise from the first elements (or top row). Our procedure is an algorithm that describes how to build up π and π' step by step from the sequence of pairs in (ii).

The second elements j from each term $\binom{i}{j}$ of our sequence are to be inserted in their appropriate place in the top row of π (i.e., so that nonincrease along rows is maintained). If the appropriate place of insertion for j already contains a smaller integer j_1, then j takes the place of j_1 and the smaller integer j_1 is inserted in the second row. If the appropriate place in the second row for j_1 is taken by j_2, then j_2 is inserted in the third row, and so on. Once the sequence of *bumpings* caused by the insertion of j is concluded, then i (the first element of the pair $\binom{i}{j}$) is inserted into π' in such a way that the corresponding rows of π and π' are always of the same length.

To obtain a feel for the mechanics of the foregoing procedure, we produce step by step the (π, π') corresponding to (11.3.1) (the circled number is the new insertion, the underlined numbers are ones "bumped" by the new insertion).

③

3 ②

3 2 ②

3 2 2 ②

3 2 2 2 ①

3 2 2 2 1 ①

3 ③ 2 2 1 1
$\underline{2}$

3 3 2 2 ② 1
2 $\underline{1}$

3 3 2 2 2 1 ①
2 1

3 3 ③ 2 2 1 1
2 $\underline{2}$
$\underline{1}$

3 3 3 ③ 2 1 1
2 2 $\underline{2}$
1

3 3 3 3 2 ② 1
2 2 2 $\underline{1}$
1

3 3 3 3 2 2 ②
2 2 2 1 $\underline{1}$
1

3 3 3 3 2 2 2 ①
2 2 2 1 1
1

3 3 3 3 2 2 2 1 ①
2 2 2 1 1
1

3 3 3 3 2 2 2 1 1 ①
2 2 2 1 1
1

③

3 ③

3 3 ③

3 3 3 ③

3 3 3 3 ③

3 3 3 3 3 ③

3 3 3 3 3 3
②

3 3 3 3 3 3
2 ②

3 3 3 3 3 3 ②
2 2

3 3 3 3 3 3 2
2 2
①

3 3 3 3 3 3 2
2 2 ①
1

3 3 3 3 3 3 2
2 2 1 ①
1

3 3 3 3 3 3 2
2 2 1 1 ①
1

3 3 3 3 3 3 2 ①
2 2 1 1 1
1

3 3 3 3 3 3 2 1 ①
2 2 1 1 1
1

3 3 3 3 3 3 2 1 1 ①
2 2 1 1 1
1

Inspection of the foregoing procedure shows immediately that the π and π' have corresponding rows of equal length and the correspondence between the appearances of i in π (resp. π') and the ith column (resp. row) sum is clear. Furthermore (using mathematical induction on the number of steps completed in the procedure), we see that bumping can only result in entries moving downward and not to the right, and thus strict decrease along columns is maintained in π at each step. Since insertions into π' are successively made from a nonincreasing sequence, we see that nonincrease must hold along rows and columns; moreover, strict decrease must obtain in the columns of π' since if $\genfrac{}{}{0pt}{}{i}{j_1}\ \genfrac{}{}{0pt}{}{i}{j_2}$ are two successive terms in our sequence, the number of bumpings induced by j_1 is greater than or equal to the number induced by j_2 (hence, insertions of i move successively to the right).

The main question, however, is: Does the foregoing procedure establish a one-to-one correspondence between the elements defined in (ii) and those defined in (iii)? Let us show that our procedure is uniquely reversible and consequently does establish a one-to-one correspondence. First we point out that the remark made at the end of the preceding paragraph allows us to pick out the last insertion made in π' (it is the rightmost appearance of the smallest integer appearing). If this integer, say i, occurs on the first row of π', then the corresponding entry in π is the last insertion in π. If i occurs on some other row, then the corresponding entry in π was the end of a bumping chain and we can clearly reverse the bumping process to determine what entry was inserted in π. Thus the procedure yields a bijection, and Theorem 11.4 is established. ■

Theorem 11.5. *If the matrix A in* (i) *of Theorem* 11.4 *corresponds to* (π, π') *in* (iii), *then the transposed matrix A^T corresponds to* (π', π).

Proof. Let us define the *class* of $\genfrac{}{}{0pt}{}{i}{j}$ in the sequence of pairs corresponding to A as the position in the first row of π where j is initially inserted. Returning to our example (11.3.1), we may list the classes of each pair directly below the pair as follows:

$$\begin{array}{rcccccccccccccccc} & 3 & 3 & 3 & 3 & 3 & 3 & 2 & 2 & 2 & 1 & 1 & 1 & 1 & 1 & 1 & 1 \\ & 3 & 2 & 2 & 2 & 1 & 1 & 3 & 2 & 1 & 3 & 3 & 2 & 2 & 1 & 1 & 1 \\ \hline \text{class} & 1 & 2 & 3 & 4 & 5 & 6 & 2 & 5 & 7 & 3 & 4 & 6 & 7 & 8 & 9 & 10 \end{array}$$

Now the entries in the first row of π are made up in order of the last entries in each class: namely, 3 3 3 3 2 2 2 1 1 1.

The important thing about class membership is that it is possible to define it in a manner that does *not* depend on the lexicographic ordering of the sequence. Namely, $\genfrac{}{}{0pt}{}{i}{j}$ is in class t if and only if t is the largest subset of pairs

$$i_1 i_2 \cdots i_{t-1} i \qquad\qquad j_1 \geqslant j_2 \geqslant \cdots \geqslant j_{t-1} \geqslant j,$$
$$\text{such that}$$
$$j_1 j_2 \cdots j_{t-1} j \qquad\qquad i_1 \geqslant i_2 \geqslant \cdots \geqslant i_{t-1} \geqslant i.$$

Thus from this last observation we see that the class of ${}^{i}_{j}$ in our sequence is the same as that of ${}^{j}_{i}$ in the sequence of reversed ordered pairs, which clearly corresponds to A^T. Hence the first row of the first plane partition corresponding to A^T is the first row of π' and the first row of the second plane partition corresponding to A^T is the first row of π.

Deleting from the ordered pairs sequence the entries that produce the first rows of the plane partitions, we may follow exactly the same procedure to establish the interchange of the second rows, and so on. Thus we see that (π', π) corresponds to A^T. ∎

COROLLARY 11.6. *The correspondence described in Theorem 11.4 produces a one-to-one correspondence between symmetric matrices A of nonnegative integers and plane partitions π with strict decrease along columns. Furthermore, the number of appearances of i in π equals the ith row sum (or ith column sum) of A.*

Proof. We invoke Theorem 11.5, noting that A is symmetric if and only if $A = A^T$. Hence, $A \leftrightarrow (\pi, \pi)$. ∎

This corollary has many interesting applications; we present one of the most striking next.

THEOREM 11.7. *Let S be a set of positive integers. The number of plane partitions of n with strict decrease along columns and with each summand in S is the coefficient of q^n in*

$$\prod_{i \in S} (1 - q^i)^{-1} \prod_{\substack{i,j \in S \\ i<j}} (1 - q^{i+j})^{-1}.$$

Proof. By Corollary 11.6, the desired generating function is

$$\sum_A \prod_{i,j \geqslant 1} q^{i a_{ij}}$$

where the sum is over all symmetric matrices A of nonnegative integers subject to the condition that $a_{ij} = 0$ if $i \notin S$ or $j \notin S$. Hence

$$\sum_A \prod_{i,j \geqslant 1} q^{i a_{ij}} = \sum_A \left(\prod_{i=1}^{\infty} q^{i a_{ii}} \right) \left(\prod_{\substack{i<j \\ i,j \in S}} q^{i a_{ij} + j a_{ji}} \right)$$

$$= \left(\prod_{i \in S} \sum_{a_{ii}=0}^{\infty} q^{ia_{ii}} \right) \left(\prod_{\substack{i<j \\ i,j \in S}} \sum_{a_{ij}=0}^{\infty} q^{(i+j)a_{ij}} \right)$$

$$= \prod_{i \in S} (1 - q^i)^{-1} \prod_{\substack{i<j \\ i,j \in S}} (1 - q^{i+j})^{-1}. \quad \blacksquare$$

In the examples there is discussed the fact that certain two-variable generating functions arise in a very natural way relative to plane partitions through Corollary 11.6. For example, the coefficient of $a^m q^n$ in

$$\prod_{n=1}^{\infty} (1 - aq^n)^{-n}$$

is the number of plane partitions of n for which the sum of the diagonal parts is m.

11.4 Higher-Dimensional Partitions

P. A. MacMahon at one time conjectured that if $M_k(n)$ denotes the number of k-dimensional partitions of n (see the definition in Section 11.1), then

$$\sum_{n=0}^{\infty} M_k(n) q^n = \prod_{i=1}^{\infty} (1 - q^i)^{-\binom{k+i-2}{k-1}}. \tag{11.4.1}$$

The conjecture is certainly true for $k = 1$ and 2. MacMahon eventually came to doubt the truth of (11.4.1) in general; however, its falsehood for all $k > 2$ was only established in the late 1960s. The following theorem provides a simple (though laborious) computational method for disproving MacMahon's conjecture; this method is easily extended to any problem involving k-dimensional partitions of small integers.

THEOREM 11.8. *Let*

$$\sum_{n=0}^{\infty} \mu_k(n) q^n = \prod_{i=1}^{\infty} (1 - q^i)^{-\binom{k+i-2}{k-1}}. \tag{11.4.2}$$

Then

$$\mu_k(0) = M_k(0) = 1, \tag{11.4.3}$$

$$\mu_k(1) = M_k(1) = 1, \tag{11.4.4}$$

$$\mu_k(2) = M_k(2) = k + 1, \tag{11.4.5}$$

$$\mu_k(3) = M_k(3) = 1 + 2k + \binom{k}{2}, \tag{11.4.6}$$

$$\mu_k(4) = M_k(4) = 1 + 4k + 4\binom{k}{2} + \binom{k}{3}, \tag{11.4.7}$$

$$\mu_k(5) = M_k(5) = 1 + 6k + 11\binom{k}{2} + 7\binom{k}{3} + \binom{k}{4}, \tag{11.4.8}$$

$$\mu_k(6) = M_k(6) + \binom{k}{4} + \binom{k}{3} = 1 + 10k + 27\binom{k}{2} + 29\binom{k}{3} + 12\binom{k}{4} + \binom{k}{5}. \tag{11.4.9}$$

Proof. We begin by treating $\mu_k(n)$:

$$\begin{aligned}
\sum_{n=0}^{\infty} \mu_k(n)q^n &= (1-q)^{-1}(1-q^2)^{-k}(1-q^3)^{-\binom{k+1}{2}} \\
&\quad \times (1-q^4)^{-\binom{k+2}{3}}(1-q^5)^{-\binom{k+3}{4}}(1-q^6)^{-\binom{k+4}{5}}\cdots \\
&= (1 + q + q^2 + q^3 + q^4 + q^5 + q^6 + \cdots) \\
&\quad \times \left(1 + kq^2 + \binom{k+1}{2}q^4 + \binom{k+2}{3}q^6 + \cdots\right) \\
&\quad \times \left(1 + \binom{k+1}{2}q^3 + \frac{1}{2}\binom{k+1}{2}\left(\binom{k+1}{2} + 1\right)q^6 + \cdots\right) \\
&\quad \times \left(1 + \binom{k+2}{3}q^4 + \cdots\right) \\
&\quad \times \left(1 + \binom{k+3}{4}q^5 + \cdots\right) \\
&\quad \times \left(1 + \binom{k+4}{5}q^6 + \cdots\right) \\
&\quad \vdots \\
&= 1 + q + (k+1)q^2 + \left(1 + 2k + \binom{k}{2}\right)q^3 \\
&\quad + \left(1 + 4k + 4\binom{k}{2} + \binom{k}{3}\right)q^4 \\
&\quad + \left(1 + 6k + 11\binom{k}{2} + 7\binom{k}{3} + \binom{k}{4}\right)q^5 \\
&\quad + \left(1 + 10k + 27\binom{k}{2} + 29\binom{k}{3} + 12\binom{k}{4} + \binom{k}{5}\right)q^6 \\
&\quad + \cdots,
\end{aligned} \tag{11.4.10}$$

Table 11.1

Ordinary partition of 6	Arrangements in higher-dimensional space	Total choices for placing arrangement
6	6 at origin	1
5 + 1	5—1→x_j-axis	k
4 + 2	4—2→x_j-axis	k
4 + 1 + 1	4—1—1→x_j-axis	k
	4—1→x_j-axis; 4—1→x_ℓ-axis	$\binom{k}{2}$
3 + 3	3—3→x_j-axis	k
3 + 2 + 1	3—2—1→x_j-axis	k
	3—2→x_j-axis; 3—1→x_ℓ-axis	$2\binom{k}{2}$
3 + 1 + 1 + 1	3—1—1—1→x_j-axis	k

Table 11.1 (continued)

Ordinary partition of 6	Arrangements in higher-dimensional space	Total choices for placing arrangement
	3 — 1 — 1 → x_j-axis; 3 — 1 → x_ℓ-axis	$2\binom{k}{2}$
	3 — 1 → x_j-axis; 3 — 1 → x_ℓ-axis; 3 — 1 → x_h-axis	$\binom{k}{3}$
	3 — 1 → x_j-axis; 3 — 1 → x_ℓ-axis; 1	$\binom{k}{2}$
2 + 2 + 2	2 — 2 — 2 → x_j-axis	k
	2 — 2 → x_j-axis; 2 — 2 → x_ℓ-axis	$\binom{k}{2}$
2 + 2 + 1 + 1	2 — 2 — 1 — 1 → x_j-axis	k
	2 — 2 — 1 → x_j-axis; 2 — 1 → x_ℓ-axis	$2\binom{k}{2}$
	2 — 1 — 1 → x_j-axis; 2 — 2 → x_ℓ-axis	$2\binom{k}{2}$

Table 11.1 (continued)

Ordinary partition of 6	Arrangements in higher-dimensional space	Total choices for placing arrangement
	2 — 2 → x_j-axis; 1 → x_ℓ-axis; 1 → x_h-axis	$3\binom{k}{3}$
	2 — 2 → x_j-axis; 2, 1 — 1; 1 → x_ℓ-axis	$2\binom{k}{2}$
$2+1+1+1+1$	2 — 1 — 1 — 1 — 1 → x_j-axis	k
	2 — 1 — 1 — 1 → x_j-axis; 1 → x_ℓ-axis	$2\binom{k}{2}$
	2 — 1 — 1 → x_j-axis; 1 — 1 → x_ℓ-axis	$\binom{k}{2}$
	2 — 1 — 1 → x_j-axis; 1 → x_ℓ-axis; 1 → x_h-axis	$3\binom{k}{3}$

Table 11.1 (continued)

Ordinary partition of 6	Arrangements in higher-dimensional space	Total choices for placing arrangement
	2, 1, 1, 1, 1, 1; x_j -axis, x_ℓ -axis, x_h -axis, x_i -axis	$\binom{k}{4}$
	2, 1, 1, 1, 1, 1; x_j -axis, x_ℓ -axis	$2\binom{k}{2}$
	2, 1, 1, 1, 1, 1; x_j-axis, x_h-axis, x_ℓ-axis	$3\binom{k}{3}$
$1+1+1+1+1+1+1$		
	1, 1, 1, 1, 1, 1; x_j -axis	k
	1, 1, 1, 1, 1, 1; x_j -axis, x_ℓ -axis	$2\binom{k}{2}$

Table 11.1 (continued)

Ordinary partition of 6	Arrangements in higher-dimensional space	Total choices for placing arrangement
	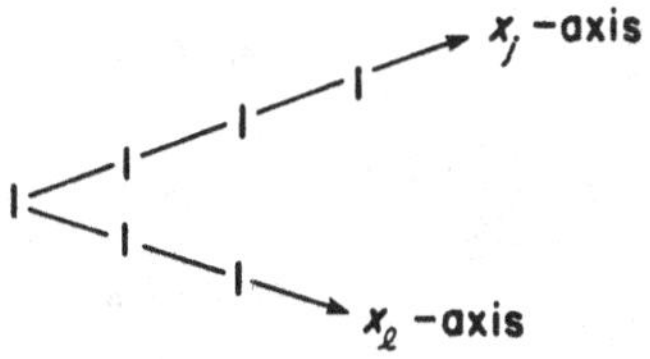 x_j-axis, x_ℓ-axis	$2\binom{k}{2}$
	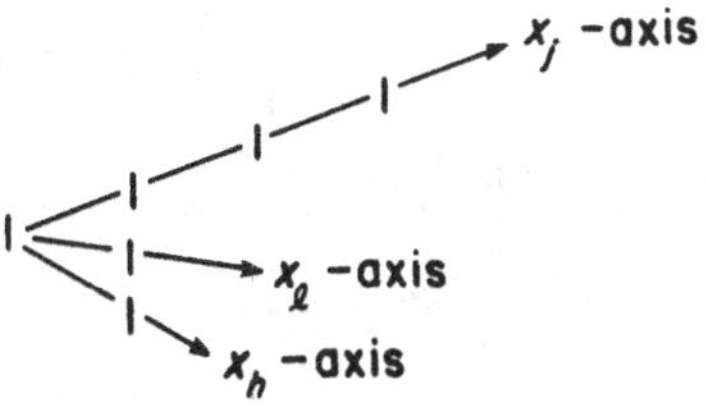 x_j-axis, x_ℓ-axis, x_h-axis	$3\binom{k}{3}$
	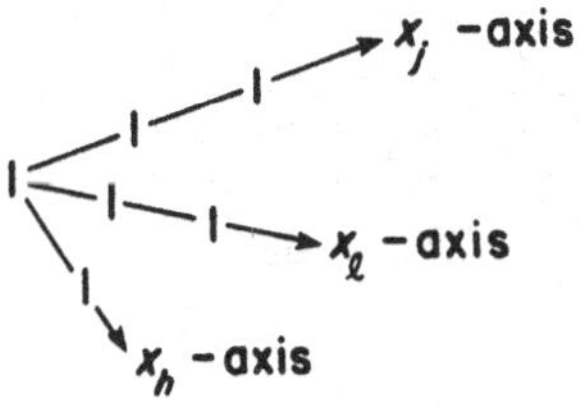 x_j-axis, x_ℓ-axis, x_h-axis	$3\binom{k}{3}$
	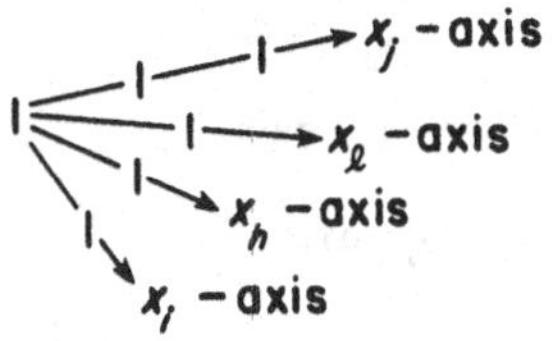 x_j-axis, x_ℓ-axis, x_h-axis, x_i-axis	$4\binom{k}{4}$
	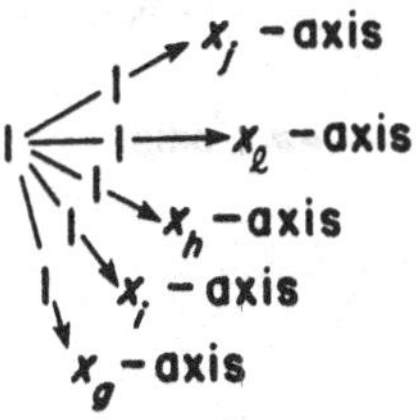 x_j-axis, x_ℓ-axis, x_h-axis, x_i-axis, x_g-axis	$\binom{k}{5}$
	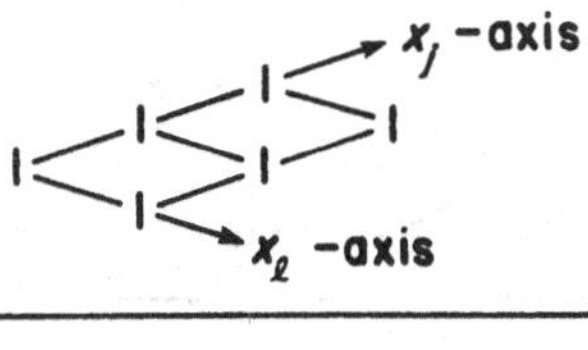 x_j-axis, x_ℓ-axis	$2\binom{k}{2}$

Table 11.1 (continued)

Ordinary partition of 6	Arrangements in higher-dimensional space	Total choices for placing arrangement
	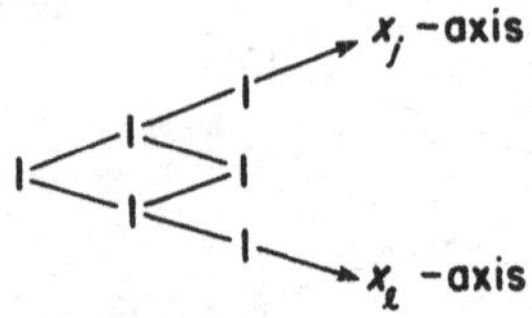 x_j -axis x_ℓ -axis	$\binom{k}{2}$
	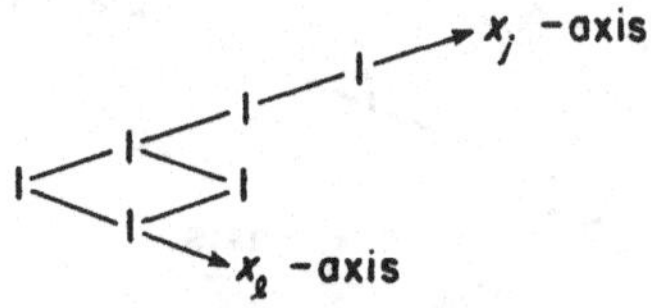 x_j -axis x_ℓ -axis	$2\binom{k}{2}$
	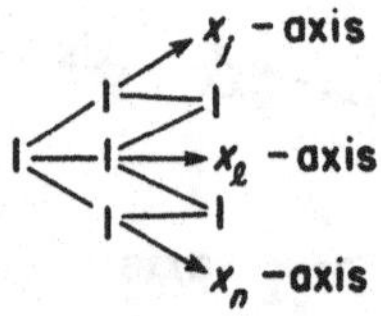 x_j -axis x_ℓ -axis x_n -axis	$3\binom{k}{3}$
	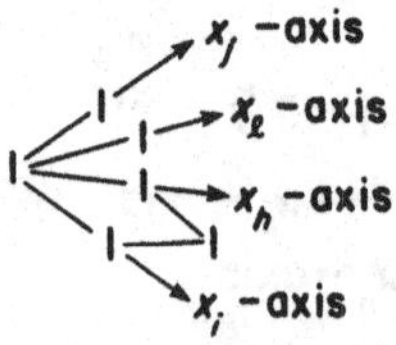 x_j -axis x_ℓ -axis x_h -axis x_i -axis	$6\binom{k}{4}$
	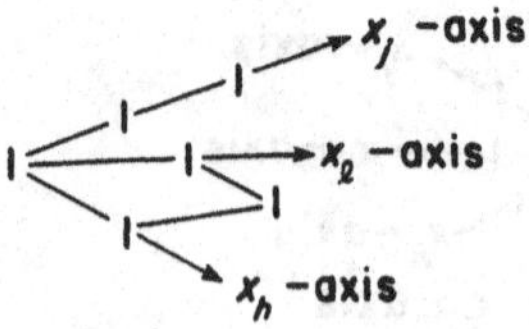 x_j -axis x_ℓ -axis x_h -axis	$3\binom{k}{3}$
	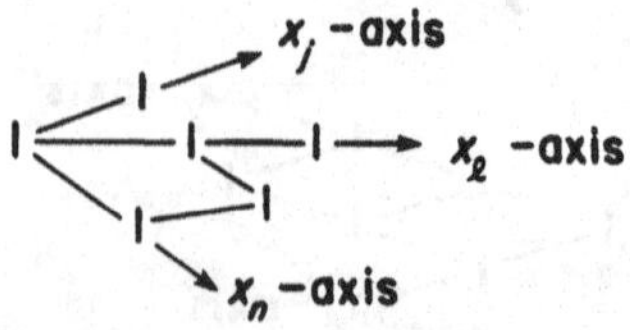 x_j -axis x_ℓ -axis x_n -axis	$6\binom{k}{3}$

and we may now determine $\mu_k(n)$ for $0 \leqslant n \leqslant 6$ by comparing coefficients in the extremes of (11.4.10).

To compute $M_k(n)$ we observe that each k-dimensional partition is made up of an ordinary (one-dimensional) partition of n appropriately distributed in the first quadrant of k-dimensional space. Thus to evaluate $M_k(2)$ we see that the ordinary partition 2 can be placed in k-dimensional space in only one way; $1 + 1$, on the other hand, can be distributed in k ways: one 1 is placed at the origin and the second is placed on any of the k-coordinate axes. Hence, $M_k(2) = 1 + k$. The remaining six expressions for $M_k(n)$ are derived in the same way. Since $M_k(6)$ is crucial for Corollary 11.9, we evaluate it using Table 11.1, which is self-explanatory. To obtain all possible k-dimensional partitions of 6, we add up all the entries in the right-hand column:

$$M_k(6) = 1 + 10k + 27\binom{k}{2} + 28\binom{k}{3} + 11\binom{k}{4} + \binom{k}{5},$$

which is the result asserted in (11.4.9). ∎

Corollary 11.9. *Equation (11.4.1) is valid for $k = 1$ or 2 and false for all $k > 2$.*

Proof. The case in which $k = 1$ is a special case of Theorem 1.1. The case in which $k = 2$ is Eq. (11.2.15). Equation (11.4.1) is equivalent to the assertion that $\mu_k(n) = M_k(n)$ for all n; however

$$\mu_k(6) - M_k(6) = \binom{k}{3} + \binom{k}{4} > 0 \quad \text{for} \quad k \geqslant 3.$$

Hence (11.4.1) is false for $k > 2$. ∎

Given the disappointing nature of the foregoing results, we can well expect disappointment on other questions concerning higher-dimensional partitions.

Observing that $M_2(3) = 6 = 3\cdot 2$, $M_2(6) = 48 = 3 \times 16$, $M_2(9) = 282 = 3 \times 94$, $M_2(12) = 1479 = 3 \times 493$, $M_2(15) = 6879 = 3 \times 2293$, we might be tempted to conjecture that $3|M_2(3n)$ for all n. Our hopes are dashed, however, by the following:

Theorem 11.10. *If $k + 1$ is a prime, then*

$$M_k((k+1)n) \equiv 0 \pmod{k+1}, \qquad 1 \leqslant n < (k+1)^k, \quad (11.4.11)$$

$$M_k((k+1)^{k+1}) \equiv 1 \pmod{k+1}. \quad (11.4.12)$$

Proof. We may represent each k-dimensional partition by a $(k+1)$-dimensional graphical representation or Ferrers graph (merely place a column of $n_{j_1 j_2 \cdots j_k}$ nodes on the line parallel to the x_{k+1} axis passing through the point $(j_1, j_2, \ldots, j_k, 0)$. The group G of $k + 1$ transformations $T^0, T^1, T^2, \ldots, T^k$ where

$$T: (x_1, x_2, \ldots, x_{k+1}) \to (x_{k+1}, x_1, x_2, \ldots, x_k)$$

always either produces from a given Ferrers graph $k + 1$ different Ferrers graphs or else leaves the initial Ferrers graph fixed. This is because G is a cyclic group of prime order and so any nonidentity in G generates the full group. If a Ferrers graph is left fixed by G, then since the only fixed points under any $T^i \in G$ are points on the diagonal $(x, x, \ldots, x)$, we see that the first Ferrers graph that partitions a multiple of $k + 1$ and is invariant under G is just the $(k + 1)$-dimensional hypercube with $k + 1$ nodes on each edge. Therefore, the k-dimensional partitions of $(k + 1)n$ can be divided into disjoint sets of $(k + 1)$ elements each (namely, the orbits of G) whenever $0 < n < (k + 1)^k$. Thus (11.4.11) is valid.

On the other hand, the only $(k + 1)$-dimensional Ferrers graph of $(k + 1)^{k+1}$ that has $k + 1$ nodes on the diagonal is the hypercube just described. Since all the other k-dimensional partitions of $(k + 1)^{k+1}$ may be separated into sets of $k + 1$ elements each, we see that

$$M_k((k + 1)^{k+1}) \equiv 1 \pmod{k + 1}. \qquad \blacksquare$$

Examples

1. An examination of the structure of the successive ranks of a partition (see Section 9.3) shows that there is a one-to-one correspondence between each ordinary partition $\pi = (\lambda_1, \ldots, \lambda_k)$ of n and ordered pairs of partitions (π_1, π_2) where $\pi_1 = (\lambda_1', \lambda_2', \ldots, \lambda_r')$, $\pi_2 = (\lambda_1'', \ldots, \lambda_r'')$, $\lambda_1 = \lambda_1'$, $k - 1 = \lambda_1''$, $\sum (\lambda_i' + \lambda_i'') = n$, $\lambda_i' - \lambda_i''$ is the ith successive rank of π.

2. The correspondence in Example 1 may be utilized to establish a one-to-one correspondence between plane partitions π of n with at most r rows and each part $\leqslant m$ and ordered pairs (π', π'') of plane partitions, each with strict decrease along columns such that each part of π' is at most m and each part of π'' is at most r.

3. Example 2 and Theorem 11.4 may be used to prove Corollary 11.3 by the use of the technique that appears in Theorem 11.7.

4. The Ferrers graphs of plane partitions (a concept introduced in Section 11.4) may be used to define six conjugates of a plane partition π (these arise from the 3! possible permutations of the coordinate axes related to the three-dimensional Ferrers graph of π). We consider one of these conjugations, say τ, for which $\tau\pi$ is the plane partition in which each row of $\tau\pi$ is the conjugate (in the ordinary partition sense) of the corresponding row of π. The *trace* of a plane partition π is the sum $\sum n_{ii}$ of the diagonal entries in π; the *conjugate trace* of π is the number of entries n_{ij} of π such that $n_{ij} \geqslant i$. The trace of π and the conjugate trace of $\tau\pi$ are identical.

5. The proof in Example 3 may be extended and combined with Example 4 to show that the coefficient of $z^m q^n$ in

$$\prod_{n\geqslant 1}(1 - zq^n)^{-n}$$

is the number of plane partitions of n whose trace is m.

6. The asymptotic work of Chapter 6, namely, Theorem 6.2, is applicable to many of the plane partition generating functions. In particular, the number of plane partitions of n, $p_{\infty,\infty}(n, \infty)$, satisfies

$$p_{\infty,\infty}(n, \infty) \sim (\zeta(3)2^{-11})^{1/36}n^{-25/36}\exp(3\cdot 2^{-2/3}\zeta(3)^{1/3}n^{2/3} + 2c)$$

where

$$\zeta(s) = \sum_{n\geqslant 1} n^{-s}, \qquad c = \int_0^\infty \frac{y \log y}{e^{2\pi y} - 1}\,dy.$$

The relevant functions for Theorem 6.2 are $D(s) = \zeta(s - 1)$, $g(\tau) = e^{-\tau}/((1 - e^{-\tau})^2)$.

Similar results can be obtained for numerous special cases of Theorem 11.7.

7*. Let us consider those r-dimensional partitions whose nonzero entries $n_{i_1 i_2 \cdots i_r}$ occur precisely at the points $(i_1, i_2, \ldots, i_r)$ that define the graphical representation of a fixed $(r - 1)$-dimensional partition π of N. Let $a_m(s_1, \ldots, s_v)$ denote the number of r-dimensional partitions of m with strict decrease holding in Eq. (11.1.1) whenever any of $i_{s_1} < j_{s_1}, i_{s_2} < j_{s_2}, \ldots, i_{s_v} < j_{s_v}$ holds, where $\{s_1, s_2, \ldots, s_v\}$ is a fixed subset of $\{1, 2, \ldots, r\}$. Let $\{t_1, t_2, \ldots, t_{r-v}\}$ be the complement of $\{s_1, s_2, \ldots, s_v\}$ in $\{1, 2, \ldots, r\}$. If

$$A(q) = \sum_{m\geqslant 0} a_m(s_1, s_2, \ldots, s_r)q^m,$$

and

$$B_0(q) = \sum_{m\geqslant 0} a_m(t_1, t_2, \ldots, t_{r-v})q^m,$$

then

$$B_0(q) = (-1)^{p(\pi)}q^{p(\pi)}A(1/q)$$

where $p(\pi)$ is the number of $(r - 1)$-dimensional partitions whose graphical representation is contained within the graphical representation of π.

Notes

This chapter has only briefly touched a very active area of research. The topic of higher-dimensional partitions, as such, originated with MacMahon (1916). However, Young tableaux (which are essentially equivalent to plane partitions with strict decrease along columns) were originated earlier by Rev. Alfred Young in his work on invariant theory. Since then, Young tableaux have played an important role in the representation theory of the symmetric group (see Rutherford, 1948); they also occur in algebraic geometry (see Lascoux, 1974a, b), and in many combinatorial problems (Kreweras,

1965, 1967). Stanley (1971) presents an extensive and readable account of current work. The material in Section 11.2 is from Carlitz (1967); his techniques were extended by Knuth and Bender (1972) to other plane partition problems. Hodge and Pedoe (1952) considered the case in which $q = 1$ for a problem in algebraic geometry. The material in Section 11.3 was initially considered by Robinson (1938) and Schensted (1961) (see also Schützenberger, 1963); we have presented Knuth's (1970) extension of the Schensted construction. The application to plane partitions is given by Knuth and Bender (1972). Recent research on plane partitions was surely inspired by the pathbreaking papers of B. Gordon and L. Houten (1968) and B. Gordon (1971).

Theorem 11.8 and its corollary were given by Atkin et al. (1967). Tietze (1940, 1941) presented all the formulas for $M_k(j)$, $1 \leqslant j \leqslant 5$; however, he was apparently unaware of MacMahon's conjecture. Wright (1968) has developed another technique for the disproof of many instances of MacMahon's conjecture. Theorem 11.10 is due to Andrews (1971); the technique arises from Wright (1968). The possibility of extending Rogers–Ramanujan type identities to plane partitions was considered briefly by Chaundy (1931); however, Gordon (1962) points out that Chaundy's derivation is in error.

Reviews of recent work related to the material in this chapter occur in LeVeque (1974), Section P64.

Examples 1–3. Knuth and Bender (1972).

Examples 4–5. Stanley (1973).

Example 6. Wright (1931).

Example 7. Stanley (1972); Stanley's work on (P, ω)-partitions allows him to deduce the result in this example easily from a general reciprocity theorem.

References

Andrews, G. E. (1971). "On a conjecture of Guinand for the plane partition function," *Proc. Edinburgh Math. Soc.* (2) **17**, 275–276.

Atkin, A. O. L., Bratley, P., MacDonald, I. G., and McKay, J. K. S. (1967). "Some computations for m-dimensional partitions," *Proc. Cambridge Phil. Soc.* **63**, 1097–1100.

Carlitz, L. (1967). "Rectangular arrays and plane partitions," *Acta Arith.* **13**, 22–47.

Chaundy, T. W. (1931). "Partition-generating functions," *Quart. J. Math. Oxford Ser.* **2**, 234–240.

Gordon, B. (1962). "Two new representations of the partition function," *Proc. Amer. Math. Soc.* **13**, 869–873.

Gordon, B. (1971). "Notes on plane partitions. V," *J. Combinatorial Theory* **B11**, 157–168.

Gordon, B. and Houten, L. (1968). "Notes on plane partitions. I, II," *J. Combinatorial Theory* **4**, 72–80, 81–99.

Hodge, W. V. D., and Pedoe, D. (1952). *Methods of Algebraic Geometry*, Vol. 2. Cambridge Univ. Press, London and New York.

Knuth, D. E. (1970). "Permutations matrices and generalized Young tableaux," *Pacific J. Math.* **34**, 709–727.

Knuth, D. E., and Bender, E. A. (1972). "Enumeration of plane partitions," *J. Combinatorial Theory* **13**, 40–54.

Kreweras, G. (1965). "Sur une classe de problèmes de dénombrement liés au treillis des partitions des entiers," *Cahiers du B.U.R.O.* No. 6, Paris.

Kreweras, G. (1967). "Traitment simultané du 'Problème de Young' et 'Problème Simon Newcomb'," *Cahiers du B.U.R.O.* No. 10, Paris.

Lascoux, A. (1974a). "Polynômes symétriques et coefficients d'intersection de cycles de Schubert," *C. R. Acad. Sci. Paris* **279**, 201–204.

Lascoux, A. (1974b). "Tableaux de Young et fonctions de Schur Littlewood," *Séminaire Delange–Pisot–Poitou* No. 4, 1–7.

LeVeque, W. J. (1974). *Reviews in Number Theory*, Vol. 4. Amer. Math. Soc., Providence, R.I.

MacMahon, P. A. (1916). *Combinatory Analysis*, Vol. 2. Cambridge Univ. Press, London and New York (reprinted by Chelsea, New York, 1960).

Robinson, G. deB. (1938). "On the representations of the symmetric group," *Amer. J. Math.* **60**, 745–760.

Rutherford, D. E. (1948). *Substitutional Analysis.* Edinburgh Univ. Press, Edinburgh (reprinted by Hafner, New York, 1968).

Schensted, C. (1961). "Longest increasing and decreasing subsequences," *Canadian J. Math.* **13**, 179–191.

Schützenberger, M.-P. (1963). "Quelques remarques sur une construction de Schensted," *Math. Scand.* **12**, 117–128.

Stanley, R. P. (1971). "Theory and application of plane partitions, I, II," *Studies in Appl. Math.* **50**, 167–188, 259–279.

Stanley, R. P. (1972). "Ordered structures and partitions," *Mem. Amer. Math. Soc.* **119**.

Stanley, R. P. (1973). "The conjugate trace and trace of a plane partition," *J. Combinatorial Theory* **A14**, 53–65.

Tietze, H. (1940–1941). "Systeme von Partitionen und Gitterpunktfiguren, I–IX," *S.-B. Math.-Natur. Abt. Bayer. Akad. Wiss.* 23–54, 69–131, 133–145, 147–166; (1940); 1–37, 39–55, 165–170, 171–186, 187–191 (1941).

Tietze, H. (1941). "Über die Anzahl komprimierter Gitterpunktmengen von gegebener Punktezahl," *Math. Z.* **47**, 352–356.

Wright, E. M. (1931). "Asymptotic partition formulae, I: Plane partitions," *Quart. J. Math. Oxford Ser.* **2**, 177–189.

Wright, E. M. (1968). "Rotatable partitions," *J. London Math. Soc.* **43**, 501–505.

CHAPTER 12

Vector or Multipartite Partitions

12.1 Introduction

As we saw in Chapter 4, problems often arise concerning the additive decomposition of nonzero vectors with nonnegative integral coordinates (also called *multipartite numbers*). Our work in Chapter 4 dealt with applications of "compositions" of multipartite numbers. Now we shall consider partitions of multipartite numbers. As usual, compositions take order of summands into account; partitions do not. The elementary theory of multipartite partitions resembles greatly the elementary theory of ordinary partitions presented in Chapter 1 related to infinite products. Unfortunately there is not known any truly simple type of infinite series that is as useful in treating multipartite partitions as the basic hypergeometric series (introduced in Chapter 2) are in treating ordinary partitions. Thus we only know of comparatively simple identities related to multipartite partition functions (see Section 12.2), and we do not have any device like Euler's pentagonal number theorem (Corollary 1.7) for the rapid calculation of any multipartite partition functions. The most useful theorems for actual evaluation of multipartite partition functions are presented in Theorem 12.3.

L. Carlitz and others have considered problems of restricted multipartite partitions. In particular, they look at k-partite partitions

$$(n_1,\ldots, n_k) = \sum_{j=1}^{r} (m_{1j}, m_{2j},\ldots, m_{kj})$$

subject to the "decreasing" condition

$$\min(m_{1j},\ldots, m_{kj}) \geqslant \max(m_{1,j+1},\ldots, m_{k,j+1}).$$

We shall examine such questions in Section 12.4.

12.2 Multipartite Generating Functions

We recall from Section 4.3 that by $P(\mathbf{n}) = P(n_1, n_2,\ldots, n_r)$ we denote the number of partitions of the "r-partite" or "multipartite" number $(n_1, n_2,\ldots, n_r)$

ENCYCLOPEDIA OF MATHEMATICS and Its Applications, Gian-Carlo Rota (ed.). 2, George E. Andrews, The Theory of Partitions

(i.e., an ordered r-tuple of nonnegative integers not all zero). That is, $P(\mathbf{n})$ is the number of distinct representations of n as a sum of multipartite numbers:

$$\mathbf{n} = \boldsymbol{\xi}^{(1)} + \boldsymbol{\xi}^{(2)} + \cdots + \boldsymbol{\xi}^{(s)} \tag{12.2.1}$$

subject to lexicographic ordering $\boldsymbol{\xi}^{(i)} \geqslant \boldsymbol{\xi}^{(i+1)}$ of the parts:

$$\boldsymbol{\xi}^{(i)} = (\xi_1^{(i)}, \ldots, \xi_r^{(i)}) > (\xi_1^{(i+1)}, \ldots, \xi_r^{(i+1)}) = \boldsymbol{\xi}^{(i+1)}$$

provided $\xi_j^{(i)} > \xi_j^{(i+1)}$ where j is the least integer such that $\xi_j^{(i)} \neq \xi_j^{(i+1)}$. If the number of parts in the partition is restricted to be at most j, we write $P_{\leqslant}(\mathbf{n}; j)$.

For simplicity in the treatment of the related generating functions we define $Q(\mathbf{n})$ as the number of partitions of $\mathbf{n}$ into distinct parts where $(0, 0, \ldots, 0)$ may be a part, and $Q(\mathbf{n}; j)$ is the number of such partitions with j parts. If we desire that $(\mathbf{0})$ be excluded as a part, we shall write $Q_+(\mathbf{n})$ or $Q_+(\mathbf{n}; j)$. The treatment in Chapter 1 extends directly to the study of $P(\mathbf{n})$ and $Q(\mathbf{n})$. In fact, obvious alterations of Theorem 1.1 allow us to prove that

$$\sum_{n_1, \ldots, n_r \geqslant 0} P(\mathbf{n}) x_1^{n_1} \cdots x_r^{n_r} = \prod_{\substack{n_1 \geqslant 0 \cdots n_r \geqslant 0, \\ \text{not all zero}}} (1 - x_1^{n_1} x_2^{n_2} \cdots x_r^{n_r})^{-1}, \tag{12.2.2}$$

$$\sum_{n_1, \ldots, n_r \geqslant 0} Q(\mathbf{n}) x_1^{n_1} \cdots x_r^{n_r} = \prod_{n_1 \geqslant 0 \cdots n_r \geqslant 0} (1 + x_1^{n_1} x_2^{n_2} \cdots x_r^{n_r}); \tag{12.2.3}$$

here for absolute convergence we need only require that $|x_i| < 1$ for $1 \leqslant i \leqslant r$.

It is now fairly obvious that all properties of ordinary (or unipartite) partitions that depend solely on infinite products can be extended without much effort to properties of multipartite numbers. For example, we have Cheema's extension of Euler's theorem:

THEOREM 12.1. *For every* $\mathbf{n}$, $Q_+(\mathbf{n})$ *equals* $\mathcal{O}(\mathbf{n})$, *the number of partitions of* $\mathbf{n}$ *(see (12.2.1)) in which each part* $(\xi_1^{(i)}, \ldots, \xi_r^{(i)})$ *has at least one odd component.*

Proof. As in the proof of Corollary 1.2,

$$\begin{aligned} 1 + \sum_{(\mathbf{n}) > (\mathbf{0})} Q_+(\mathbf{n}) x_1^{n_1} \cdots x_r^{n_r} &= \prod_{(\mathbf{n}) > (\mathbf{0})} (1 + x_1^{n_1} x_2^{n_2} \cdots x_r^{n_r}) \\ &= \prod_{(\mathbf{n}) > (\mathbf{0})} \frac{(1 - x_1^{2n_1} x_2^{2n_2} \cdots x_r^{2n_r})}{(1 - x_1^{n_1} x_2^{n_2} \cdots x_r^{n_r})} \\ &= \prod_{\substack{(\mathbf{n}) > (\mathbf{0}) \\ \text{at least one } n_i \text{ odd}}} (1 - x_1^{n_1} x_2^{n_2} \cdots x_r^{n_r})^{-1} \\ &= \sum_{(\mathbf{n}) \geqslant (\mathbf{0})} \mathcal{O}(\mathbf{n}) x_1^{n_1} \cdots x_r^{n_r}; \end{aligned}$$

hence, $Q_+(\mathbf{n}) = \mathcal{O}(\mathbf{n})$ for every $(\mathbf{n})$. ∎

In surveying our work on ordinary partitions we see that the properties of partition ideals of order 1 presented in Section 8.3 were almost entirely related to the fact that the relevant generating functions were infinite products. We can, therefore, develop a corresponding theory for multipartite partitions. For example, we can obtain Subbarao's generalization of Euler pairs for multipartite partitions:

THEOREM 12.2. *Let S_1 and S_2 be sets of positive integers. The number of partitions of* $\mathbf{n}$ *into parts* $\boldsymbol{\xi}^{(i)} = (\xi_1^{(i)},\ldots,\xi_r^{(i)})$ *in which* $\xi_j^{(i)} \in S_1$ *for all i and* $1 \leqslant j \leqslant r$ *and where no part repeats more than* $t-1$ *times always equals the number of partitions of* $\mathbf{n}$ *into parts* $\boldsymbol{\xi}^{(i)}$ *with* $\xi_j^{(i)} \in S_1$, $1 \leqslant j \leqslant r$, *and some* $\xi_{j_0}^{(i)} \in S_2$ *for all i if and only if* $tS_1 \subseteq S_1$ *and* $S_2 = S_1 - tS_1$.

12.3 Bell Polynomials and Formulas for Multipartite Partition Functions

To obtain useful formulas for computation, we shall utilize the well-known Bell polynomials. Although this procedure will not produce any result even approximating Corollary 1.7, it does produce effective computational formulas, especially since there exist extensive tables of the Bell polynomials.

The Bell polynomials (first extensively studied by E. T. Bell) arise as an aid to the task of taking the nth derivative of a composite function. Namely, we hope to find a formula for the nth derivative of

$$h(t) = f(g(t)).$$

If we denote

$$\frac{d^n h}{dt^n} = h_n, \qquad \frac{d^n f}{dg^n} = f_n, \qquad \frac{d^n g}{dt^n} = g_n,$$

then we see that

$$\begin{aligned} h_1 &= f_1 \\ h_2 &= f_1 g_2 + f_2 g_1^{\,2} \\ h_3 &= f_1 g_3 + 3f_2 g_2 g_1 + f_3 g_1^{\,3} \\ &\;\vdots \end{aligned}$$

It is a simple matter to establish by mathematical induction that

$$\begin{aligned} h_n = f_1\alpha_{n1}(g_1,\ldots,g_n) + f_2\alpha_{n2}(g_1,\ldots,g_n) + \cdots \\ + f_n\alpha_{nn}(g_1,\ldots,g_n) \end{aligned} \tag{12.3.1}$$

where $\alpha_{ni}(g_1,\ldots,g_n)$ is a homogeneous polynomial of degree i in $g_1,\ldots,g_n$.

In light of this last observation, we see that the study of h_n may be reduced to the study of the Bell polynomials:

$$Y_n(g_1, g_2, \ldots, g_n) = \alpha_{n1}(g_1, \ldots, g_n) + \alpha_{n2}(g_1, \ldots, g_n) + \cdots + \alpha_{nn}(g_1, \ldots, g_n). \tag{12.3.2}$$

Note that Y_n is a polynomial in n variables and the fact that g_i was originally an ith derivative is not necessary in the consideration of (12.3.2). Finally, we note that the choice of the $f(t)$ as e^t in (12.3.1) produces the simple formula

$$Y_n(y_1, y_2, y_3, \ldots, y_n) = e^{-y} \frac{d^n e^y}{dt}. \tag{12.3.3}$$

This formula is important for two reasons: *First*, it provides a nice recurrence for the Bell polynomials (we now write $D = d/dt$):

$$\begin{aligned} Y_{n+1}(g_1, g_2, \ldots, g_{k+1}) &= e^{-g}D^n(De^g) \\ &= e^{-g}D^n(g_1 e^g) \\ &= \sum_{k=0}^{n} \binom{n}{k} (e^{-g}D^{n-k}e^g)D^k g_1 \\ &= \sum_{k=0}^{n} \binom{n}{k} Y_{n-k}(g_1, \ldots, g_{n-k})g_{k+1}. \end{aligned} \tag{12.3.4}$$

Second, we obtain from (12.3.4) a concise expression for the generating function of the Bell polynomials:

$$\mathscr{B}(u) = \sum_{n=0}^{\infty} \frac{Y_n u^n}{n!}. \tag{12.3.5}$$

The desired expression is

$$\log \mathscr{B}(u) = \sum_{n=1}^{\infty} \frac{u^n g_n}{n!}. \tag{12.3.6}$$

To verify (12.3.6) we need only differentiate with respect to u and observe that a comparison of the coefficients of u^n in the resulting equation produces an identity equivalent to (12.3.4).

If we exponentiate (12.3.6) and expand the infinite product of exponential functions on the resulting right-hand side, we obtain the following explicit formula for the Bell polynomials:

$$Y_n(g_1, \ldots, g_n) = \sum_{(k) \vdash n} \frac{n!}{k_1! \cdots k_n!} \left(\frac{g_1}{1!}\right)^{k_1} \left(\frac{g_2}{2!}\right)^{k_2} \cdots \left(\frac{g_n}{n!}\right)^{k_n}. \tag{12.3.7}$$

The Bell polynomials are useful in many problems in combinatorics; we shall restrict our consideration to their application in multipartite partition problems.

Let

$$\mathscr{P}_j(x_1,\ldots,x_r) = \mathscr{P}_j = 1 + \sum_{(\mathbf{n})>0} P_{\leqslant}(\mathbf{n};j)x_1^{n_1}\cdots x_r^{n_r}; \tag{12.3.8}$$

$$\mathscr{Q}_j(x_1,\ldots,x_r) = \mathscr{Q}_j = 1 + \sum_{(\mathbf{n})>0} Q(\mathbf{n};j)x_1^{n_1}\cdots n_r^{n_r}; \tag{12.3.9}$$

$$F(u) = 1 + \sum_{j=1} \mathscr{P}_j u^i; \tag{12.3.10}$$

$$G(u) = 1 + \sum_{j=1}^{\infty} \mathscr{Q}_j u^j. \tag{12.3.11}$$

THEOREM 12.3

$$\mathscr{P}_j = Y_j(0!\beta_r(1), 1!\beta_r(2), 2!\beta_r(3), 3!\beta_r(4),\ldots,(j-1)!\beta_r(j))/j!, \tag{12.3.12}$$

$$(-1)^j\mathscr{Q}_j = Y_j(-0!\beta_r(1), -1!\beta_r(2), -2!\beta_r(3),\ldots,-(j-1)!\beta_r(j))/j!, \tag{12.3.13}$$

where $\beta_i(m) = \prod_{i=1}^{j}(1-x_i^m)^{-1}$.

Proof. The arguments establishing Eqs. (12.2.2) and (12.2.3) are easily refined to show that

$$F(u) = \prod_{\mathbf{n}\geqslant 0}(1 - ux_1^{n_1}x_2^{n_2}\cdots x_r^{n_r})^{-1} \tag{12.3.14}$$

and

$$G(u) = \prod_{\mathbf{n}\geqslant 0}(1 + ux_1^{n_1}x_2^{n_2}\cdots x_r^{n_r}). \tag{12.3.15}$$

Therefore

$$\begin{aligned}
\log F(u) &= -\sum_{\mathbf{n}\geqslant 0}\log(1 - ux_1^{n_1}x_2^{n_2}\cdots x_r^{n_r}) \\
&= \sum_{m=1}^{\infty}\sum_{\mathbf{n}\geqslant 0}\frac{u^m x_1^{n_1 m}x_2^{n_2 m}\cdots x_r^{n_r m}}{m} \\
&= \sum_{m=1}^{\infty}\frac{u^m}{m}(1-x_1^m)^{-1}(1-x_2^m)^{-1}\cdots(1-x_r^m)^{-1} \\
&= \sum_{m=1}\frac{u^m}{m!}(m-1)!\beta_r(m).
\end{aligned} \tag{12.3.16}$$

Equation (12.3.14) now follows from the fact that (12.3.16) is a special case of (12.3.6), which is equivalent to (12.3.5). Next

$$\log G(-u) = -\log F(u)$$
$$= -\sum_{m=1}^{\infty} \frac{u^m}{m!}(m-1)!\beta_r(m), \tag{12.3.17}$$

and this yields (12.3.13). ■

To illustrate the usefulness of Theorem 12.3, we calculate $\mathscr{P}_2$: By (12.3.12)

$$\mathscr{P}_2 = \tfrac{1}{2}Y_2(\beta_r(1), \beta_r(2))$$
$$= \tfrac{1}{2}(\beta_r(2) + \beta_r(1)^2)$$
$$= \frac{1}{2}\left[\prod_{i=1}^{r}(1 - x_i^2)^{-1} + \prod_{i=1}^{r}(1 - x_i^2)\right]$$
$$= \frac{1}{2}\frac{1 + \prod_{i=1}^{r}(1 + x_i)}{\prod_{i=1}^{r}(1 - x_i)^2}$$
$$= \frac{1}{2}\frac{\prod_{i=1}^{r}(1 - x_i) + \prod_{i=1}^{r}(1 + x_i)}{\prod_{i=1}^{r}(1 - x_i)(1 - x_i^2)}. \tag{12.3.18}$$

From here it is a reasonably simple matter to compute the actual series expansion.

12.4 Restricted Bipartite Partitions

To conclude this chapter we shall consider certain restricted partitions of bipartite numbers for which considerations similar to those used in Section 11.2 seem to work surprisingly well.

We shall look at $\pi(n, m)$, the number of partitions of (n, m) into "steadily decreasing" parts, that is, partitions of the form

$$(n, m) = (n_1, m_1) + (n_2, m_2) + \cdots + (n_r, m_r) \tag{12.4.1}$$

where $\min(n_i, m_i) \geqslant \max(n_{i+1}, m_{i+1})$, $i \leqslant i < r$. Our main result (due to L. Carlitz) is Corollary 12.5, which is a partition identity different from the simple extensions of partition ideals of order 1 considered in Section 12.2.

Theorem 12.4. *For* $|x| < 1$, $|y| < 1$,

$$\sum_{n,m\geqslant 0} \pi(n, m)x^n y^m = \prod_{n=1}^{\infty}(1 - x^n y^{n-1})^{-1}(1 - x^{n-1}y^n)^{-1}(1 - x^{2n}y^{2n})^{-1}. \tag{12.4.2}$$

Proof. Let $\pi(a, b|n, m)$ denote the number of partitions of (n, m) of the type (12.4.1) subject to the additional restriction $\min(a, b) \geqslant \max(n_1, m_1)$, and write

$$\xi_{ab} = \sum_{r,s \geqslant 0} \pi(a, b|r, s) x^r y^s.$$

Then clearly

$$\xi_{nm} = \sum_{r,s=0}^{\min(n,m)} x^r y^s \xi_{rs} \tag{12.4.3}$$

where we have merely classified each partition with largest part $\max(r, s) \leqslant \min(n, m)$.

Hence if

$$F_1(u) = \sum_{r=0}^{\infty} u^r \xi_{rr},$$

then by (12.4.3) (note that $\xi_{nm} = \xi_{nn}$ if $m \geqslant n$, $\xi_{nm} = \xi_{mm}$ if $n \geqslant m$)

$$\begin{aligned}
F_1(u) &= \sum_{n=0}^{\infty} u^n \sum_{r,s=0}^{n} x^r y^s \xi_{rs} \\
&= \sum_{r,s=0}^{\infty} x^r y^s \xi_{rs} \sum_{n=\max(r,s)}^{\infty} u^n \\
&= \sum_{r \geqslant s} x^r y^s \xi_{ss} \sum_{n=r}^{\infty} u^n + \sum_{r \leqslant s} x^r y^s \xi_{rr} \sum_{n=s}^{\infty} u^n - \sum_{r=0}^{\infty} x^r y^r \xi_{rr} \sum_{n=r}^{\infty} u^n \\
&= (1-u)^{-1} \left\{ \sum_{r \geqslant s} (xu)^r y^s \xi_{ss} + \sum_{r \leqslant s} x^r (yu)^s \xi_{rr} - \sum_{r=0}^{\infty} (xyu)^r \xi_{rr} \right\} \\
&= (1-u)^{-1} \left\{ (1-xu)^{-1} \sum_{s=0}^{\infty} (xyu)^s \xi_{ss} \right. \\
&\qquad \left. + (1-yu)^{-1} \sum_{r=0}^{\infty} (xyu)^r \xi_{rr} - \sum_{r=0}^{\infty} (xyu)^r \xi_{rr} \right\} \\
&= \frac{(1-xyu^2)}{(1-u)(1-xu)(1-yu)} F_1(uxy).
\end{aligned} \tag{12.4.4}$$

Iteration of (12.4.4) yields directly

$$F_1(u) = \prod_{n=0}^{\infty} \frac{(1 - x^{2n+1} y^{2n+1} u^2)}{(1 - x^n y^n u)(1 - x^{n+1} y^n u)(1 - x^n y^{n+1} u)}. \tag{12.4.5}$$

Finally (by Abel's lemma):

$$\sum_{n,m=0}^{\infty} \pi(n, m)x^n y^m = \lim_{r\to\infty} \xi_{rr}$$

$$= \lim_{u\to 1^-} (1 - u)F_1(u)$$

$$= \prod_{n=1}^{\infty} \frac{(1 - x^{2n-1}y^{2n-1})}{(1 - x^n y^n)(1 - x^{n-1}y^n)(1 - x^n y^{n-1})}$$

$$= \prod_{n=1}^{\infty} \frac{1}{(1 - x^{2n}y^{2n})(1 - x^{n-1}y^n)(1 - x^n y^{n-1})}. \qquad \blacksquare$$

Since the infinite product in (12.4.2) is clearly the generating function for $\pi_1(n, m)$, the number of bipartite partitions of (n, m) in which all parts are of the forms $(2a, 2a)$ or $(a - 1, a)$ or $(a, a - 1)$, we obtain the following result immediately.

COROLLARY 12.5. *For all n and m,* $\pi(n, m) = \pi_1(n, m)$. ■

Examples

1. The formula of Cayley (referred to after Example 2 in Chapter 5)

$$\frac{1}{(1-q)(1-q^2)\cdots(1-q^i)} = \sum \frac{1}{1^{p_1}2^{p_2}3^{p_3}\cdots i^{p_i}p_1!p_2!\cdots p_i!} \times \frac{1}{(1-q)^{p_1}(1-q^2)^{p_2}\cdots(1-q^i)^{p_i}},$$

where the summation runs over all partitions $(1^{p_1}2^{p_2}3^{p_3}\cdots)$ of i, is an immediate corollary of Theorem 12.3.

2. The *summatory maximum* of $n_1, n_2, \ldots, n_r$ (written $\text{smax}(n_1, n_2, \ldots, n_r)$) is defined by $\text{smax}(n_1, n_2, \ldots, n_r) = n_1 + n_2 + \cdots + n_r - (r-1)\min(n_1, n_2, \ldots, n_r)$. The identity $\text{smax}(n_1, n_2, \ldots, n_r) = \max(n_1, n_2, \ldots, n_r)$ is valid in general only if $r = 1$ of 2.

3. The proof of Theorem 12.4 may be extended to prove that the number of partitions of $(n_1, \ldots, n_r)$ in which the minimum coordinate of each part is at least as large as the summatory maximum of the next part equals the number of partitions of $(n_1, \ldots, n_r)$ in which each part has one of the $2r - 1$ forms $(a + 1, a, \ldots, a), (a, a + 1, a, \ldots, a), \ldots, (a, a, \ldots, a, a + 1), (ar + 2, ar + 2, \ldots, ar + 2), (ar + 3, ar + 3, \ldots, ar + 3), \ldots, (ar + r, ar + r, \ldots, ar + r)$ $(a \geqslant 0)$. The proof relies on the crucial observation that

$$\frac{1}{1 - xu} + \frac{1}{1 - yu} - 1 = \frac{1 - xyu^2}{(1 - xu)(1 - yu)}$$

is a special case of

$$- \prod_{i=1}^{r}\left(\frac{1}{1 - x_i u} - 1\right) + \frac{1}{(1 - ux_1)\cdots(1 - ux_r)}$$

$$= \frac{1 - x_1 x_2 \cdots x_r u^r}{(1 - ux_1)(1 - ux_2)\cdots(1 - ux_r)}.$$

4. From (12.3.18) it follows easily that the number of partitions of $(n_1 n_2 \cdots n_r)$ into at most two parts is $[\frac{1}{2}(n_1 + 1)(n_2 + 1)\cdots(n_r + 1) + \frac{1}{2}]$. We note that such formulas for any $P(\mathbf{n}; j)$ are directly computable from Theorem 12.3.

5. Application of Eq. (12.3.6) to Example 5 of Chapter 11 shows that the generating function for plane partitions with trace n is

$$\frac{1}{n!} Y_n\left(\frac{0!q}{(1 - q)^2}, \frac{1!q^2}{(1 - q^2)^2}, \frac{2!q^3}{(1 - q^3)^2}, \ldots, \frac{(n - 1)!q^n}{(1 - q^n)^2}\right).$$

6. From Example 5 we see that the generating function for plane partitions with trace 2 is

$$\frac{q^2(1 + q^2)}{(1 - q)^2(1 - q^2)^2}.$$

7. The number of partitions of $(n_1, \ldots, n_r)$ is the same as the number of factorizations (order discounted) of the integer $N = p_1^{n_1} p_2^{n_2} \cdots p_r^{n_r}$ where the p_i are primes. Thus there are four factorizations of 12: $12, 6\cdot 2, 4\cdot 3, 3\cdot 2\cdot 2$; and there are four partitions of (2, 1): (2, 1), (1, 1) + (1, 0), (2, 0) + (0, 1), (1, 0) + (1, 0) + (0, 1).

8*. It is not difficult to show that

$$\left\{\prod_{i=1}^{r}(1 - x_i)(1 - x_i^2)\cdots(1 - x_i^n)\right\} \mathscr{P}_n(x_1, x_2, \ldots, x_r)$$

is a polynomial in $x_1, x_2, \ldots, x_r$; this follows from Theorem 12.3. Somewhat surprising is B. Gordon's theorem that the coefficients of this polynomial are all nonnegative.

Notes

MacMahon (1915–1916, 1917) was the first person to investigate multipartitite partitions in detail. An extensive account of recent work is given by Cheema and Motzkin (1971). Theorem 12.1 is due to Cheema (1964) and Theorem 12.2 is due to Subbarao (1971). The Bell polynomials (studied extensively by Bell, 1934) are treated in greater detail by Riordan (1958) and Comtet (1974); Eq. (12.3.7) is known as Faa di Bruno's formula. N. J. Fine has done some very interesting work connected with sums like (12.3.7). Theorem 12.3 goes back to MacMahon (1917); it has also been rediscovered by Wright (1956). Theorem 12.4 is by Carlitz (1963a, b); related work is done by Carlitz and Roselle (1966), Roselle (1966a, b), and Andrews (1976). Reviews related to the material in this chapter are found in Section P64 of LeVeque (1974).

Example 1. MacMahon (1915–1916).
Examples 2, 3. Andrews (1976).
Example 8. Gordon (1963).

With regard to Example 8, we must mention the work of L. Solomon (1977). He interprets (12.3.10) and (12.3.11) as Poincaré series of certain graded vector spaces associated with the symmetric group S_n on n letters, and he is able to show that the polynomial in Example 8 is also a certain Poincaré series. This interpretation of these polynomials immediately implies Gordon's theorem. Numerous other interesting relations between multipartitite partitions and group theory are explored in Solomon's paper.

References

Andrews, G. E. (1976). "An extension of Carlitz's bipartition identity," to appear.

Bell, E. T. (1934). "Exponential polynomials," *Ann. of Math.* **35**, 258–277.

Carlitz, L. (1963a). "Some generating functions," *Duke Math. J.* **30**, 191–201.

Carlitz, L. (1963b). "A problem in partitions," *Duke Math. J.* **30**, 203–213.

Carlitz, L., and Roselle, D. P. (1966). "Restricted bipartite partitions," *Pacific J. Math.* **19**, 221–228.

Cheema, M. S. (1964). "Vector partitions and combinatorial identities," *Math. Comp.* **18**, 414–420.

Cheema, M. S., and Motzkin, T. S. (1971). "Multipartitions and multipermutations," *Proc. Symp. Pure Math.* **19**, 39–70.

Comtet, L. (1974). *Advanced Combinatorics*. D. Reidel, Dordrecht.

Fine, N. J. (1959). "Sums over partitions," *Rep. Inst. Theory of Numbers (Boulder)* pp. 86–94.

Gordon, B. (1963). "Two theorems on multipartite partitions," *J. London Math. Soc.* **38**, 459–464.

LeVeque, W. J. (1974). *Reviews in Number Theory*, Vol. 4. Amer. Math. Soc., Providence, R.I.

MacMahon, P. A. (1915–1916). *Combinatory Analysis*, Vols. 1 and 2. Cambridge Univ. Press, London and New York (reprinted by Chelsea, New York, 1960).

MacMahon, P. A. (1917). "Memoir on the theory of partitions of numbers, VII," *Philos. Trans. Roy. Soc. London* **A217**, 81–113.

Riordan, J. (1958). *An Introduction to Combinatorial Analysis*. Wiley, New York.

Robertson, M. M. (1962). "Partitions of large multipartites," *Amer. J. Math.* **84**, 16–34.

Roselle, D. P. (1966a). "Generalized Eulerian functions and a problem in partitions," *Duke Math. J.* **33**, 293–304.

Roselle, D. P. (1966b). "Restricted k-partite partitions," *Math. Nachr.* **32**, 139–148.

Roselle, D. P. (1974). "Coefficients associated with the expansion of certain products," *Proc. Amer. Math. Soc.* **45**, 144–150.

Solomon, L. (1977). "Partition identities related to symmetric polynomials in several sets of variables," to appear.

Subbarao, M. V. (1971). "Partition theorems for Euler pairs," *Proc. Amer. Math. Soc.* **28**, 330–336.

Wright, E. M. (1956). "Partitions of multipartite numbers," *Proc. Amer. Math. Soc.* **7**, 880–890.

CHAPTER 13

Partitions in Combinatorics

13.1 Introduction

The subject matter for this chapter could well fill a book by itself. Our justification for an abbreviated treatment lies in the fact that our expressed goal was to treat the term "partitions" primarily as "partitions of numbers." Not surprisingly, partitions of numbers often are closely related to partition problems in combinatorics. For this reason, we shall look at a few topics where partitions of numbers play an important role: finite vector spaces, partitions of sets, and symmetric functions. The first topic mentioned has been studied extensively in recent years; we shall present a simple but fundamental result of Knuth that relates partitions to the combinatorics of both finite vector spaces and finite sets. Partitions of sets have been neatly related to symmetric functions recently by P. Doubilet and we shall introduce this work in Sections 13.3 and 13.4.

13.2 Partitions and Finite Vector Spaces

We recall from Chapter 3 (Theorem 3.1) that the Gaussian polynomial

$$\begin{bmatrix} N+M \\ M \end{bmatrix} = \frac{(q)_{N+M}}{(q)_N(q)_M}$$

is the generating function for $p(N, M, n)$, the number of partitions of n into at most M parts each not exceeding N. Surprisingly (at first), these polynomials arise in the study of finite vector spaces:

THEOREM 13.1. *Let $V_n(q)$ denote the vector space of dimension n over the finite field $GF(q)$ of q (= a prime power) elements. Then there are $\left[\begin{smallmatrix} n \\ m \end{smallmatrix}\right]$ subspaces of $V_n(q)$ of dimension m.*

Remark. We present two proofs: The first is fast and elementary. The second shows explicitly the relationship between partitions and finite vector spaces.

ENCYCLOPEDIA OF MATHEMATICS and Its Applications, Gian-Carlo Rota (ed.). 2, George E. Andrews, The Theory of Partitions

First proof of Theorem 13.1. To determine the m-dimensional subspaces of $V_n(q)$, we first determine all possible m-tuples $(v_1, v_2, \ldots, v_m)$ of m linearly independent vectors. Such m-tuples may be chosen as follows: v_1 may be any of the $q^n - 1$ nonzero elements of $V_n(q)$; v_2 may be any of the $q^n - q$ vectors lying outside the subspace spanned by v_1; v_3 may be any of the $q^n - q^2$ vectors lying outside the subspace spanned by $\{v_1, v_2\}$, and so on. Hence the m-tuple $(v_1, v_2, \ldots, v_m)$ may be chosen in

$$(q^n - 1)(q^n - q)(q^n - q^2)\cdots(q^n - q^{m-1})$$

ways.

Now each such m-tuple $(v_1, v_2, \ldots, v_m)$ spans an m-dimensional subspace of $V_n(q)$; however, several may span the same subspace. In fact, precisely

$$(q^m - 1)(q^m - q)\cdots(q^m - q^{m-1})$$

span the same subspace since this number is the number of ways of choosing a subset of m linearly independent elements from $V_m(q)$. Hence the number of m-dimensional subspaces of $V_n(q)$ is

$$\frac{(q^n - 1)(q^n - q)\cdots(q^n - q^{m-1})}{(q^m - 1)(q^m - q)\cdots(q^m - q^{m-1})}$$

$$= \frac{q^{\binom{m}{2}}(-1)^m(1 - q^n)(1 - q^{n-1})\cdots(1 - q^{n-m+1})}{q^{\binom{m}{2}}(-1)^m(1 - q^m)(1 - q^{m-1})\cdots(1 - q)}$$

$$= \begin{bmatrix} n \\ m \end{bmatrix}.$$

■

Second proof of Theorem 13.1. We choose a fixed basis for $V_n(q)$, say $u_1, u_2, \ldots, u_n$. Then we know from linear algebra that each subspace of dimension m has a canonical basis $v_1, v_2, \ldots, v_m$ given by

$$v_i = \sum_{j=1}^{n} C_{ij}u_j \equiv (C_{i1}, C_{i2}, \ldots, C_{in}) \tag{13.2.1}$$

where $C_{ir_i} = 1$, $C_{ij} = 0$ for $j > r_i$, $C_{sr_i} = 0$ for $s < i$, for $1 \leqslant i \leqslant m$, $n \geqslant r_1 > r_2 > \cdots > r_k \geqslant 1$. For clarity let us consider $n = 9$, $m = 5$, $r_1 = 8$, $r_2 = 7$, $r_3 = 5$, $r_4 = 3$, $r_5 = 2$:

$$U_1 = (C_{11}, 0, 0, C_{14}, 0, C_{16}, 0, 1, 0),$$

$$U_2 = (C_{21}, 0, 0, C_{24}, 0, C_{26}, 1, 0, 0),$$

$$U_3 = (C_{31}, 0, 0, C_{34}, 1, 0, 0, 0, 0),$$

$$U_4 = (C_{41}, 0, 1, 0, 0, 0, 0, 0, 0),$$

$$U_5 = (C_{51}, 1, 0, 0, 0, 0, 0, 0, 0).$$

Now the positions of the undetermined C_{ij} in the foregoing array correspond to the graphical representation or Ferrers graph

$$\begin{array}{ccc} \cdot & \cdot & \cdot \\ \cdot & \cdot & \cdot \\ \cdot & \cdot & \\ \cdot & & \\ \cdot & & \end{array}$$

of the partition 3 + 3 + 2 + 1 + 1 of 10. Thus since each undetermined C_{ij} may be chosen in q ways, there are q^{10} subspaces of $V_9(q)$ with this form of canonical basis. Indeed we see that to each partition of r with at most m parts and largest part at most $n - m$ we may by the foregoing procedure produce q^r different subspaces of dimension m. Therefore the total number of subspaces of dimension m is

$$\sum_{r=0}^{\infty} p(n - m, m, r)q^r = \begin{bmatrix} (n - m) + m \\ m \end{bmatrix} = \begin{bmatrix} n \\ m \end{bmatrix},$$

by Theorem 3.1. ∎

There are many other theorems relating Gaussian polynomials to finite vector spaces; we mention a few in the examples.

13.3 Partitions of Sets

Let us begin with the simple question: How many ways can a set of n elements be split up into a set of disjoint subsets, that is, how many "partitions" are there of $\mathbf{n} = \{1, 2, \ldots, n\}$?

The answer to this question is called the nth *Bell number* B_n (after Eric Temple Bell). Table 13.1 gives the first few Bell numbers and the related partitions.

Table 13.1

n	Partitions of $\{1, 2, \ldots, n\}$	B_n
0	ϕ	1
1	{1}	1
2	{1, 2}; {1}, {2}	2
3	{1, 2, 3}; {1, 2}, {3}; {1, 3}, {2}; {2, 3}, {1}; {1}, {2}, {3}	5

We remark that if $\pi = \{\beta_1, \beta_2, \ldots, \beta_m\}$ where each $\beta_i \subset \{1, 2, \ldots, n\} = \mathbf{n}$, then π is a *partition* of $\mathbf{n}$ provided $\bigcup_{i=1}^m \beta_i = \mathbf{n}$ and $\beta_i \cap \beta_j = \phi$ (the empty set) whenever $i \neq j$. The β_i are called the *blocks* of π. Notice that there is a unique partition (in the sense of Chapter 1) of the number n associated with each π, namely, the partition $\lambda(\pi)$ whose parts are $|\beta_1|, |\beta_2|, |\beta_3|, \ldots, |\beta_m|$ where $|\beta_i|$ is the number of elements of the block β_i.

For simplicity of notation we make the following conventions concerning the partition $\lambda = (n_1 n_2 \cdots n_m) = (1^{r_1} 2^{r_2} 3^{r_3} \cdots)$ of the number n:

$$\lambda! = n_1! n_2! n_3! \cdots n_m! = 1!^{r_1} 2!^{r_2} 3!^{r_3} \cdots,$$

$$|\lambda| = r_1! r_2! r_3! \cdots,$$

$$\text{sign } \lambda = (-1)^{r_2 + 2r_3 + 3r_4 + \cdots} = (-1)^{n - r_1 - r_2 - r_3 - \cdots} = (-1)^{\sigma(\lambda) - \#(\lambda)}.$$

THEOREM 13.2. For each partition λ of n given by $(1^{\lambda_1} 2^{\lambda_2} 3^{\lambda_3} \cdots m^{\lambda_m})$ there are exactly

$$\frac{n!}{\lambda_1! \lambda_2! \cdots \lambda_m! (1!)^{\lambda_1} (2!)^{\lambda_2} \cdots (m!)^{\lambda_m}} = \frac{n!}{|\lambda| \lambda!}$$

partitions π of $\mathbf{n}$ for which $\lambda(\pi) = \lambda$.

Proof. We begin with the standard interpretation of the multinomial coefficient, namely, that there are

$$\frac{n!}{n_1! n_2! \cdots n_j!} = \binom{n}{n_1, n_2, \ldots, n_j} \tag{13.3.1}$$

ordered j-tuples of subsets $(S_1, \ldots, S_j)$ that are mutually disjoint, whose union is $\{1, 2, \ldots, n\}$, and where $|S_i| = n_i$. To see this we note that the set S_1 can be chosen in $\binom{n}{n_1}$ ways and the remaining $j - 1$ sets can then be chosen in

$$\binom{n - n_1}{n_2, \ldots, n_j}$$

ways. Consequently

$$\binom{n}{n_1, n_2, \ldots, n_j} = \binom{n}{n_1} \binom{n - n_1}{n_2, n_3, \ldots, n_j}$$

$$= \binom{n}{n_1} \binom{n - n_1}{n_2} \binom{n - n_1 - n_2}{n_3, n_4, \ldots, n_j}$$

$$\vdots$$

$$= \binom{n}{n_1}\binom{n-n_1}{n_2}\cdots\binom{n-n_1-n_2-\cdots-n_{j-1}}{n_j}$$

$$= \frac{n!}{n_1!n_2!\cdots n_j!}. \tag{13.3.2}$$

Hence if there are λ_1 singleton subsets, λ_2 two-element subsets, and so on, then the number of ordered $(\lambda_1 + \lambda_2 + \cdots + \lambda_m)$-tuples of subsets partitioning **n** is

$$\frac{n!}{(1!)^{\lambda_1}(2!)^{\lambda_2}\cdots(m!)^{\lambda_m}}. \tag{13.3.3}$$

Several of these "ordered" partitions may correspond to the same partition of n; in fact, the one-element subsets may be permuted $\lambda_1!$ ways, the two-element subsets $\lambda_2!$ ways, and so on; consequently, the number of partitions π of **n** with $\lambda(\pi) = \lambda$ is just the number in (13.3.3) divided by $\lambda_1!\lambda_2!\cdots\lambda_m!$, and this is precisely the assertion of this theorem. ■

THEOREM 13.3. *The following formulas hold for the Bell numbers*:

$$B_{n+1} = \sum_{k=0}^{n}\binom{n}{k}B_k; \tag{13.3.4}$$

$$B_n = Y_n(1, 1, \ldots, 1) \tag{13.3.5}$$

(*where* $Y_n(g_1, g_2, \ldots, g_n)$ *is the Bell polynomial defined in* (12.3.2));

$$\sum_{n=0}^{\infty}\frac{B_n x^n}{n!} = \exp(e^x - 1). \tag{13.3.6}$$

Proof. We shall prove (13.3.4) using a simple combinatorial argument. The other identities follow directly from properties of the Bell polynomials derived in Chapter 12.

There are B_{n+1} partitions of $\mathbf{n} + 1 = \{1, 2, \ldots, n + 1\}$. Now $n + 1$ lies in a block of size $k + 1$ where $0 \leqslant k \leqslant n$, and there are clearly $\binom{n}{k}$ choices for this block. Once this block is chosen, the remaining set of $n - k$ numbers may be partitioned in B_{n-k} ways. Hence, summing over all admissible k, we see that

$$B_{n+1} = \sum_{k=0}^{n}\binom{n}{k}B_{n-k} = \sum_{k=0}^{n}\binom{n}{k}B_k,$$

which is (13.3.4).

Equation (13.3.4) together with $B_0 = 1$ uniquely determines the Bell numbers. By (12.3.4) we see that $Y_n(1, 1, \ldots, 1)$ satisfies the same recurrence, and since $Y_0(1, 1, \ldots, 1) = 1$, it follows that $B_n = Y_n(1, 1, \ldots, 1)$ for all n. Hence (13.3.5) is valid.

Finally by (12.3.6),

$$\sum_{n=0}^{\infty} \frac{B_n x^n}{n!} = \sum_{n=0}^{\infty} \frac{Y_n(1, 1, \ldots, 1)x^n}{n!}$$

$$= \exp\left(\sum_{n=1}^{\infty} \frac{x^n}{n!}\right) = \exp(e^x - 1),$$

which is (13.3.6). ■

For our applications to symmetric functions in Section 13.4 we must briefly consider the lattice L_n of partitions of the set $\{1, 2, \ldots, n\} = \mathbf{n}$. Our ultimate object is to obtain Theorem 13.9, which provides an explicit computational formula for the Möbius function of this lattice.

First we define a *partial order* on L_n by saying $\pi_1 \leqslant \pi_2$ whenever each block of π_2 is contained in a block of π_1. It is a simple matter to show that under this partial order L_n forms a lattice. The *union* (or *join*) $\pi_1 \vee \pi_2$ is the coarsest common refinement of π_1 and π_2, namely, i and j are in the same block of $\pi_1 \vee \pi_2$ if and only if they are in the same block in π_1 and in the same block in π_2. The *intersection* (or *meet*) $\pi_1 \wedge \pi_2$ is, on the other hand, defined by the condition that i and j are in the same block if and only if there exists a chain $i = i_0, i_1, i_2, \ldots, i_r = j$ such that for $0 \leqslant m < r$, i_m and i_{m+1} are in the same block of π_1 or π_2.

The *incidence algebra* is the algebra of real-valued functions of two variables whose domain is $L_n \times L_n$ where addition is standard addition of functions, and multiplication is defined by the convolution:

$$(fg)(\pi_1, \pi_2) = \sum_{\pi_1 \leqslant \pi \leqslant \pi_2} f(\pi_1, \pi)g(\pi, \pi_2). \tag{13.3.7}$$

The unit of this algebra is clearly the *Kronecker δ-function*: $\delta(\pi_1, \pi_2) = 1$ if $\pi_1 = \pi_2$ and $\delta(\pi_1, \pi_2) = 0$ if $\pi_1 \neq \pi_2$. Another fundamental function is the *zeta function*: $\zeta(\pi_1, \pi_2) = 1$ if $\pi_1 \leqslant \pi_2$, $\zeta(\pi_1, \pi_2) = 0$ if $\pi_1 \nleqslant \pi_2$.

The following two results are phrased for the lattice of partitions L_n; however, they are valid in a much more general setting where L_n is replaced by any locally finite partially ordered set.

LEMMA 13.4. *The zeta function* $\zeta(\pi_1, \pi_2)$ *possesses an inverse in the incidence algebra* (*to be called the* Möbius function $\mu(\pi_1, \pi_2)$).

Proof. We proceed by induction on the number of elements in the segment

$$[\pi_1, \pi_2] = \{\pi | \pi_1 \leqslant \pi \leqslant \pi_2\}.$$

If the segment has only one element, then clearly $\pi_1 = \pi_2$ and $\mu(\pi_1, \pi_1) = 1$. If $\mu(\pi_1, \pi_2)$ has now been defined whenever $[\pi_1, \pi_2]$ has less than k elements, we see that if $[\bar{\pi}_1, \bar{\pi}_2]$ has k elements, then

$$\mu(\bar{\pi}_1, \bar{\pi}_2) = -\sum_{\bar{\pi}_1 \leqslant \pi < \bar{\pi}_2} \mu(\bar{\pi}_1, \pi),$$

and so $\mu(\bar{\pi}_1, \bar{\pi}_2)$ is uniquely determined. ■

THEOREM 13.5 (Möbius inversion). *Let $f(\pi)$ and $g(\pi)$ be real-valued functions with domain L_n. Then*

$$g(\pi_0) = \sum_{\pi \leqslant \pi_0} f(\pi) \qquad \text{for all} \quad \pi_0 \in L_n \tag{13.3.8}$$

if and only if

$$f(\pi_0) = \sum_{\pi \leqslant \pi_0} g(\pi)\mu(\pi, \pi_0) \qquad \text{for all} \quad \pi_0 \in L_n. \tag{13.3.9}$$

Furthermore

$$g(\pi_0) = \sum_{\pi \geqslant \pi_0} f(\pi) \qquad \text{for all} \quad \pi_0 \in L_n \tag{13.3.10}$$

if and only if

$$f(\pi_0) = \sum_{\pi \geqslant \pi_0} \mu(\pi_0, \pi)g(\pi) \qquad \text{for all} \quad \pi_0 \in L_n. \tag{13.3.11}$$

Remark. Other names for the Möbius inversion process are "the generalized inclusion–exclusion principle" and "generalized sieve methods." When the lattice under consideration is not L_n but the lattice of subsets of a finite set ordered by "inclusion," the Möbius inversion process is precisely classical inclusion–exclusion.

Proof. Each of the four implications asserted in the theorem is proved in the same manner. We shall therefore only prove that (13.3.8) implies (13.3.9). Assuming (13.3.8), we see that

$$\begin{aligned}
\sum_{\pi \leqslant \pi_0} g(\pi)\mu(\pi, \pi_0) &= \sum_{\pi \leqslant \pi_0} \sum_{\pi' \leqslant \pi} f(\pi')\mu(\pi, \pi_0) \\
&= \sum_{\pi' \leqslant \pi_0} f(\pi') \sum_{\pi' \leqslant \pi \leqslant \pi_0} \mu(\pi, \pi_0) \\
&= \sum_{\pi' \leqslant \pi_0} f(\pi')(\zeta\mu)(\pi', \pi_0) \\
&= \sum_{\pi' \leqslant \pi_0} f(\pi')\delta(\pi', \pi_0) = f(\pi_0).
\end{aligned}$$

■

Our next result provides the necessary groundwork for the computation of $\mu(\mathbf{0}, I)$ where $\mathbf{0}$ is the minimal element of L_n ($\mathbf{0} = [\{1, 2, \ldots, n\}]$ and I is the maximal element of L_n ($I = [\{1\}, \{2\}, \ldots, \{n\}]$).

DEFINITION 13.1. Let ϕ be a function from $\mathbf{n}$ to a set C. We form a partition π by putting i and j in the same block if and only if $\phi(i) = \phi(j)$; π is called the kernel of ϕ, and we write $\operatorname{Ker} \phi = \pi$.

LEMMA 13.6. *Let $\pi \in L_n$, and let C be a set with X elements. Let $f(\pi)$ denote the number of functions ϕ from $\mathbf{n}$ to C with* $\operatorname{Ker} \phi = \pi$, *then*

$$f(\pi) = X(X-1)\cdots(X - \nu(\pi) + 1) \equiv (X)_{\nu(\pi)} \qquad (13.3.12)$$

where $\nu(\pi)$ is the number of blocks of π.

Proof. The elements of the first block may map onto any one of the X elements of C; the elements of the second block may map onto any one of the remaining $X - 1$ elements of C, and so on. Hence, the total number of functions admissible is

$$X(X-1)(X-2)\cdots(X - \nu(\pi) + 1) = (X)_{\nu(\pi)}. \qquad \blacksquare$$

COROLLARY 13.7. *For each $\pi_0 \in L_n$*

$$\sum_{\pi \leqslant \pi_0} (X)_{\nu(\pi)} = X^{\nu(\pi_0)}; \qquad (13.3.13)$$

$$\sum_{\pi \leqslant \pi_0} X^{\nu(\pi)} \mu(\pi, \pi_0) = (X)_{\nu(\pi_0)}. \qquad (13.3.14)$$

Proof. Equation (13.3.14) follows from (13.3.13) by Theorem 13.5. To see (13.3.13) we count in two ways the number of functions from $\mathbf{n}$ to C that remain constant on the blocks of π_0. Clearly there are $X^{\nu(\pi_0)}$ such functions. On the other hand, we may split these functions up according to their kernel π. Now the only admissible kernels π are those partitions which are "refined" by π_0, that is, those π such that $\pi \leqslant \pi_0$. Hence, by Lemma 13.6, the number of functions from $\{1, 2, \ldots, n\}$ to C that remain constant on blocks of π is $(X)_{\nu(\pi)}$; therefore

$$\sum_{\pi \leqslant \pi_0} (X)_{\nu(\pi)} = X^{\nu(\pi_0)},$$

which establishes (13.3.13) and completes our proof. $\blacksquare$

We now introduce the well-known *Stirling numbers of the first kind* $s(n, k)$, and *Stirling numbers of the second kind* $S(n, k)$:

$$(X)_n = \sum_{k=0}^{n} s(n, k) X^k; \qquad (13.3.15)$$

$$X^n = \sum_{k=0}^{n} S(n, k)(X)_k. \tag{13.3.16}$$

THEOREM 13.8.

$$\sum_{\substack{\pi \in L_n \\ v(\pi)=k}} 1 = S(n, k); \tag{13.3.17}$$

$$\sum_{\substack{\pi \in L_n \\ v(\pi)=k}} \mu(\pi, I) = s(n, k); \tag{13.3.18}$$

$$\mu(\mathbf{0}, I) = (-1)^{n-1}(n-1)!. \tag{13.3.19}$$

Proof. From (13.3.13) with $\pi_0 = I$, and thus $v(\pi_0) = n$, we see that

$$X^n = \sum_{\pi \in L_n} (X)_{v(\pi)} = \sum_{k=0}^{n} (X)_k \left(\sum_{\substack{\pi \in L_n \\ v(\pi)=k}} 1 \right), \tag{13.3.20}$$

but the $S(n, k)$ are clearly uniquely determined by (13.3.16) and so since (13.3.20) is of precisely the same form, we conclude that (13.3.17) is valid.

Equation (13.3.18) follows from (13.3.14) and (13.3.15) in exactly the same way.

Finally, since there is only one single block partition of $\mathbf{n} = \{1, 2, \ldots, n\}$ (namely, $\mathbf{0}$), we see that by (13.3.18)

$$\begin{aligned} \mu(\mathbf{0}, I) &= s(n, 1) \\ &= \text{coefficient of } X \text{ in } X(X-1)\cdots(X-n+1) \\ &= (-1)^{n-1}(n-1)!, \end{aligned}$$

which is (13.3.19). ■

THEOREM 13.9. *Suppose* $\pi_1, \pi_2 \in L_n$, $\pi_1 \leqslant \pi_2$. *We define the partition* $\lambda(\pi_1, \pi_2) = (1^{r_1}2^{r_2}3^{r_3}\cdots m^{r_m})$, *where* r_i *is the number of blocks of* π_1 *made up of exactly* i *blocks of* π_2. *Then*

$$\mu(\pi_1, \pi_2) = (-1)^{\sum_{\substack{i=1 \\ r_i>0}}^{m} (r_i - 1)} \cdot \prod_{i=1}^{m} (i-1)!^{r_i}. \tag{13.3.21}$$

Proof. Inspection of the segment $[\pi_1, \pi_2]$ reveals that it is isomorphic to the direct product of r_1 copies of L_1, r_2 copies of $L_2, \ldots,$ and r_m copies of L_m. It is not difficult to see that the Möbius function on the direct product of partially ordered sets is just the product of the Möbius functions. Hence we obtain (13.3.21). ■

13.4 The Combinatorics of Symmetric Functions

If we refine the considerations of Section 13.3 concerning functions whose domain is a set of n elements, we are immediately led to the elementary theory of symmetric functions. This approach to symmetric functions not only provides us with a simple proof of the fundamental theorem of symmetric functions (Theorem 13.12), but, more importantly, yields quite computable formulas for the coefficients in many classical symmetric function identities (see Theorems 13.11 and 13.12).

We now consider the set F of functions from $\mathbf{n} = \{1, 2, 3, \ldots, n\}$ to a set C that is countably infinite:

$$C = \{x_1, x_2, x_3, \ldots\}.$$

The elements of C are to be viewed as the indeterminates for the ring of polynomials with countably many variables over the reals.

DEFINITION 13.2. For each $f \in F$, the generating function $\gamma(f)$ is defined by

$$\gamma(f) = \prod_{d \in \mathbf{n}} f(d) = \prod_{i=1}^{\infty} x_i^{|f^{-1}(x_i)|},$$

and if $T \subset F$, we define

$$\gamma(T) = \sum_{f \in T} \gamma(f).$$

Our initial work concerns three special subsets of F:

$$\mathscr{K}_\pi = \{f \in F | \operatorname{Ker} f = \pi\}, \tag{13.4.1}$$

$$\mathscr{S}_\pi = \{f \in F | \operatorname{Ker} f \leqslant \pi\}, \tag{13.4.2}$$

$$\mathscr{A}_\pi = \{f \in F | \operatorname{Ker} f \vee \pi = I\}, \tag{13.4.3}$$

and we let $k_\pi = \gamma(\mathscr{K}_\pi)$, $s_\pi = \gamma(\mathscr{S}_\pi)$, $a_\pi = \gamma(\mathscr{A}_\pi)$.

Notice that the subscript π on each of k_π, s_π, and a_π refers to a partition of the set $\mathbf{n}$. The following definition concerns subscripts that are partitions of integers.

DEFINITION 13.3. For each integer n and each partition $\lambda = (\lambda_1 \lambda_2 \lambda_3 \cdots \lambda_m)$ of n we define

$$k_\lambda = \sum x_{i_1}^{\lambda_1} x_{i_2}^{\lambda_2} \cdots x_{i_m}^{\lambda_m},$$

$$a_n = \sum x_{i_1} x_{i_2} \cdots x_{i_n}, \qquad a_\lambda = a_{\lambda_1} a_{\lambda_2} \cdots a_{\lambda_m},$$

$$s_n = \sum x_i^n, \qquad s_\lambda = s_{\lambda_1} s_{\lambda_2} \cdots s_{\lambda_m}.$$

The next theorem clarifies the relationship between functions whose subscript is a partition of **n** and those whose subscript is a partition of an integer.

THEOREM 13.10. *For each partition* π *of* **n**,

$$k_\pi = |\lambda(\pi)| k_{\lambda(\pi)}; \tag{13.4.4}$$

$$s_\pi = s_{\lambda(\pi)}; \tag{13.4.5}$$

$$a_\pi = \lambda(\pi)! a_{\lambda(\pi)}. \tag{13.4.6}$$

Proof. We write

$$\lambda(\pi) = (1^{r_1}2^{r_2}3^{r_3}\cdots) = (n_1, n_2, \ldots, n_m).$$

First we examine k_π: If $f \in k_\pi$, then $\operatorname{Ker} f = \pi$, and so

$$\gamma(f) = x_{i_1}^{n_1} x_{i_2}^{n_2} \cdots x_{i_m}^{n_m}.$$

Furthermore it is clear that $r_1! r_2! r_3! \cdots = |\lambda(\pi)|$ elements of k_π have exactly this generating function. Consequently

$$\begin{aligned} k_\pi &= \sum |\lambda(\pi)| x_{i_1}^{n_1} \cdots x_{i_m}^{n_m} \\ &= |\lambda(\pi)| k_{\lambda(\pi)}. \end{aligned}$$

Next we treat $\mathscr{S}_\pi$: From (13.4.2) we see that $f \in \mathscr{S}_\pi$ if and only if its kernel is less refined than π, that is, if and only if f is constant on blocks of π. Hence if the blocks of π are $B_1, B_2, B_3, \ldots, B_m$, then

$$\begin{aligned} s_\pi &= \gamma(\mathscr{S}_\pi) \\ &= \sum_{f \in \mathscr{S}_\pi} \gamma(f) \\ &= \sum_{f \in \mathscr{S}_\pi} \gamma(f|_{B_1}) \gamma(f|_{B_2}) \cdots \gamma(f|_{B_m}) \\ &= \prod_{i=1}^{m} \left(\sum_{\substack{\text{constant} \\ \text{functions} \\ f: B_i \to C}} \gamma(f) \right) \\ &= \prod_{i=1}^{m} (x_1^{|B_i|} + x_2^{|B_i|} + x_3^{|B_i|} + \cdots) \\ &= \prod_{i=1}^{m} s_{|B_i|} = s_{\lambda(\pi)}. \end{aligned}$$

Finally we consider $\mathscr{A}_\pi$: For each $f \in \mathscr{A}_\pi$, we have $\operatorname{Ker} f \vee \pi = I$; this means that no two numbers in the same block of f can be in the same block of π as well. The latter formulation is equivalent to saying that f is one-to-one on the blocks of π. Therefore if the blocks of π are $B_1, B_2, B_3, \ldots, B_m$, then

$$\begin{aligned} a_\pi &= \gamma(\mathscr{A}_\pi) \\ &= \sum_{f \in \mathscr{A}_\pi} \gamma(f|_{B_1})\gamma(f|_{B_2})\cdots\gamma(f|_{B_m}) \\ &= \prod_{i=1}^{m} \left(\sum_{\substack{\text{one-to-one} \\ \text{functions} \\ f: B_i \to C}} \gamma(f) \right). \end{aligned}$$

Now each $\gamma(f)$ in this last expression is of the form $x_{j_1}x_{j_2}\cdots x_{j_{|B_i|}}$, and each such term arises from $|B_i|!$ such functions. Hence

$$\begin{aligned} a_\pi &= \prod_{i=1}^{m} \left(\sum |B_i|! x_{j_1}x_{j_2}\cdots x_{j_{|B_i|}}\right) \\ &= |B_1|!|B_2|!\cdots|B_m|! a_{|B_1|}a_{|B_2|}\cdots a_{|B_m|} \\ &= \lambda(\pi)! a_{\lambda(\pi)}. \end{aligned}$$

■

Our next result will be seen to imply easily the fundamental theorem on symmetric functions (Theorem 13.12).

THEOREM 13.11. *For each partition π of* $\mathbf{n}$,

$$k_\pi = \sum_{\sigma \le \pi} \mu(\sigma, \pi) s_\sigma, \tag{13.4.7}$$

$$k_\pi = \sum_{\sigma \le \pi} \frac{\mu(\sigma, \pi)}{\mu(\sigma, I)} \sum_{\tau \ge \sigma} \mu(\sigma, \tau) a_\tau. \tag{13.4.8}$$

Proof. We begin by observing from (13.4.1), (13.4.2), and (13.4.3) that

$$s_\pi = \sum_{\sigma \le \pi} k_\sigma; \tag{13.4.9}$$

$$a_\pi = \sum_{\sigma : \sigma \vee \pi = I} k_\sigma. \tag{13.4.10}$$

Equation (13.4.9) can be easily transformed via Möbius inversion (Eq. (13.3.8)):

$$k_\pi = \sum_{\sigma \le \pi} s_\sigma \mu(\sigma, \pi),$$

which is (13.4.7).

The second portion of this theorem is somewhat harder, requiring a double application of Möbius inversion: By (13.4.10)

$$a_\pi = \sum_\sigma \delta(\sigma \vee \pi, I)k_\sigma$$

$$= \sum_\sigma \left(\sum_{\tau \geqslant \sigma \vee \pi} \mu(\tau, I) \right) k_\sigma$$

$$= \sum_{\tau \geqslant \pi} \mu(\tau, I) \sum_{\sigma \leqslant \tau} k_\sigma.$$

Hence by Möbius inversion

$$\mu(\pi, I) \sum_{\sigma \leqslant \pi} k_\sigma = \sum_{\tau \geqslant \pi} \mu(\pi, \tau) a_\tau,$$

or

$$\sum_{\sigma \leqslant \pi} k_\sigma = \sum_{\tau \geqslant \pi} \frac{\mu(\pi, \tau)}{\mu(\pi, I)} a_\tau, \tag{13.4.11}$$

which is permissible since $\mu(\pi, I) \neq 0$. Applying Möbius inversion to (13.4.11), we see that

$$k_\pi = \sum_{\sigma \leqslant \pi} \mu(\sigma, \pi) \sum_{\tau \geqslant \sigma} \frac{\mu(\sigma, \tau)}{\mu(\sigma, I)} a_\tau,$$

which is (13.4.8). ■

THEOREM 13.12 (Fundamental theorem on symmetric functions). *Every polynomial symmetric function in the variables $x_1, x_2, x_3, \ldots$ is actually a polynomial in $a_1, a_2, a_3, \ldots$.*

Proof. Let $p(x_1, x_2, x_3, \ldots)$ be a polynomial symmetric in the x_i. Then

$$p(x_1, x_2, x_3, \ldots)$$

$$= \sum_{N_i \geqslant n_i \geqslant 0} c_{n_1 n_2 \cdots n_m} \sum x_{i_1}^{n_1} x_{i_2}^{n_2} \cdots x_{i_m}^{n_m}$$

$$= \sum_{N_i \geqslant n_i \geqslant 0} c_{n_1 n_2 \cdots n_m} k_{(n_1, n_2, \ldots, n_m)}$$

$$= \sum_{N_i \geqslant n_i \geqslant 0} c_{n_1 n_2 \cdots n_m} \frac{1}{|\lambda(\pi)|} k_\pi$$

(where each π is some partition of some $\mathbf{n}$ ($n \leqslant N_1 + \cdots + N_m$) with m blocks of size $n_1, n_2, \ldots, n_m$)

$$= \sum_{N_i \geq n_i \geq 0} c_{n_1 n_2 \cdots n_m} \frac{1}{|\lambda(\pi)|} \left(\sum_{\sigma \leq \pi} \frac{\mu(\sigma, \pi)}{\mu(\sigma, I)} \sum_{\tau \geq \sigma} \mu(\sigma, \tau) a_\tau \right)$$

$$= \sum_{N_i \geq n_i \geq 0} c_{n_1 n_2 \cdots n_m} \frac{1}{|\lambda(\pi)|} \left(\sum_{\sigma \leq \pi} \frac{\mu(\sigma, \pi)}{\mu(\sigma, I)} \sum_{\tau \geq \sigma.} \mu(\sigma, \tau) \lambda(\tau)! a_{\lambda(\tau)} \right)$$

and since the $a_{\lambda(\tau)}$ are products of the a_i, we have the desired result. ■

Examples

1. If we consider a vector space $V_n(q)$ over $GF(q)$ of dimension n, and count linear transformations f of $V_n(q)$ into a space $\mathscr{X}$ over $GF(q)$ of X elements subject to the condition $f(V_n(q)) \cap \mathscr{Z} = (0)$, where $\mathscr{Z}$ (with Z elements) is a subspace of $\mathscr{X}$, then it is easy to show that the number of such transformations is

$$(X - Z)(X - Zq) \cdots (X - Zq^{n-1}).$$

2. If in Example 1 we require further that the linear transformations f must be such that $f(V_n(q)) \cap \mathscr{Y}$ is of dimension k, where $Z \subset Y \subset \mathscr{X}$ with Z, Y, and X elements respectively, then the number of such f is

$$\begin{bmatrix} n \\ k \end{bmatrix} (Y - Z)(Y - Zq) \cdots (Y - Zq^{k-1})(X - Y)(X - Yq) \cdots (X - Yq^{n-k-1}).$$

3. Examples 1 and 2 imply that

$$\prod_{i=0}^{n-1} (X - Zq^i) = \sum_{k=0}^{n} \begin{bmatrix} n \\ k \end{bmatrix} \prod_{j=0}^{k-1} (Y - Zq^j) \prod_{h=0}^{n-k-1} (X - Yq^h).$$

4. Example 3 may also be deduced from Corollary 2.4.
5. The linear operator L on polynomials in X over the complex numbers given by

$$L((X - 1)(X - q) \cdots (X - q^{n-1})) = 1$$

satisfies

$$L(X^n) = \sum_{k=0}^{n} \begin{bmatrix} n \\ k \end{bmatrix}.$$

6. Let $V_m(q)$ be an m-dimensional subspace of $V_{m+n}(q)$. By counting the h-dimensional subspaces of $V_{m+n}(q)$ that intersect $V_m(q)$ in a k-dimensional space, we may prove that

$$\sum_{k=0}^{h} \begin{bmatrix} n \\ k \end{bmatrix} \begin{bmatrix} m \\ h - k \end{bmatrix} q^{k(m-h+k)} = \begin{bmatrix} m + n \\ h \end{bmatrix}.$$

7. For any sequence of integers $a_1, a_2, \ldots, a_n$ such that $a_{i+1} - a_i = 0$ or 1 for $1 \leq i < n$, it is true that

$$\begin{bmatrix} N \\ n \end{bmatrix} = \sum_{k=0}^{n} q^{(n-k)(a_{k+1}-a_k)} \begin{bmatrix} a_k \\ k \end{bmatrix} \begin{bmatrix} n - a_{k+1} \\ n - k \end{bmatrix}.$$

8. From (13.3.16) it is easy to deduce that

$$S(n+1, k) = S(n, k-1) + kS(n, k).$$

9. From (13.3.16) and the trivial identity

$$X^{n+1} = X \cdot (1 + (X - 1))^n,$$

it is easy to show that

$$S(n+1, k) = \sum_{j=0}^{n} \binom{n}{j} S(j, k-1).$$

10. From Example 8, it follows that if $S_k(t) = \sum_{n=0}^{\infty} S(n+k, k)t^n$, then

$$S_k(t) = \frac{S_{k-1}(t)}{1 - kt}.$$

Hence

$$S_k(t) = \frac{1}{(1-t)(1-2t)\cdots(1-kt)}.$$

11. For each partition λ of n, we consider $a_\lambda = \sum_{\mu \vdash n} C_{\lambda\mu} k_\mu$. From Theorem 13.11 we can show that $C_{\mu\lambda} = C_{2\mu}$.

12. For every pair of positive integers n and k, each partition of $n!k$ into parts each not exceeding n is actually a sum of k partitions of $n!$. For example: there are 37 partitions of $18 = 3!3$ into parts each $\leqslant 3$:

$$(3^6) = (3^2) + (3^2) + (3^2) \qquad (2^9) = (2^3) + (2^3) + (2^3)$$

$$(123^5) = (123) + (3^2) + (3^2) \qquad (1^3 2^3 3^3) = (1^3 3) + (2^3) + (3^2)$$

$$(1^3 3^5) = (1^3 3) + (3^2) + (3^2) \qquad (1^5 2^2 3^3) = (1^3 3) + (1^2 2^2) + (3^2)$$

$$(2^3 3^4) = (2^3) + (3^2) + (3^2) \qquad (1^7 23^3) = (1^6) + (123) + (3^2)$$

$$(1^2 2^2 3^4) = (1^2 2^2) + (3^2) + (3^2) \qquad (1^9 3^3) = (1^6) + (1^3 3) + (3^2)$$

$$(1^4 23^4) = (1^4 2) + (3^2) + (3^2) \qquad (2^6 3^2) = (2^3) + (2^3) + (3^2)$$

$$(1^6 3^4) = (1^6) + (3^2) + (3^2) \qquad (1^2 2^5 3^2) = (1^2 2^2) + (2^3) + (3^2)$$

$$(12^4 3^3) = (123) + (2^3) + (3^2) \qquad (1^4 2^4 3^2) = (1^4 2) + (2^3) + (3^2)$$

$$(1^8 2^2 3^2) = (1^6) + (1^2 2^2) + (3^2) \qquad (1^6 2^3 3^2) = (1^6) + (2^3) + (3^2)$$

$$(1^{10} 23^2) = (1^6) + (1^4 2) + (3^2) \qquad (1^2 2^8) = (1^2 2^2) + (2^3) + (2^3)$$

$$(1^{12} 3^2) = (1^6) + (1^6) + (3^2) \qquad (1^4 2^7) = (1^4 2) + (2^3) + (2^3)$$

$$(12^7 3) = (123) + (2^3) + (2^3) \qquad (1^6 2^6) = (1^6) + (2^3) + (2^3)$$

$$(1^3 2^6 3) = (1^3 3) + (2^3) + (2^3) \qquad (1^8 2^5) = (1^6) + (1^2 2^2) + (2^3)$$

$$(1^5 2^5 3) = (1^3 3) + (1^2 2) + (2^3) \qquad (1^{10} 2^4) = (1^6) + (1^4 2) + (2^3)$$

$$(1^7 2^4 3) = (1^6) + (123) + (2^3) \qquad (1^{12} 2^3) = (1^6) + (1^6) + (2^3)$$

$$(1^9 2^3 3) = (1^6) + (1^3 3) + (2^3) \qquad (1^{14} 2^2) = (1^6) + (1^6) + (1^2 2^2)$$

$$(1^{11} 2^2 3) = (1^6) + (1^4 2) + (123) \qquad (1^{16} 2) = (1^6) + (1^6) + (1^4 2)$$

$$(1^{13} 23) = (1^6) + (1^6) + (123) \qquad (1^{18}) = (1^6) + (1^6) + (1^6)$$

$$(1^{15} 3) = (1^6) + (1^6) + (1^3 3)$$

Notes

The material in Section 13.2 was taken from D. E. Knuth (1971). The relationship between q-series and finite vector spaces has an extensive literature; the beginnings of this subject may be traced to Jordan (1870), Landsberg (1893), and Dickson (1901). Among the recent works in this area we mention Carlitz (1954), Carlitz and Hodges (1955a, b), Hodges (1964, 1965), Fulton (1969), Rota and Goldman (1969, 1970), and Andrews (1971). Reviews of recent work in this area are given in Section T35 of LeVeque (1974).

The literature on the partitions of finite sets is extensive; see Rota (1964a). We have taken most of our work from Rota (1964a), Rota and Frucht (1965), and Doubilet (1972).

Apart from the examples we have considered there are many other applications of partitions in pure and applied mathematics. Some of the applications to particle physics are given in Temperley (1952) and Bohr and Kalckar (1937). Partitions as applied to the representation theory of the symmetric group have already been mentioned in Chapter 11; see also Littlewood (1950), Robinson (1961), and Rutherford (1947).

The partitions of n also possess a lattice structure relative to the partial ordering: $(\lambda_1\lambda_2\cdots\lambda_s) \geqslant (\lambda_1'\lambda_2'\cdots\lambda_s')$ when $\sum_{j=1}^{i}\lambda_j \geqslant \sum_{j=1}^{i}\lambda_j'$ for each i, $1 \leqslant i \leqslant s$. This lattice has been extensively treated by Brylawski (1973) and has been applied by Snapper (1971) to problems in group theory.

Attention should also be drawn to Stanley (1972), who has extended the partition analysis of MacMahon (1916) to numerous combinatorial problems through the use of (P, ω)-partitions.

Examples 1, 2. Rota and Goldman (1970).

Example 5. Rota and Goldman (1969).

Examples 6, 7. Bender (1971). Example 6 is the q-analog of the Chu–Vandermonde summation and is a special case of Corollary 2.4. Example 7 is an extremely surprising and unusual generalization of Example 6.

Example 11. Doubilet (1972).

Example 12. Knutson (1972).

References

Andrews, G. E. (1971). "On the foundations of combinatorial theory, V: Eulerian differential operators," *Studies in Appl. Math.* **50**, 345–375.

Auluck, F. C. (1951). "On some new types of partitions associated with generalized Ferrars graphs," *Proc. Cambridge Phil. Soc.* **47**, 679–686.

Auluck, F. C., and Kothari, D. S. (1946). "Statistical mechanics and the partition of numbers," *Proc. Cambridge Phil. Soc.* **42**, 272–277.

Bender, E. A. (1971). "A generalized q-binomial Vandermonde convolution," *Discrete Math.* **1**, 115–119.

Bohr, N., and Kalckar, F. (1937). "On the transmutation of atomic nuclei by impact of material particles," *Kgl. Danske Vidensk Selskab Mat. Fys. Medd.* **14**, No. 10.

Brylawski, T. (1973). "The lattice of integer partitions," *Discrete Math.* **6**, 201–219.

Carlitz, L. (1954). "Representations by quadratic forms in a finite field," *Duke Math. J.* **21**, 123–137.

Carlitz, L., and Hodges, J. H. (1955a). "Representations by Hermitian forms in a finite field," *Duke Math. J.* **22**, 393–405.

Carlitz, L., and Hodges, J. H. (1955b). "Distribution of bordered symmetric, skew and hermitian matrices in a finite field," *J. reine angew. Math.* **195**, 192–201.

Dickson, L. E. (1901). *Linear Groups with an Exposition of the Galois Field Theory.* Teubner, Leipzig (reprinted by Dover, New York, 1958).

Doubilet, P. (1972). "On the foundations of combinatorial theory, VII: Symmetric functions through the theory of distribution and occupancy," *Studies in Appl. Math.* **51**, 377–396.

Fulton, J. D. (1969). "Symmetric involutory matrices over finite fields and modular rings of integers," *Duke Math. J.* **36**, 401–407.

Hodges, J. H. (1964). "Simultaneous pairs of linear and quadratic matrix equations over a finite field," *Math. Z.* **84**, 38–44.

Hodges, J. H. (1965). "A symmetric matrix equation over a finite field," *Math. Nachr.* **30**, 221–228.

Jordan, C. (1870). *Traité des substitutions et des équations algébrique.* Gauthier-Villars, Paris.

Knuth, D. E. (1971). "Subspaces, subsets, and partitions," *J. Combinatorial Theory* **A-10**, 178–180.

Knutson, D. (1972). "A lemma on partitions," *Amer. Math. Monthly* **79**, 1111–1112.

Landsberg, G. (1893). "Über eine Anzahlbestimmung und eine damit zusammenhängende Reihe," *J. reine angew. Math.* **111**, 87–88.

LeVeque, W. J. (1974). *Reviews in Number Theory*, Vol. 5. Amer. Math. Soc., Providence, R.I.

Littlewood, D. E. (1950). *The Theory of Group Characters and Matrix Representations of Groups*, 2nd ed. Oxford Univ. Press, London and New York.

MacMahon, P. A. (1916). *Combinatory Analysis*, Vol. 2. Cambridge Univ. Press, London and New York (reprinted by Chelsea, New York, 1960).

Robinson, G. de B. (1961). *Representation Theory of the Symmetric Group.* Univ. of Toronto Press, Toronto.

Rota, G.-C. (1964a). "The number of partitions of a set," *Amer. Math. Monthly* **71**, 498–504.

Rota, G.-C. (1964b). "On the foundations of combinatorial theory, I: Theory of Möbius functions," *Z. Wahrscheinlichkeitstheorie und verw. Gebiete* **2**, 340–368.

Rota, G.-C., and Frucht, R. (1965). "Polinomios de Bell y particiones de conjuntos finitos," *Scientia: Rev. Ci. Técnica (Chile)* No. 126, 5–10.

Rota, G.-C., and Goldman, J. (1969). "The number of subspaces of a vector space," *Recent Progress in Combinatorics* (W. Tutte, ed.), pp. 75–83. Academic Press, New York.

Rota, G.-C., and Goldman, J. (1970). "On the foundations of combinatorial theory, IV: Finite vector spaces and Eulerian generating functions," *Studies in Appl. Math.* **49**, 239–258.

Rutherford, D. E. (1947). *Substitutional Analysis.* Edinburgh Univ. Press, Edinburgh (reprinted by Hafner, New York, 1968).

Snapper, E. (1971). "Group characters and nonnegative integral matrices," *J. Algebra* **19**, 520–535.

Stanley, R. P. (1972). "Ordered structures and partitions," *Mem. Amer. Math. Soc.* **119**.

Temperley, H. N. V. (1952). "Statistical mechanics and the partition of numbers, II: The form of crystal surfaces," *Proc. Cambridge Phil. Soc.* **48**, 683–697.

Wright, E. M. (1968, 1971, 1972). "Stacks, I, II, III," *Quart. J. Math.* **19**, 313–320; **22**, 107–116; **23**, 153–158.

CHAPTER 14

Computations for Partitions

14.1 Introduction

In most of the applied and many of the theoretical aspects of partitions we are interested in actually enumerating or perhaps completely exhibiting a set of partitions of n subject to certain conditions. There are certain elementary algorithms to apply that generally do the job if n is small; these we discuss in the next section.

The parity of $p(n)$ has been of interest over the years, and we describe in Section 14.3 the most effective algorithm known for determining $p(n)$ modulo 2, as well as other algorithms that derive from generating functions.

Computations for higher-dimensional partitions are considered in Section 14.4; here we include the work of D. E. Knuth, which is the only quite general simplification known. Sections 14.5–14.7 give reasonably short tables related to well-known partition functions. In Section 14.8, we present a guide to more extensive tables.

14.2 Elementary Algorithms

The simplest way to enumerate all the partitions of n is to write them in lexicographic order. To proceed from $(\lambda_1, \lambda_2, \ldots, \lambda_s)$ to the next partition in this order we note that

if $\lambda_s > 1$, the next partition is $(\lambda_1, \lambda_2, \ldots, \lambda_{s-1}, \lambda_s - 1, 1)$; (14.2.1)

if $\lambda_{s-r} = c > 1$ but $\lambda_{s-r+1} = \lambda_{s-r+2} = \cdots = \lambda_s = 1$, then the next partition is obtained by replacing $\lambda_{s-r}, \lambda_{s-r+1}, \ldots, \lambda_s$ by $(c-1), (c-1), \ldots, (c-1), d$ where $0 < d \leqslant c - 1$, and the number α of appearances of $(c-1)$ is so chosen that $\alpha(c-1) + d = c + r = \lambda_{s-r} + \lambda_{s-r+1} + \cdots + \lambda_s$. (14.2.2)

The following computer program in FORTRAN IV utilizes the foregoing algorithm to compute all the partitions of each integer not exceeding 20:

ENCYCLOPEDIA OF MATHEMATICS and Its Applications, Gian-Carlo Rota (ed.). 2, George E. Andrews, The Theory of Partitions

```
      DIMENSIØN IP(20)
      DO 80 N=1,20
      WRITE (6,5) N
    5 FØRMAT (I10)
      IP(1)=N
      DØ 10 L=2,20
   10 IP(L)=0
   15 CØNTINUE
      WRITE (6,20) IP(1), IP(2), . . ., IP(20)
C   COMMENT IP(J) is the Jth part of the partition of N just computed.
   20 FORMAT (20I3)
      J=1
   25 CØNTINUE
      IF (IP(J)-1) 35,35,30
   30 J=J+1
      GØ TØ 25
   35 CØNTINUE
      IF (J-1) 75,75,40
   40 M=J-2
      K=N
      IF (M) 45,55,45
   45 DØ 50 L=1,M
   50 K=K-IP(L)
   55 IQ=IP(J-1)
      JQ=K/(IP(J-1)-1)
   60 DØ 65 LQ=1,JQ
   65 IP(J-2+LQ)=IQ-1
      IP(J-1+JQ)=K-JQ*(IQ-1)
      MQ=J+JQ
      DØ 70 NQ=MQ,N
   70 IP(NQ)=0
      GØ TØ 15
   75 CØNTINUE
   80 CØNTINUE
      STØP
      END
```

The printout begins with

$$
\begin{array}{ccccc}
 & & 1 & & \\
1 & 0 & 0 & \cdots & 0 \\
 & & 2 & & \\
2 & 0 & 0 & \cdots & 0 \\
1 & 1 & 0 & \cdots & 0 \\
 & & 3 & & \\
3 & 0 & 0 & \cdots & 0 \\
2 & 1 & 0 & \cdots & 0 \\
1 & 1 & 1 & \cdots & 0
\end{array}
$$

It is a simple matter to put in a subroutine that will examine whether each partition $(\lambda_1, \ldots, \lambda_s) = (\mathrm{IP}(1), \mathrm{IP}(2), \ldots, \mathrm{IP}(S))$ fulfills various prescribed conditions.

There is also a related algorithm due to Hindenburg that may be restricted to the partitions of n into exactly m parts. The order now is reverse lexicographic: the parts are written in *ascending* order and $(\Lambda_1 \cdots \Lambda_m)$ appears before $(\Lambda_1' \cdots \Lambda_m')$ in the list if for some j: $\Lambda_1 = \Lambda_1', \Lambda_2 = \Lambda_2', \ldots, \Lambda_{j-1} = \Lambda_{j-1}', \Lambda_j < \Lambda_j'$.

The algorithm begins with the first element in this ordering: $(1, 1, \ldots, 1, n - m + 1)$, and it passes from $(\Lambda_1, \ldots, \Lambda_m)$ to the next partition as follows:

find the largest j such that $\Lambda_m - \Lambda_j \geqslant 2$; (14.2.3)

replace $\Lambda_j, \Lambda_{j+1}, \ldots, \Lambda_m$ by $(\Lambda_j + 1, \Lambda_j + 1, \ldots, \Lambda_j + 1, \Lambda_m')$ where Λ_m' is so chosen that the number being partitioned is still n. For example, when $m = 5$, $n = 12$ (14.2.4)

$$
\begin{array}{ccccc}
1 & 1 & 1 & 1 & 8 \\
1 & 1 & 1 & 2 & 7 \\
1 & 1 & 1 & 3 & 6 \\
1 & 1 & 1 & 4 & 5 \\
1 & 1 & 2 & 2 & 6 \\
1 & 1 & 2 & 3 & 5 \\
1 & 1 & 2 & 4 & 4 \\
1 & 1 & 3 & 3 & 4 \\
1 & 2 & 2 & 2 & 5 \\
1 & 2 & 2 & 3 & 4 \\
1 & 2 & 3 & 3 & 3 \\
2 & 2 & 2 & 2 & 4 \\
2 & 2 & 2 & 3 & 3
\end{array}
$$

Again we have an easily programmable algorithm.

14.3 Algorithms from Generating Functions

One value of our work on generating functions and partition identities lies in the fact that the time of computation for certain partition functions may be greatly decreased through algorithms derived from these theoretical considerations.

For example, Corollary 1.8: $p(n) - p(n-1) - p(n-2) + p(n-5) + p(n-7) - \cdots = 0$, provides us with all values of $p(n)$ for $n \leqslant N$ after on the order of $\frac{2}{3}(6n^3)^{\frac{1}{2}}$ operations. This is quite efficient compared to the algorithm that arises from Example 2 in Chapter 6, where the $r(n)$ in $\prod_{n \geqslant 1} (1 - q^n)^{-a_n} = \sum_{n \geqslant 0} r(n)q^n$ may be computed from the recurrence

$$nr(n) = \sum_{\substack{h,j \geqslant 1 \\ h \cdot j \leqslant n}} r(n - hj)ja_j,$$

which requires on the order of $n^2 \log n$ operations.

Exact formulas such as those in Chapter 5 also provide useful means for computation.

The partition identities of Chapters 7–9 are also useful in this regard. For example, in Theorem 8.5 we see that $A_{k,i}(n)$ can be computed in on the order of $cn^{3/2}$ (c a constant that depends on k) steps from

$$\sum_{n \geqslant 0} A_{k,i}(n)q^n = \frac{1}{(q)_\infty} \sum_{n=0} (-1)^n q^{\frac{1}{2}(2k+1)n(n+1) - in}(1 - q^{(2n+1)i}).$$

On the other hand, if we compute $B_{k,i}(n)$ from

$$\sum_{m \geqslant 0} b_{k,i}(m, n) = B_{k,i}(n),$$

$$b_{k,i}(m, n) = b_{k,i-1}(m, n) + b_{k,k-i+1}(m - i + 1, n - m),$$

$$b_{k,0}(m, n) = 0,$$

$$b_{k,i}(m, n) = \begin{cases} 1 & \text{if } m = n = 0, \\ 0 & \text{if } m \leqslant 0 \text{ or } n \leqslant 0 \text{ but } m^2 + n^2 \neq 0, \end{cases}$$

then we must use essentially dn^2 steps (d a constant that depends on k).

In general, when we undertake computations of a particular partition function, consideration of and transformations of the related generating function should always be carefully examined in view of the possible time-saving that can be effected.

We close this section with the following theorem, due to MacMahon, which provides an even better algorithm than Corollary 1.8 for computing the parity of $p(n)$:

THEOREM 14.1

$$p(4n) \equiv p(n) + p(n-7) + p(n-9) + \cdots + p(n-\alpha_i) + \cdots \pmod 2,$$

$$p(4n+1) \equiv p(n) + p(n-5) + p(n-11) + \cdots + p(n-\beta_i) + \cdots \pmod 2,$$

$$p(4n+3) \equiv p(n) + p(n-3) + p(n-13) + \cdots + p(n-\gamma_i) + \cdots \pmod 2,$$

$$p(4n+6) \equiv p(n) + p(n-1) + p(n-15) + \cdots + p(n-\delta_i) + \cdots \pmod 2,$$

where the right-hand series are terminated just before the arguments become negative and $\alpha_i = i(8i \mp 1)$, $\beta_i = i(8i \mp 3)$, $\gamma_i = i(8i \mp 5)$, $\delta_i = i(8i \mp 7)$.

Proof.

$$\begin{aligned}
\sum_{n \geq 0} p(n)q^n &= \prod_{n=1}^{\infty} (1-q^n)^{-1} \\
&\equiv \prod_{n=1} (1+q^n)^{-1} && \pmod 2 \\
&= \prod_{n=1}^{\infty} (1-q^{2n-1}) && \text{(Eq. (1.2.5))} \\
&\equiv \prod_{n=1}^{\infty} (1+q^{4n-1})(1+q^{4n-3}) && \pmod 2 \\
&= \prod_{n=1}^{\infty} (1-q^{4n})^{-1} \prod_{n=1}^{\infty} (1-q^{4n})(1+q^{4n-1})(1+q^{4n-3}) \\
&= \left(\sum_{n \geq 0} p(n)q^{4n}\right) \sum_{n=0}^{\infty} q^{\frac{1}{2}n(n+1)} && \text{(by Theorem 2.8).}
\end{aligned}$$

The recurrences stated in Theorem 14.1 are now easily obtained by comparing coefficients of q^{4n}, q^{4n+1}, q^{4n+3} and q^{4n+6} in the foregoing congruence. ■

14.4 Computations for Higher-Dimensional Partitions

As long as we are treating problems in which the related generating function is of some well-known form, for example, an infinite product, then it is possible to proceed as we have indicated in Section 14.3. Unfortunately, such good fortune is not available when we consider partitions of dimension d with $d > 2$, as we have seen in Chapter 11. Thus, computation of even three-dimensional partitions becomes a cumbersome problem. The only useful tool known is a device of MacMahon that is closely related to our proof of Theorem 3.7. MacMahon's technique was seen by D. E. Knuth to apply to a quite general setting, and we shall present his generalization.

We shall consider a set P partially ordered by $\textcircled{\leqslant}$ and an order reversing map ω from P to the set of nonnegative integers; that is, if $x \textcircled{\leqslant} y$ in P, then $\omega(x) \geqslant \omega(y)$.

DEFINITION 14.1. We shall call the order reversing map ω a *labeling* if additionally only finitely many x have $\omega(x) > 0$.

The following lemma isolates those elements of P with positive labels; this immediately reduces our consideration from all of P (which is generally infinite) to a finite subset of P.

LEMMA 14.2. *Let the partial ordering $\textcircled{\leqslant}$ be extended to a linear ordering $\leqslant$ of P. There is a bijection between the labelings of P and pairs of sequences*

$$n_1 \geqslant n_2 \geqslant \cdots \geqslant n_m \qquad \textit{(positive integers)},$$

$$x_1, x_2, \ldots, x_m \qquad \textit{(elements of } P), \qquad (14.4.1)$$

subject to the conditions that (i) $x \in P$ *and* $x \textcircled{\leqslant} x_j$ *implies* $x = x_i$ *for some* $i < j$; (ii) $x_i > x_{i+1}$ *implies* $n_i > n_{i+1}$, *for* $1 \leqslant i < m$.

Proof. The correspondence is obvious once we exhibit the appropriate construction: Write down all x with a given largest label in increasing order relative to $\leqslant$; then write the x with the next largest label, again in increasing order relative to $\leqslant$; and so on. Once all positive labels are exhausted, write the label of each x listed directly above that x. Inspection shows that this procedure produces the desired sequences and is uniquely reversible. ■

DEFINITION 14.2. We shall say that a labeling ω of P is a *P-partition* of n if $\sum_{x \in P} \omega(x) = n$.

Now by Lemma 14.2, every P-partition corresponds uniquely to a pair of sequences

$$n_1 \geqslant n_2 \geqslant n_3 \geqslant \cdots \qquad \text{(an infinite sequence of nonnegative integers)},$$

$$x_1, x_2, \ldots, x_m \qquad \text{(distinct elements of } P), \qquad (14.4.2)$$

where $m \geqslant 0$; (i) $x \in P$ and $x \leqslant x_j$ implies $x = x_i$ for some $i < j$, (ii) $x_i > x_{i+1}$ implies $n_i > n_{i+1}$ for $1 \leqslant i < m$; (iii) if $m > 0$, then $n_m > n_{m+1}$; (iv) if $m > 0$, then there exists $x \in P$ such that $x < x_m$ and $x \neq x_i$ for $1 \leqslant i \leqslant m$. Finite sequences $\{x_i\}$ satisfying (i) and (iv) are called *topological sequences*. To see this correspondence, we merely attach infinitely many zeros to the sequence n_i and we delete x_m if it is the least element of $P - \{x_1, \ldots, x_{m-1}\}$.

For example, let P be the set of points in the plane with nonnegative coordinates and with the partial ordering $(i, j) \textcircled{\leqslant} (i', j')$ if $i \leqslant i'$ and $j \leqslant j'$. This partial ordering may be extended to a linear ordering through the lexicographic order $(i, j) \leqslant (i', j')$ if $i < i'$ or $i = i'$ and $j \leqslant j'$. In this case, P-partitions as characterized in (14.4.1) correspond precisely to plane partitions (written upside down). Thus the plane partition

$$\begin{array}{l} 1 \\ 12 \\ 431 \\ 5332 \end{array}$$

corresponds to the (P, ω)-partition

$$\begin{array}{cccccccccc} 5 & 4 & 3 & 3 & 3 & 2 & 2 & 1 & 1 & 1 \\ (0,0) & (0,1) & (1,0) & (1,1) & (2,0) & (1,2) & (3,0) & (2,1) & (0,2) & (0,3) \end{array}$$

The corresponding sequence (14.4.2) is

$$\begin{array}{cccccccccccc} 5 & 4 & 3 & 3 & 3 & 2 & 2 & 1 & 1 & 1 & 0 & 0 \ \ldots \\ (0,0) & (0,1) & (1,0) & (1,1) & (2,0) & (1,2) & (3,0) & (2,1) & & & & \end{array}$$

Clearly there is a one-to-one correspondence between the (14.4.1) sequences and the (14.4.2) sequences.

DEFINITION 14.3. The *index* of a topological sequence $x_1, x_2, \ldots, x_m$ is the sum of all j such that $x_j > x_{j+1}$ with the convention that $x_m > x_{m+1}$.

Thus, returning to our example, we see that the topological sequence

$$(0,0) \quad (0,1) \quad (1,0) \quad (1,1) \quad (2,0)\,(1,2)\,(3,0) \quad (2,1)$$

has *index* $5 + 7 + 8 = 20$.

Given a topological sequence $x_1, x_2, \ldots, x_m$, every sequence $n_1 \geqslant n_2 \geqslant \cdots$ of nonnegative integers that satisfies conditions (ii) and (iii) listed after (14.4.2) corresponds uniquely to an infinite sequence $p_1 \geqslant p_2 \geqslant p_3 \geqslant \cdots$ obtained by subtracting 1 from each of $n_1, \ldots, n_j$ for each j such that $x_j > x_{j+1}$ or $j = m$. Returning to our example, we see that the sequence $p_1 \geqslant p_2 \geqslant p_3 \geqslant \cdots$ in this case is

$$2, 1, 0, 0, 0, 0, 0, 0, 0, 0, \ldots.$$

The remarkable fact is that the sum of the p_i plus the index equals the sum of the n_i and this is the only relationship between the three sequences $\{x_1, \ldots, x_m\}$, $\{n_i\}$, $\{p_i\}$. These observations establish the following result.

THEOREM 14.3. *Let P be an infinite partially ordered set. Then there is a one-to-one between correspondence P-partitions of n and ordered pairs $(\{x_i\}, \{p_i\})$ where $\{x_i\}$ is a topological sequence and $\{p_i\}$ is a partition of $n - k$, with k the index of $\{x_i\}$.* ■

COROLLARY 14.4. *Let P be an infinite partially ordered set; let s(n) be the number of P-partitions of n, and let t(k) be the number of topological sequences of P having index k. Then*

$$\sum_{n\geq 0} s(n)q^n = \left(\sum_{n\geq 0} t(n)q^n\right)\left(\sum_{n\geq 0} p(n)q^n\right)$$
$$= (q)_\infty^{-1} \sum_{n\geq 0} t(n)q^n. \qquad \blacksquare$$

Generally topological sequences are easier to find (and index) than arbitrary (P, ω)-partitions. For example, we may construct topological sequences of points in the first quadrant as follows: First choose a set of, say, m points that correspond to the graphical representation of some partition. Then number these points using the first m integers, so that there is strict increase along rows and columns, making sure that m is not in the first column. If, say,

$$\begin{matrix} \cdot & & \\ \cdot & \cdot & \cdot \end{matrix}$$

is chosen, then the corresponding topological sequences are

$$\begin{matrix} 3 & & \\ 1 & 2 & 4 \end{matrix} \qquad \text{and} \qquad \begin{matrix} 2 & & \\ 1 & 3 & 4, \end{matrix}$$

meaning $x_1 = (0, 0)$, $x_2 = (1, 0)$, $x_3 = (0, 1)$, $x_4 = (2, 0)$, and $x_1 = (0, 0)$, $x_2 = (0, 1)$, $x_3 = (1, 0)$, $x_4 = (2, 0)$, respectively. In this instance our topological sequences are essentially standard Young tableaux (to obtain standard Young tableaux, reverse the sequence), with the special requirement that m not be in the first column.

14.5 Brief Tables of Partition Functions

Table 14.1 lists $p(n)$, the number of partitions of n; $p(\mathcal{O}, n)(= p(\mathcal{D}, n)$ the number of partitions of n with odd parts; $A_{2,2}(n)(= B_{2,2}(n))$, the number of partitions of n into parts congruent to 1 or 4 modulo 5, and $A_{2,1}(n)$ $(= B_{2,1}(n))$, the number of partitions of n into parts congruent to 2 or 3 modulo 5.

14.6 Table of the Plane Partition Function

Table 14.2, for $M_2(n)$, the number of plane partitions of n, was easily computed from the simple recurrence (Chapter 6, Example 2)

$$nM_2(n) = \sum_{j=1}^{n} M_2(n-j)\sigma_2(j)$$

where $\sigma_2(j)$ is the sum of the squares of the divisors of j.

Table 14.1

n	$p(n)$	$p(\mathcal{O}, n)$	$A_{2,2}(n)$	$A_{2,1}(n)$
1	1	1	1	0
2	2	1	1	1
3	3	2	1	1
4	5	2	2	1
5	7	3	2	1
6	11	4	3	2
7	15	5	3	2
8	22	6	4	3
9	30	8	5	3
10	42	10	6	4
11	56	12	7	4
12	77	15	9	6
13	101	18	10	6
14	135	22	12	8
15	176	27	14	9
16	231	32	17	11
17	297	38	19	12
18	385	46	23	15
19	490	54	26	16
20	627	64	31	20
21	792	76	35	22
22	1002	89	41	26
23	1255	104	46	29
24	1575	122	54	35
25	1958	142	61	38
26	2436	165	70	45
27	3010	192	79	50
28	3718	222	91	58
29	4565	256	102	64
30	5604	296	117	75
31	6842	340	131	82
32	8349	390	149	95
33	10143	448	167	105
34	12310	512	189	120
35	14883	585	211	133
36	17977	668	239	152
37	21637	760	266	167
38	26015	864	299	190
39	31185	982	333	210
40	37338	1113	374	237

Table 14.1 (continued)

n	$p(n)$	$p(\mathcal{O}, n)$	$A_{2,2}(n)$	$A_{2,1}(n)$
41	44583	1260	415	261
42	53174	1426	465	295
43	63261	1610	515	324
44	75175	1816	575	364
45	89134	2048	637	401
46	105558	2304	709	448
47	124754	2590	783	493
48	147273	2910	871	551
49	173525	3264	961	604
50	204226	3658	1065	673
51	239943	4097	1174	739
52	281589	4582	1299	820
53	329931	5120	1429	899
54	386155	5718	1579	997
55	451276	6378	1735	1091
56	526823	7108	1913	1207
57	614154	7917	2100	1321
58	715220	8808	2311	1457
59	831820	9792	2533	1593
60	966467	10880	2785	1756
61	1121505	12076	3049	1916
62	1300156	13394	3345	2108
63	1505499	14848	3659	2301
64	1741630	16444	4010	2525
65	2012558	18200	4380	2753
66	2323520	20132	4794	3019
67	2679689	22250	5231	3287
68	3087735	24576	5717	3599
69	3554345	27130	6233	3917
70	4087968	29927	6804	4281
71	4697205	32992	7409	4655
72	5392783	36352	8080	5084
73	6185689	40026	8790	5521
74	7089500	44046	9573	6021
75	8118264	48446	10406	6537
76	9289091	53250	11322	7118
77	10619863	58499	12294	7721
78	12132164	64234	13363	8401
79	13848650	70488	14498	9103
80	15796476	77312	15742	9894

Table 14.1 (continued)

n	$p(n)$	$p(\mathcal{O}, n)$	$A_{2,2}(n)$	$A_{2,1}(n)$
81	18004327	84756	17066	10715
82	20506255	92864	18512	11631
83	23338469	101698	20050	12587
84	26543660	111322	21732	13653
85	30167357	121792	23519	14761
86	34262962	133184	25466	15995
87	38887673	145578	27540	17285
88	44108109	159046	29796	18710
89	49995925	173682	32196	20203
90	56634173	189586	34806	21854
91	64112359	206848	37582	23579
92	72533807	225585	40594	25483
93	82010177	245920	43802	27480
94	92669720	267968	47276	29671
95	104651419	291874	50974	31975
96	118114304	317788	54979	34502
97	133230930	345856	59239	37153
98	150198136	376256	63843	40058
99	169229875	409174	68747	43114
100	190569292	444793	74040	46447

14.7 Table of Gaussian Polynomials

Since these polynomials

$$\begin{bmatrix} n \\ r \end{bmatrix} = \frac{(1 - q^n)\cdots(1 - q^{n-r+1})}{(1 - q^r)\cdots(1 - q)}$$

have been so important throughout our work, we include self-explanatory Table 14.3, complete for $n \leqslant 12$. We note that since

$$\begin{bmatrix} n \\ r \end{bmatrix} = \begin{bmatrix} n \\ n - r \end{bmatrix},$$

we need only consider $0 \leqslant r \leqslant n/2$. Furthermore

$$\begin{bmatrix} n \\ 0 \end{bmatrix} = 1,$$

$$\begin{bmatrix} n \\ 1 \end{bmatrix} = 1 + q + q^2 + \cdots + q^{n-1};$$

hence we consider only $2 \leqslant r \leqslant n/2$.

Table 14.2

n	$M_2(n)$	n	$M_2(n)$	n	$M_2(n)$
1	1	41	409383981	81	23498 9042219523
2	3	42	593001267	82	31738 2398602028
3	6	43	856667495	83	42817 1324714100
4	13	44	1234363833	84	57697 8362008262
5	24	45	1774079109	85	77663 3557947931
6	48	46	2543535902	86	104422 6582040722
7	86	47	3537993036	87	140249 8445554353
8	160	48	5191304973	88	188168 0993051045
9	282	49	7391026522	89	252192 5777635221
10	500	50	1 0499640707	90	337650 8618954817
11	859	51	1 4883573114	91	451605 3506209319
12	1479	52	2 1053676445	92	603409 5681424573
13	2485	53	2 9720561230	93	805440 1761691018
14	4167	54	4 1871334614	94	1074059 4259633282
15	6879	55	5 8874385349	95	1430879 4779412286
16	11297	56	8 2623976486	96	1904421 9737399823
17	18334	57	11 5737404664	97	2532294 5742768655
18	29601	58	16 1825846160	98	3364043 3065994170
19	47330	59	22 5863047430	99	4464887 4867815365
20	75278	60	31 4689799781	100	5920606 6030052023
21	118794	61	43 7699333376		
22	186475	62	60 7771804065		
23	290783	63	84 2541287719		
24	451194	64	116 6117605448		
25	696033	65	161 1415838202		
26	1068745	66	222 3312543970		
27	1632658	67	306 2906627106		
28	2483234	68	421 3276093961		
29	3759612	69	578 7232662336		
30	5668963	70	793 7771067795		
31	8512309	71	1087 2114256046		
32	12733429	72	1487 0591703377		
33	18974973	73	2031 1959869076		
34	28175955	74	2770 7354427729		
35	41691046	75	3774 5732428153		
36	61484961	76	5135 4666391773		
37	90379784	77	6978 1543686979		
38	132441995	78	9470 1959299374		
39	193487501	79	12836 4196860773		
40	281846923	80	17378 1688194937		

Table 14.3

$\begin{bmatrix} n \\ r \end{bmatrix}$	1	q	q^2	q^3	q^4	q^5	q^6	q^7	q^8	q^9	q^{10}	q^{11}	q^{12}	q^{13}	q^{14}	q^{15}	q^{16}	q^{17}	q^{18}
$\begin{bmatrix} 4 \\ 2 \end{bmatrix}$	1	1	2	1	1														
$\begin{bmatrix} 5 \\ 2 \end{bmatrix}$	1	1	2	2	2	1	1												
$\begin{bmatrix} 6 \\ 2 \end{bmatrix}$	1	1	2	2	3	2	2	1	1										
$\begin{bmatrix} 6 \\ 3 \end{bmatrix}$	1	1	2	3	3	3	3	2	1	1									
$\begin{bmatrix} 7 \\ 2 \end{bmatrix}$	1	1	2	2	3	3	3	2	2	1	1								
$\begin{bmatrix} 7 \\ 3 \end{bmatrix}$	1	1	2	3	4	4	5	4	4	3	2	1	1						
$\begin{bmatrix} 8 \\ 2 \end{bmatrix}$	1	1	2	2	3	3	4	3	3	2	2	1	1						
$\begin{bmatrix} 8 \\ 3 \end{bmatrix}$	1	1	2	3	4	5	6	6	6	6	5	4	3	2	1	1			
$\begin{bmatrix} 8 \\ 4 \end{bmatrix}$	1	1	2	3	5	5	7	7	8	7	7	5	5	3	2	1	1		
$\begin{bmatrix} 9 \\ 2 \end{bmatrix}$	1	1	2	2	3	3	4	4	4	3	3	2	2	1	1				
$\begin{bmatrix} 9 \\ 3 \end{bmatrix}$	1	1	2	3	4	5	7	7	8	8	8	7	7	5	4	3	2	1	1
$\begin{bmatrix} 9 \\ 4 \end{bmatrix}$	1	1	2	3	5	6	8	9	11	11	12	11	11	9	8	6	5	3	2
$\begin{bmatrix} 10 \\ 2 \end{bmatrix}$	1	1	2	2	3	3	4	4	5	4	4	3	3	2	2	1	1		
$\begin{bmatrix} 10 \\ 3 \end{bmatrix}$	1	1	2	3	4	5	7	8	9	10	10	10	10	9	8	7	5	4	3
$\begin{bmatrix} 10 \\ 4 \end{bmatrix}$	1	1	2	3	5	6	9	10	13	14	16	16	18	16	16	14	13	10	9
$\begin{bmatrix} 10 \\ 5 \end{bmatrix}$	1	1	2	3	5	7	9	11	14	16	18	19	20	20	19	18	16	14	11

Table 14.3 (continued)

$\begin{bmatrix} n \\ r \end{bmatrix}$	1	q	q^2	q^3	q^4	q^5	q^6	q^7	q^8	q^9	q^{10}	q^{11}	q^{12}	q^{13}	q^{14}	q^{15}	q^{16}	q^{17}	q^{18}
$\begin{bmatrix} 11 \\ 2 \end{bmatrix}$	1	1	2	2	3	3	4	4	5	5	5	4	4	3	3	2	2	1	1
$\begin{bmatrix} 11 \\ 3 \end{bmatrix}$	1	1	2	3	4	5	7	8	10	11	12	12	13	12	12	11	10	8	7
$\begin{bmatrix} 11 \\ 4 \end{bmatrix}$	1	1	2	3	5	6	9	11	14	16	19	20	23	23	24	23	23	20	19
$\begin{bmatrix} 11 \\ 5 \end{bmatrix}$	1	1	2	3	5	7	10	12	16	19	23	25	29	30	32	32	32	30	29
$\begin{bmatrix} 12 \\ 2 \end{bmatrix}$	1	1	2	2	3	3	4	4	5	5	6	5	5	4	4	3	3	2	2
$\begin{bmatrix} 12 \\ 3 \end{bmatrix}$	1	1	2	3	4	5	7	8	10	12	13	14	15	15	15	15	14	13	12
$\begin{bmatrix} 12 \\ 4 \end{bmatrix}$	1	1	2	3	5	6	9	11	15	17	21	23	27	28	31	31	33	31	31
$\begin{bmatrix} 12 \\ 5 \end{bmatrix}$	1	1	2	3	5	7	10	13	17	21	26	30	35	39	43	46	48	49	49
$\begin{bmatrix} 12 \\ 6 \end{bmatrix}$	1	1	2	3	5	7	11	13	18	22	28	32	39	42	48	51	55	55	58

14.8 Guide to Tables

Gupta, Gwyther, and Miller (1958) present the most extensive readily available table of partitions. This table lists $p(n)$ for $n \leqslant 600$, and extensive tables are given for the number of partitions of n into at most m parts. Their work greatly extends a previous table by Gupta (1939). Barton, David, and Kendall (1966) present short tables of partitions and bipartitions. N. J. A. Sloane (1973) lists numerous partition sequences in his fascinating *Handbook of Integer Sequences.* The papers by Andrews, Cheema, Churchhouse, and Burnell and Houten in the book *Computers in Number Theory* (Academic Press, New York, 1971) describe some recent uses of computers in partition problems.

Notes

The algorithm of Hindenburg in Section 14.2 is taken from Dickson's *History of the Theory of Numbers* (1920, Vol. 2, p. 106); the entire Chapter 3 of that book presents a detailed early history of partitions. The history from 1940 to 1972 may be found in LeVeque's (1974) *Reviews in Number Theory*, Chapter P, Vol. 4, while in Vol. 6, Section Z30, reviews of various tables of partitions may be found.

Theorem 14.1 is from MacMahon (1921); MacMahon's work was greatly extended by Parkin and Shanks (1967). The material in Section 14.4 is taken from Knuth (1970); Knuth includes a computer program for finding the topological sequences related to solid partitions. The P-partition concept has been generalized extensively by Stanley (1972) to (P, ω)-partitions, and he has shown that numerous partition and permutation problems can be treated through the use of this powerful method.

The excellent book by Nijenhuis and Wilf (1975) presents a number of algorithms for the computation of compositions and partitions; Chapters 5 and 6 of their text treat compositions while Chapters 9 and 10 are related to partitions.

References

Andrews, G. E. (1971). "The use of computers in search of identities of the Rogers–Ramanujan type," *Computers in Number Theory* (A. O. L. Atkin and B. J. Birch, eds.), pp. 377–387. Academic Press, New York.

Barton, D. E., David, F. N., and Kendall, M. G. (1966). *Symmetric Functions and Allied Tables*. Cambridge Univ. Press, London and New York.

Burnell, D., and Houten, L. (1971). "Multiplanar partitions," *Computers in Number Theory* (A. O. L. Atkin and B. J. Birch, eds.), pp. 401–404. Academic Press, New York.

Cheema, M. S. (1971). "Computers in the theory of partitions," *Computers in Number Theory* (A. O. L. Atkin and B. J. Birch, eds.), pp. 389–395. Academic Press, New York.

Churchhouse, R. F. (1971). "Binary partitions," *Computers in Number Theory* (A. O. L. Atkin and B. J. Birch, eds.), pp. 397–400. Academic Press, New York.

Dickson, L. E. (1920). *History of the Theory of Numbers*, Vol. 2. Carnegie Institution of Washington, Washington, D.C. (reprinted by Chelsea, New York, 1952, 1966).

Gupta, H. (1939). *Tables of Partitions*. Indian Math. Soc., Madras.

Gupta, H., Gwyther, A. E., and Miller, J. C. P. (1958). *Tables of Partitions* (Roy. Soc. Math. Tables, Vol. 4).

Knuth, D. E. (1970). "A note on partitions," *Math. Comp.* **24**, 955–961.

LeVeque, W. J. (1974). *Reviews in Number Theory*, Vols. 4 and 6. Amer. Math. Soc., Providence, R.I.

MacMahon, P. A. (1921). "Note on the parity of the number which enumerates the partitions of a number," *Proc. Cambridge Phil. Soc.* **20**, 281–283.

Nijenhuis, A., and Wilf, H. S. (1975). *Combinatorial Algorithms*. Academic Press, New York.

Parkin, T. R., and Shanks, D. (1967). "On the distribution of parity in the partition function," *Math. Comp.* **21**, 466–480.

Sloane, N. J. A. (1973). *A Handbook of Integer Sequences*. Academic Press, New York.

Stanley, R. P. (1972). "Ordered structures and partitions," *Mem. Amer. Math. Soc.* **119**.

Index for Definitions of Symbols

Author Index

Numbers set in *italics* designate pages on which complete literature citations are given.

Subject Index